Advances in Experimental Medicine and Biology

Volume 1103

More information about this series at http://www.springer.com/series/5584

Mari Dezawa
Editor

Muse Cells

Endogenous Reparative Pluripotent Stem Cells

Springer

Editor
Mari Dezawa
Tohoku University Graduate School of Medicine
Sendai, Miyagi, Japan

ISSN 0065-2598 ISSN 2214-8019 (electronic)
Advances in Experimental Medicine and Biology
ISBN 978-4-431-56845-2 ISBN 978-4-431-56847-6 (eBook)
https://doi.org/10.1007/978-4-431-56847-6

Library of Congress Control Number: 2018962258

This Springer imprint is published by the registered company Springer Japan KK part of Springer Nature.
The registered company address is: Shiroyama Trust Tower, 4-3-1 Toranomon, Minato-ku, Tokyo 105-6005, Japan

Foreword

The past decade (2008–2017) has ushered in a new era in biology that should be called the age of adult stem cells. The previous decade (1997–2007) was the age of the embryonic stem cell (ESC) that culminated with the discovery of the induced pluripotent stem cell (iPSC) in 2006. This book describes the progress made in understanding adult stem cells and particularly the Muse cell.

Before 1998, most people thought that stem cells come from only two sources: the bone marrow and embryos. Bone marrow cells are the source of hematopoietic stem cells that make our blood cells, and ESC of course made all of us. Pluripotent stem cells, able to make all the germ cell layers, were thought to come only from embryos.

Frightened by visions of embryo farms to make cells and organs for transplantation, many countries passed legislation to forbid harvesting and even studies of ESC. A little thought should have convinced all of us that organisms must have evolved adult pluripotent stem cells. To repair even a minor wound, dozens of cell types are required.

Over 400 different kinds of cells inhabit our bodies. If all these cell types were necessary to repair our tissues, where are they kept? Do we have depots of cells lined up and ready to travel to all parts to fix injuries? Such depots of course do not exist. If they existed, somebody would have seen them and described them.

The only and the most efficient solution would be to have a cell that can make all the cells of the body, a cell that hides until it is needed, a cell that can self-renew and activate when needed, a cell that would seek injured tissues, and a cell that would make the cells needed to repair the tissues. This cell is of course the Muse cell.

Each chapter of this book provides timely and detailed discussions and insights into Muse cells that cannot be readily obtained from any single paper published by Mari Dezawa and her colleagues. The book pulls together data and observations from many studies and a decade of observing Muse cells in vitro and in vivo to present a compelling case for these cells being the most important cell that repairs our bodies.

Chapter 1 is entitled "The Muse Cell Discovery: Thanks to Wine and Science." It describes how the cell was discovered accidentally in 2007 by Professor Mari Dezawa. One evening in 2007, Dr. Dezawa was preparing stem cells from rat bone marrow at Kyoto University, when a friend, Professor Fujiyoshi, called to invite her to a wine party.

Dr. Dezawa fortuitously put the cells away in a solution, placed them in an incubator, and went out to the wine party. The next morning, Dr. Dezawa discovered that she had mistakenly placed the cells in a solution of trypsin rather than culture media. The cells seemed dead, but she noticed some floating debris that contained cells. She cultured these cells and grew out pluripotent stem cells.

With Dr. Fujiyoshi's help, Dr. Dezawa named them Muse cells for these *mu*ltilineage-differentiating *s*tress-*e*nduring *c*ells. Further studies showed that these Muse cells expressed the embryonic stem cell marker SSEA3 and the mesenchymal cell marker CD105. When she looked for these two markers, she found SSEA3+/CD105+ cells in many tissues.

Muse cells exhibit three critical characteristics. First, Muse cells make progenitors that outgrow and reduce the percentage of Muse cells in vitro and in vivo. Second, Muse cells remain quiescent until activation, whereupon they detach from their hiding places, become pluripotent, and enter the bloodstream. Third, Muse cells home to injured tissues and produce the cells to repair the tissues.

Chapter 2 describes how Muse cells transform from quiescent somatic cells to floating pluripotent cells. Muse cells express markers of pluripotent embryonic stem cells, e.g., Nanog, Oct3/4, and Sox32, but these pluripotent nuclear factors remain localized to the cytoplasm as long as the cells are adherent but upregulate 50–100-fold to enter the nucleus when the cells detach and become pluripotent.

Upon detachment, the cells begin to secrete serpins and 14-3-3 proteins that prevent apoptosis and repair DNA, partly accounting for the stress tolerance of the cells. While the cells are initially mesenchymal cells, they readily cross lineages to make hepatocytes, neurons, melanocytes, myocytes, endothelial, and many other cells necessary to repair tissues.

The mechanisms that prevent tumorigenesis of Muse cells are of particular interest. In adherent cultures, Muse cells proliferate as rapidly as fibroblasts but generate more non-Muse cells through asymmetric divisions, and the proportion of Muse cells gradually decline to several percent or less.

In suspension, Muse cells proliferate by symmetric division through an unusual and possibly unique mechanism. The Muse cells produce slender flat cells that enwrap the Muse cells and allow the Muse cells to proliferate to form embryoid bodies with the flat cells covering the cells until growth ceases after about 14 days. If the wrapping cells are removed, proliferation resumes.

Comparison of Muse cells with pluripotent ESC provides insights into how stem cells regulate pluripotency. ESCs exhibit stable and high rates of self-renewal, can generate whole tissues such as hematopoietic tissues, depend on leukemia-inhibiting factor and STAT3 (LIF/STAT3) and BMP4 signaling to sustain self-renewal, and use glycolysis for energy.

In contrast, Muse cells depend on fibroblast growth factor (FGF) and activin/transforming growth factor (TGF) beta for self-renewal, engage in more asymmetric divisions and less self-renewal, utilize oxidative phosphorylation and mitochondria respiration for energy, and derive their signals for what cells to make from the injured and dying cells.

Chapter 3 by Yoshihiro Kushida, Shohei Wakao, and Mari Dezawa demonstrates how Muse cells are the endogenous reparative stem cells of the body, how the cells home to injury sites and exert their reparative effects through paracrine, anti-inflammatory, anti-fibrotic, anti-immune, and anti-apoptotic mechanisms, as well as cellular replacement in the heart, liver, kidney, brain, and other organs.

Their recent discovery that sphingosine-1-phosphate (S1P) is the homing signal for Muse cells was a significant breakthrough. Injured tissues, such as stroke and myocardial infarcts, release S1P, associated with increases in circulating S1P and Muse cells in peripheral blood. In patients with acute myocardial infarcts, recovery of ejection fraction correlated with the rise of circulating Muse cells.

In animals, Muse cells migrate to and around cardiac infarcts expressing S1P. Non-Muse mesenchymal stem cells (MSC) do not. Antagonists of S1P receptor 2, expressed by Muse cells, block this homing behavior. Both Muse and non-Muse MSCs express CXCR4 receptors and migrate to bone marrow and serum in response to stromal-derived factor-1 (SDF-1).

A very high percentage (approximately 90%) of Muse cells express the human leukocyte antigen-G (HLA-G) after reaching injured tissues. HLA-G is a potent immunosuppressing factor expressed by trophoblasts in the placenta that effectively prevents immune rejection of the placenta by mothers. Only 20% of non-Muse mesenchymal/stem cells express HLA-G.

HLA-G is not the only anti-immune mechanism that Muse cells have. Like other MSCs, Muse cells express interferon gamma-induced indoleamine-2,3 dioxygenase (IDO). Muse cells inhibit T-cell proliferation, as well as IDO expression in Muse cells, suggesting Muse cells are highly evolved to shut off local but not systemic immune responses.

Muse cells express cytokines, growth, tissue growth factor-alpha and tissue growth factor-beta, HGF, vascular endothelial growth factor (VEGF), epidermal growth factor (EGF), leukemia-inhibiting factor (LIF), and corticotrophin-releasing hormone. Muse cells also express factors expressed by MSC, such as metalloprotease-1, metalloprotease-2, and metalloprotease-9, to suppress fibrosis.

Muse cells have both paracrine and pleiotropic effects on injured tissues, often specific for the tissue type. For example, upon arrival in infarcted rabbit hearts, Muse cells express GATA-4, a differentiation factor for cardiomyocytes. If this expression is silenced, the cells are not as effective in repairing the heart.

Chapter 4 is very special. It describes the protocols for preparing Muse cells, lessons hard-won over a decade of studying the cells, so that their results can be readily reproduced. The chapter contains advice such as subculturing at a ratio of 1:2 and never at 1:3 or else the ratio of Muse cells will decrease, to change media every 2–3 days, and not to use cells over 11 passages for experiments.

Several chapters delve deeply into the mechanisms by which Muse cells repair double-stranded DNA damage (Chap. 5, by Umberto et al.), how Muse cells affect the immune system (Chap. 6, by Perone et al.), and mitochondria and metabolism of Muse cells (Chap. 7, by Trosko). They compare Muse cells with other stem cells.

Chapters 8. 9, 10, 11, 12, 13, 14, 15, and 16 describe the use of Muse cells to treat acute myocardial infarcts (Chap. 8), stroke (Chap. 9), peripheral blood Muse cells (Chap. 10), chronic kidney disease (Chap. 11), liver (Chaps. 12 and 13), skin (Chap. 14), aortic aneurysm (Chap. 15), and ischemic lung (Chap. 16). These chapters illustrate the relevance of Muse cells to the major organ systems of the body.

The above list is so long that several organs and tissues are conspicuously absent, e.g., stomach and gut, bladder, muscle, spine and other bones, spinal cord, and eyes. This is not because work is not being done to study Muse cell treatment of these conditions but simply work in these areas has not matured as quickly as in the heart, lung, kidney, and brain.

Chapter 17 by Dr. Dezawa is about clinical trials. Conceptually, the clinical trials are isolating cells from bone marrow, adipose tissue, and other sources, expanded and intravenously infused into patients with myocardial infarcts, stroke, renal failure, and liver diseases. Phase 1 studies for myocardial infarcts have already started. As the chapter indicates, the list of target diseases is growing.

For anyone who seeks to understand and study Muse cells, this book is an absolute necessity. For all students of stem cell biology, this book should be one of the top ten textbooks in universities. Finally, for clinicians interested in using stem cells to treat patients, this is a must-read book. Muse cells are the most important stem cell therapy of our generation.

Distinguished Professor of Cell Biology and Neuroscience, Wise Young
Richard H. Shindell Chair in Neuroscience
W. M. Keck Center for Collaborative Neuroscience,
Rutgers, State University of New Jersey,
Piscataway, NJ, USA

Preface

Minute repairs that occur daily in every organ are key to tissue homeostasis in the living body. These highly sophisticated repair processes are often taken for granted, however, and the precise mechanisms that lead to these repairs are not well understood.

We first reported multilineage-differentiating stress-enduring (Muse) cells in 2010. They were identified as stress-tolerant endogenous pluripotent stem cells. Muse cells are non-tumorigenic, consistent with the fact that they normally reside in adult tissues, and are thus associated with fewer safety concerns. At that time, Muse cells were regarded as pluripotent stem cells residing only in mesenchymal tissues such as the bone marrow and dermis. Because Muse cells can be obtained from easily accessible tissues and are able to differentiate into various cell types, they are considered to be a feasible cell type for regenerative medicine. Due to advances in Muse cell research by several groups around the world, Muse cells are now known to distribute throughout the entire body and to function as "endogenous reparative stem cells." Although they are not the only feasible cells for regenerative medicine, they have great biological implications.

Through evolution, all creatures have developed the ability to adapt to drastic changes in the global environment. Therefore, the system involved in tissue homeostasis is presupposed to be stress-tolerant. As mentioned above, Muse cells are stress-tolerant. Both basic and clinical studies have demonstrated that Muse cells are endowed with properties that enable them to perform daily minute reparative functions in the living body that contribute to tissue homeostasis. Muse cells are suggested to stably mobilize from the bone marrow into the circulating blood and are thus supplied to every organ where they replace damaged or lost functional cells to repair and maintain tissues. In tissue that is severely damaged, such as by ischemia or injury, Muse cells promptly mobilize into the circulating blood, preferentially home to the damaged tissue, and repair the site by spontaneous differentiation into tissue-compatible cells. This series of reactions of endogenous Muse cells might not be sufficient to effectively repair tissue when the number of mobilized Muse cells is inadequate or the activity of Muse cells is lost. In such cases, there is reasonable evidence that supplying additional highly active Muse cells could result in efficient tissue repair.

For successful adaptation to various environmental changes, systems relevant to tissue homeostasis might be organized along multiple lines, although the whole scheme is still unclear. Muse cells are considered to have a key role in one of the homeostatic systems. New insight into their activities and contributions may provide a glimpse into the mysteries of the body – that is, how our body's reparative functions are strategically organized.

Due to their unique functions, Muse cells are expected to provide safe and efficient medical treatment. Muse cells have a specific receptor that responds to damage signals from the body, enabling the Muse cells to preferentially migrate to and accumulate into the damaged tissue after systemic administration. This great advantage of Muse cells makes regenerative medicine available to patients by intravenous drip, one of the simplest and least burdensome approaches for both patients and clinicians. Because Muse cells are pluripotent, they can target multiple diseases. Moreover, full utilization of the innate reparative system is safe and provides medical care compatible with the laws of nature. Muse cells may open the door to a "next-generation medical care" that is well suited to the body's natural repair system, without requiring the introduction or manipulation of artificial genes. In this sense, the medical care provided by Muse cell therapy is "reparative medicine" rather than regenerative medicine. On the basis of these unique beneficial properties of Muse cells, Life Science Institute, Inc., a subsidiary of Mitsubishi Chemical Holdings Group, began clinical trials using Muse cells to treat acute myocardial infarction in January 2018. I believe, therefore, that this book is very timely.

Many groups in several countries are performing Muse cell research. It might be difficult to grasp the whole breadth of Muse cells by only reading individually published articles. In addition, readers interested in Muse cells may be searching for a precise protocol for collecting Muse cells. For these reasons, we made special efforts to cover a comprehensive range of the Muse cell research and to include a technical chapter that provides a practical method for isolating Muse cells. As the editor, I recruited selected authors who are conducting high-quality Muse cell studies, collected all the updated knowledge ranging from basic to clinical studies, and have systematically consolidated the updated information for doctoral students, postdoctoral researchers, basic researchers, clinicians, and industrial developers. I hope that this book will provide readers a broad spectrum of knowledge regarding Muse cells, from the technical issues to the basic properties of Muse cells, their mobility in vivo, their importance in tissue homeostasis, and their feasibility for clinical applications.

Science continuously progresses. Muse cells are still largely unexplored, but the research will push forward, and new findings will accumulate. I sincerely hope that this book will inspire researchers to dig deeper to uncover the exciting potential of Muse cells to yield fruitful next-generation medical care and an exciting future.

Sendai, Japan Mari Dezawa

Contents

Chapter 1
The Muse Cell Discovery, Thanks to Wine and Science

Mari Dezawa

Abstract Multilineage-differentiating stress-enduring (Muse) cells, identified as cells positive for the pluripotent marker stage-specific embryonic antigen (SSEA-3+), were discovered as stress-tolerant pluripotent stem cells from among mesenchymal stem cells (MSCs) and fibroblasts, as well as from the adult human bone marrow mononucleated fraction. MSCs are a crude population of cells that differentiate into multiple cell types covering all three germ layers in low proportion and were thus deduced to contain a genuine pluripotent stem cell subpopulation. Muse cells constitute several percent of MSCs and 1 of ~3000 bone marrow mononucleated cells. They exhibit pluripotent gene expression as well as trilineage differentiation and self-renewal capabilities at the single-cell level, while, in contrast, MSC cells other than Muse cells do not exhibit these characteristics. These characteristics indicate that Muse cells correspond to the subpopulation of MSC cells responsible for the pluripotent aspect of MSCs. In addition to their pluripotency, Muse cells play an important role in vivo as endogenous stem cells that contribute to tissue homeostasis through daily reparative maintenance and to tissue reconstruction through their unique reparative functions following serious tissue damage. This chapter describes how my research team discovered Muse cells.

Keywords Stress · Trypsin · Mesenchymal stem cells · Transdifferentiation · Pluripotent · SSEA-3 · Bone marrow · Fibroblasts

1.1 MSCs as the Basis of the Muse Cell Discovery

Stem cells are defined as cells that are able to self-renew and differentiate [1]. According to their range of differentiation, stem cells are further categorized into totipotent stem cells, represented by fertilized eggs, which generate both embryonic

M. Dezawa (✉)
Department of Stem Cell Biology and Histology, Tohoku University Graduate School of Medicine, Sendai, Japan
e-mail: mdezawa@med.tohoku.ac.jp

M. Dezawa (ed.), *Muse Cells*, Advances in Experimental Medicine and Biology 1103, https://doi.org/10.1007/978-4-431-56847-6_1

and extraembryonic cells; pluripotent stem cells, represented by the blastocyst inner cell mass that generates cells of all three germ layers; multipotent stem cells that generate multiple cell types belonging to one or two germ layers, as seen in mesenchymal stem cells (MSCs); and unipotent stem cells, such as muscle stem cells that generate only one type of cell [1]. While each stem cell type has its advantages, pluripotent stem cells have been considered a gold standard of regenerative medicine for their ability to generate various cell types in the body, implying their applicability to a variety of diseases.

The dogma that tissue stem cells, such as neural stem cells (NSCs) and hematopoietic stem cells (HSCs), differentiate into cell types that comprise the tissue in which they reside [2, 3] and, unlike pluripotent stem cells, do not cross the boundaries between the three germ layers was challenged by studies of MSCs. MSCs are cells harvested from mesenchymal tissues such as the bone marrow (BM), adipose tissue, umbilical cord, and dental pulp as adherent crude cell populations whose morphology is similar to that of fibroblasts [4, 5]. The transdifferentiation abilities of BM-MSCs were first demonstrated by their differentiation into osteocytes, chondrocytes, and adipocytes in the presence of various cytokines and reagents [6]. This three-differentiation set is still frequently used to confirm the characteristics of harvested cells as MSCs. The three differentiation types all belong to the mesodermal lineage, but MSCs are also reported to differentiate into other cell types, mostly by gene introduction and cytokine induction and not spontaneously in vitro, for example, endothelial cells [7], cardiac muscle cells [8], skeletal muscle cells [9], hepatocytes [10], neuronal cells [11], epithelial cells [12], peripheral glial cells [13], and insulin-producing cells [14]. A series of reports suggested that MSCs are pluripotent as they are able to cross the boundaries between their mesodermal origin to ectodermal or endodermal cells. The percent of differentiated cells, however, is generally low, suggesting that only a small subpopulation of MSCs differentiate across germline boundaries [4, 5]. In connection with in vitro differentiation, in vivo differentiation of MSCs was also studied. Most MSCs, however, do not demonstrate efficient engraftment into damaged tissue. A very small number of administered MSCs, if any, are suggested to integrate into the damaged tissue and express the appropriate tissue-specific markers [15–17]. Like the in vitro experiments, the results of these in vivo experiments suggest that only a small subpopulation of MSCs have the ability to engraft into the tissue and to differentiate into tissue-specific marker-positive cells.

The above studies were conducted with crude MSC populations and not a purified cell population with distinct properties. MSCs are generally positive for the mesenchymal markers CD105, CD73, and CD90, but there is no exclusive unique marker that definitively characterizes MSCs [18]. Indeed, they are usually obtained from mesenchymal tissues by collecting adherent cells and not by collecting cells positive for a particular marker [6]. Because MSCs comprise a crude population, however, they must contain a subpopulation of cells responsible for their pluripotent-like characteristics.

MSCs are pleiotropic. These cells have been applied to a number of clinical trials targeting various diseases all over the world [19–21]. The majority of reports suggest that the beneficial effect of MSCs is delivered mainly by their paracrine and immunomodulatory effects and not by cell replacement based on transdifferentiation [22, 23]. Identification of the subpopulation of cells among MSCs that are truly pluripotent, however, would greatly benefit regenerative medicine as MSCs are established to be safe and non-tumorigenic and the ability to select out this subpopulation would allow for a more targeted treatment approach. Therefore, we began studies to address how to identify and purify these hypothetical pluripotent stem cells from among MSCs.

1.2 How Wine Contributed to the Advancement of Science

In 2003, I was working at Kyoto University, trying to establish an induction system to generate dopaminergic neurons, glial cells, and skeletal muscle cells from MSCs by gene introduction and cytokine stimulation. As mentioned earlier, MSCs are reported to transdifferentiate into various kinds of cells, and indeed, dopaminergic neurons, Schwann cells, and skeletal muscle cells were generated from MSCs by a combination of Notch gene introduction and cytokine cocktail treatment [11, 24, 25]. The induction efficiency was generally low, however, and I had a feeling that not all of the MSCs were participating in the differentiation and that only a subpopulation of the MSCs was truly responsible for the transdifferentiation I had observed.

One day in 2003, a member of my technical staff asked me whether she should discard some cultured adult rat BM-MSCs that had not yet been used in any experiments. She explained to me that several strange cell clusters had been generated by the naïve BM-MSCs cultured in a 10-cm dish and thus the cells may not be appropriate for the subsequent planned experiments. I observed the cells under a phase-contrast microscope and found that although the cells were cultured in basic culture media composed of alpha-minimum essential medium supplied with 10% fetal bovine serum, the BM-MSCs spontaneously formed cell clusters that appeared very similar to embryoid bodies formed from embryonic stem (ES) cells in suspension culture. The clusters contained pigmented-like cells and hairlike structures, reminiscent of ES cell-derived embryoid bodies (Fig. 1.1a) [26]. I asked her to not discard the BM-MSCs and to notify me if she observed such clusters again. At the same time, I collected the clusters and examined them by immunocytochemistry. Interestingly, I found cells positive for markers of each of the three primary germ layers: ectodermal (GFAP), endodermal (cytokeratin 7), and mesodermal (smooth muscle actin), suggesting the presence of pluripotent-like cells in the clusters (Fig. 1.1b).

Although the appearance was similar, the clusters differed from ES-derived embryoid bodies. The decisive difference was the proliferation speed. The proliferation speed of the MSC-derived clusters slowed down when the clusters reached a certain size and contained pigment-like cells and hairlike structures. ES-derived embryoid bodies, however, proliferate exponentially because they are tumorigenic [27]. In addition, nobody in our lab handled ES cells, and thus contamination of the cultures with ES cells was highly unlikely. While the frequency of the cluster

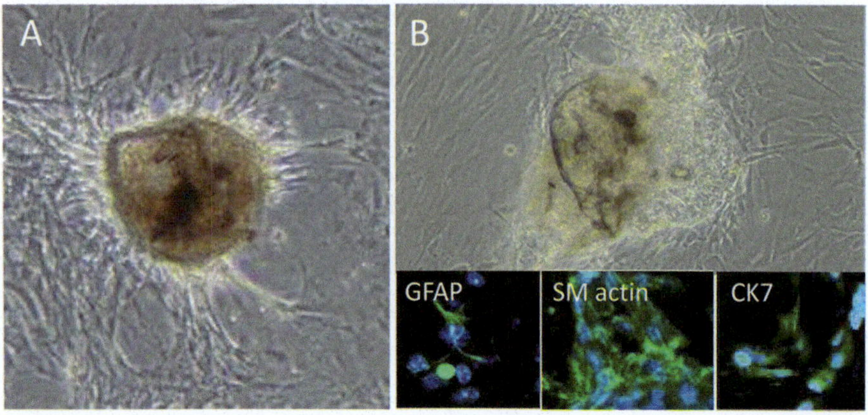

Fig. 1.1 Clusters spontaneously generated in BM-MSCs. Naturally emerged cell clusters similar to ES cell-derived embryoid body in rat BM-MSCs (**A**) and human BM-MSCs (**B**). The cluster in (**B**) contained cells positive for ectodermal (glial fibrillar acidic protein, GFAP), mesodermal (smooth muscle actin, SM actin), and endodermal (cytokeratin7, CK7) lineage cells. (Picture (**B**) was adopted with permission from Ref. [26])

appearance was low in naïve BM-MSCs, the observed characteristics of the clusters suggested that BM-MSCs indeed contained a subpopulation of non-tumorigenic pluripotent-like stem cells. If we could find a method to identify and purify those cells, it would be a great advantage for regenerative medicine.

We began to work on identifying these hypothetical pluripotent-like cells among BM-MSCs. Naïve human and mouse BM-MSCs were also found to form similar clusters with a low frequency. Next, mouse BM-MSC-derived clusters were dispersed into single cells and injected into blastocysts to examine whether or not they formed chimera. Possibly because the proliferation speed of the clusters was slower than that of ES cells, chimera formation was not observed at that time. After struggling for 4 years to identify this hypothetical group of cells, one day in the summer of 2007, I made a fortuitous mistake.

In the afternoon, I was performing a subculture. Skeletal muscle cells generated from MSCs by gene introduction and cytokine treatment were expanded by subculture for a transplantation experiment [24]. The cells were detached from the culture dish by trypsin treatment, collected by centrifugation, and about to be replated for expansion. Then, I received a phone call from my collaborator, Professor Fujiyoshi at the School of Science. He was calling to invite colleagues to a wine party that evening (Fig. 1.2). It was a hot summer afternoon, and a wine party sounded very appealing to me. I hurried to finish my work so that I could get to the wine party, and because of this, I made a simple mistake that I didn't notice until the next day.

The next morning, I was in the culture facility to check the subcultured cells. Immediately upon looking at the culture dishes, I noticed a change in the color of the culture media. The medium, which is usually light pink, had turned yellow. When the culture dishes were inspected under phase-contrast microscopy, we found

Fig. 1.2 Drink party that gave me the chance to discover Muse cells. The drinking companion, Prof. Fujiyoshi (School of Science, Kyoto Univ), Prof. Nabeshima (School of Medicine, Kyoto Univ), Ms. Tanaka (in Fujiyoshi's lab), and Mari Dezawa at Japanese restaurant Kikusui, Kyoto. The member often had drink parties in downtown Kyoto. Prof. Fujiyoshi is the person who called me to come to the wine party while I was doing subculture in 2007, summer

that all of the cells had disappeared. I was upset and couldn't understand what had happened and just blankly looked at the culture dishes that contained no cells. I carefully recalled my memory to reconstruct what I had done before leaving the lab the day before. I realized that the medium I supplied after the subculture was not the culture medium but rather the protease trypsin. Because I had rushed, I neglected to write the name of each solution on the tube, and because both solutions were light pink, I had confused trypsin for the culture medium.

Whenever I have a failed experiment, I have a habit of closely observing the sample before disposal. At that time, I observed the empty dishes for more than 30 min and then noticed that something was floating on the surface of the medium. I adjusted the microscope focus to the surface and found that a small number of cells were still alive! It was amazing that those cells were able to survive the harsh environment of the trypsin solution for more than 16 h. Trypsin solution, which contains only trypsin and HEPES buffer, contains no nutrients.

I collected the living cells by centrifugation. Skeletal muscle cells generated from MSCs comprise three types of cells: Pax7+ muscle stem cells, MyoD+ myoblasts, and terminally differentiated myogenin+ multinuclear myotubes [24]. The majority of the collected cells were Pax7+ stem cells and were not myoblasts nor myotubes. I immediately called Professor Fujiyoshi to explain what happened, and he said this mistake could be a lucky clue to identifying the hypothetical pluripotent-like cells from among BM-MSCs.

Stem cells are generally stress-tolerant, particularly in the case of tissue stem cells, and are activated from a dormant state by a stimulus such as stress to enter into the cell cycle [28]. We attempted to reproduce the mistake (but without drinking the wine this time). BM-MSCs were placed under trypsin solution for 16 h. If hypothetical pluripotent-like cells exist among MSCs, those cells would be considered true "stem cells" and should thus be tolerant to stress compared with other cells among BM-MSCs. After trypsin incubation for 16 h, the majority of the cells had died, but a small number of cells survived. The cells that remained after the 16-h trypsin incubation were subjected to single-cell suspension culture, a method occasionally used in stem cell research, and cultured for 7~10 days. Single cells began to proliferate and formed clusters similar to the embryoid bodies generated from ES cells in suspension culture (Fig. 1.3a, b) [26]. This was reminiscent of the event I encountered in 2003, the spontaneous embryoid body-like cluster formation observed in cultured BM-MSCs.

Thanks to the mistake I made by rushing to get to the wine party, I was convinced that we now had a clue to the pluripotent-like cell population in BM-MSCs. We were very careful however to identify a rational marker that would allow us to distinguish and characterize the cells. We could publish the findings based on stress treatment at that time, but we could easily foresee unstable reproducibility of the cells if we reported that the cells could only be isolated by long-term trypsin treatment. As a matter of fact, enzyme activity occasionally differs among companies and batch numbers, and tiny issues may have large effects on outcomes. For scientific reproducibility, there was a strong need to find a reasonable robust marker.

The approach was simple. We compared the cell surface markers of BM-MSCs before and after long-term trypsin treatment. The signal for many of the markers became nearly zero after treatment, but one marker, stage-specific embryonic antigen (SSEA)-3, attracted our attention. Cells positive for SSEA-3 from among

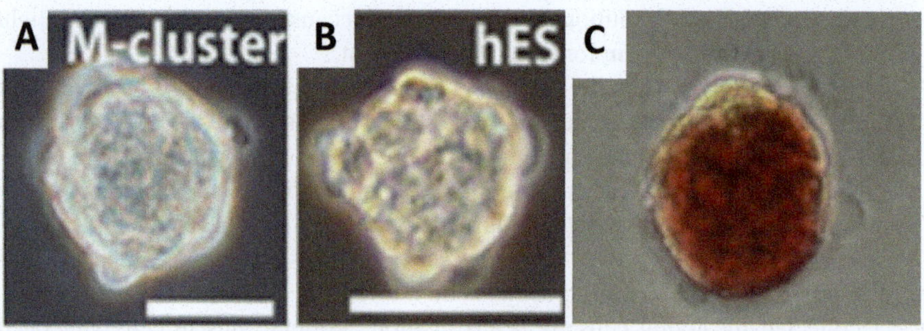

Fig. 1.3 Muse cell-derived clusters. Muse cell-derived cluster (**A**. M-cluster) and ES cell-derived cluster (**B**) formed in suspension culture. Bars = 30 μm. (**C**) Alkaline phosphatase reaction of M-cluster. (Pictures were adopted with permission from Refs. [37] and [26])

BM-MSCs were originally very low, around one to several percent, but the number became substantially higher after the trypsin treatment [26].

SSEA-3 was first reported by Shevinsky et al. in 1982 as a monoclonal antibody that recognizes globo-series sugar chains Galβ1-3GalNAcβ1-3Galα1-4Galβ1-4Glcβ1 of glycolipids and glycoproteins on the cell membrane of mouse developmentally pluripotent state cells, namely, zygote, 2~8 cell stage, morula, and epiblast stem cells (Fig. 1.4) [29]. In humans, the inner cell mass is positive for SSEA-3 [30]. In addition to these naturally existing pluripotent cells, SSEA-3 also recognizes the surface of human pluripotent stem cells, NTERA (teratoma cell line), ES cells, embryonic carcinoma cells, embryonic germ cells, and induced pluripotent stem cells [29, 31]. The function of the epitope that is recognized by SSEA-3 antibody is not yet clarified, but SSEA-3 is considered a reliable surface marker for pluripotent stem cells.

When SSEA-3+ cells were isolated from BM-MSCs and cultured in single-cell suspension, the cells successfully generated clusters similar to the embryoid bodies of ES cells, as seen in 16 h of trypsin treatment, and those clusters were positive for alkaline phosphatase reaction (Fig. 1.3c), generated cells representative of all three germ layers, and were self-renewable (Fig. 1.5), suggesting their pluripotency [26]. Professor Fujiyoshi has a talent for naming phenomena – he named the cells "multilineage-differentiating stress-enduring" (Muse) cells with a strong consideration for their stress tolerability, which provided us the opportunity to observe the cells. Currently, SSEA-3 is routinely used to isolate Muse cells from various tissue sources [32–36].

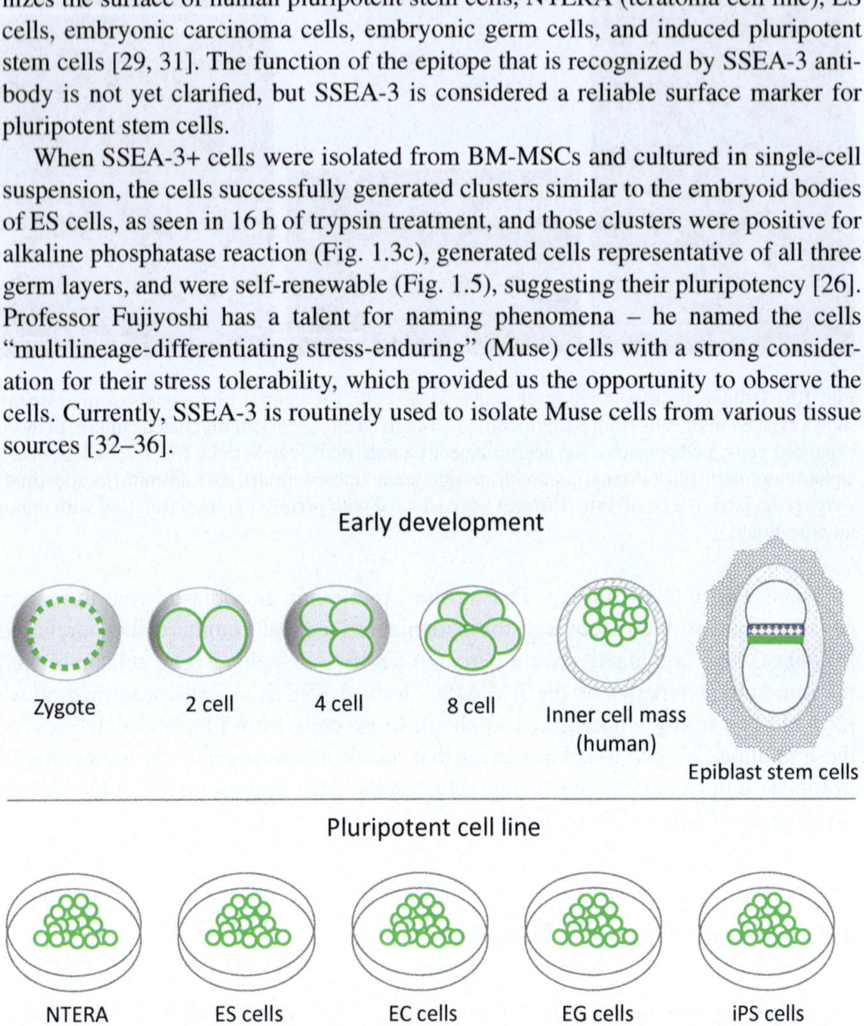

Fig. 1.4 SSEA-3 expression in pluripotent stem cells. SSEA-3 is expressed in the pluripotent stage from zygote to epiblast stem cells. SSEA-3 is also the surface marker for pluripotent cell lines, NTERA (teratoma cell line), ES cells, embryonic carcinoma cells, embryonic germ cells, and induced pluripotent stem (iPS) cells

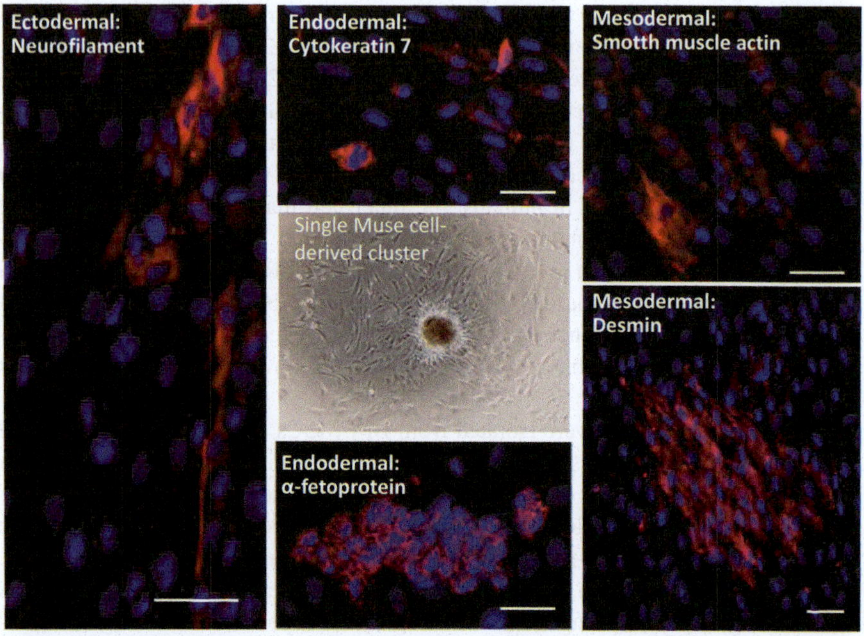

Fig. 1.5 Trilineage differentiation of single Muse cell. The central phase-contrast microscopy shows expansion of cells from a single Muse cell-derived cluster in gelatin-coated culture. In those expanded cells, cells positive for neurofilament (ectodermal), cytokeratin 7 (CK7, endodermal), alpha-fetoprotein (endodermal), smooth muscle actin (mesodermal), and desmin (mesodermal) were recognized. Bars = 50 μm. (Pictures were adopted with permission from Ref. [26] with minor modifications)

There was another finding. During the experiment, fibroblasts were used as a negative control. We discovered the undeniable fact that commercially purchased fibroblasts also contained several percent of SSEA-3+ cells. Those cells exhibited the same characteristics as the BM-MSC-derived SSEA-3+ cells described above [26]. No one in my lab expected to obtain Muse cells from fibroblasts. In view of these findings, we had to acknowledge that one of the most generally used cultured fibroblast types contains pluripotent Muse cells. This finding hinted at the unique distribution of Muse cells in vivo.

1.3 Lessons from the Discovery

Looking back at what happened during the discovery of Muse cells, I learned several important lessons from a series of experiences. As in life, discovery sometimes occurs serendipitously. Therefore, failures and mistakes should be cherished and considered as great chances. Second, humans categorize events as successes or failures, but from the standpoint of nature, there is no success or failure; there are only natural phenomena. By looking at all phenomena from a neutral point of view, we may encounter more opportunities for serendipitous discoveries.

References

1. Lovell-Badge R (2001) The future for stem cell research. Nature 414:88–91
2. Gage FH (2000) Mammalian neural stem cells. Science 287:1433–1438
3. Weissman IL, Shizuru JA (2008) The origins of the identification and isolation of hematopoietic stem cells, and their capability to induce donor-specific transplantation tolerance and treat autoimmune diseases. Blood 112:3543–3553
4. Kuroda Y, Kitada M, Wakao S, Dezawa M (2011) Bone marrow mesenchymal cells: how do they contribute to tissue repair and are they really stem cells? Arch Immunol Ther Exp 59:369–378
5. Samsonraj RM, Raghunath M, Nurcombe V, Hui JH, van Wijnen AJ, Cool SM (2017) Concise review: multifaceted characterization of human mesenchymal stem cells for use in regenerative medicine. Stem Cells Transl Med 6:2173–2185
6. Pittenger MF, Mackay AM, Beck SC, Jaiswal RK, Douglas R, Mosca JD, Moorman MA, Simonetti DW, Craig S, Marshak DR (1999) Multilineage potential of adult human mesenchymal stem cells. Science 284:143–147
7. Oswald J, Boxberger S, Jorgensen B, Feldmann S, Ehninger G, Bornhauser M, Werner C (2004) Mesenchymal stem cells can be differentiated into endothelial cells in vitro. Stem Cells 22:377–384
8. Makino S, Fukuda K, Miyoshi S, Konishi F, Kodama H, Pan J, Sano M, Takahashi T, Hori S, Abe H, Hata J, Umezawa A, Ogawa S (1999) Cardiomyocytes can be generated from marrow stromal cells in vitro. J Clin Invest 103:697–705
9. Dezawa M, Ishikawa H, Hoshino M, Itokazu Y, Nabeshima Y (2005) Potential of bone marrow stromal cells in applications for neuro-degenerative, neuro-traumatic and muscle degenerative diseases. Curr Neuropharmacol 3:257–266
10. Oyagi S, Hirose M, Kojima M, Okuyama M, Kawase M, Nakamura T, Ohgushi H, Yagi K (2006) Therapeutic effect of transplanting hgf-treated bone marrow mesenchymal cells into ccl4-injured rats. J Hepatol 44:742–748
11. Dezawa M, Kanno H, Hoshino M, Cho H, Matsumoto N, Itokazu Y, Tajima N, Yamada H, Sawada H, Ishikawa H, Mimura T, Kitada M, Suzuki Y, Ide C (2004) Specific induction of neuronal cells from bone marrow stromal cells and application for autologous transplantation. J Clin Invest 113:1701–1710
12. Spees JL, Olson SD, Ylostalo J, Lynch PJ, Smith J, Perry A, Peister A, Wang MY, Prockop DJ (2003) Differentiation, cell fusion, and nuclear fusion during ex vivo repair of epithelium by human adult stem cells from bone marrow stroma. Proc Natl Acad Sci U S A 100:2397–2402
13. Dezawa M, Takahashi I, Esaki M, Takano M, Sawada H (2001) Sciatic nerve regeneration in rats induced by transplantation of in vitro differentiated bone-marrow stromal cells. Eur J Neurosci 14:1771–1776
14. Phinney DG, Prockop DJ (2007) Concise review: mesenchymal stem/multipotent stromal cells: the state of transdifferentiation and modes of tissue repair–current views. Stem Cells 25:2896–2902
15. Petersen BE, Bowen WC, Patrene KD, Mars WM, Sullivan AK, Murase N, Boggs SS, Greenberger JS, Goff JP (1999) Bone marrow as a potential source of hepatic oval cells. Science 284:1168–1170
16. Lu C, Arai M, Misao Y, Chen X, Wang N, Onogi H, Kobayashi H, Uno Y, Takemura G, Minatoguchi S, Fujiwara T, Fujiwara H (2006) Autologous bone marrow cell transplanta-

tion improves left ventricular function in rabbit hearts with cardiomyopathy via myocardial regeneration-unrelated mechanisms. Heart Vessel 21:180–187
17. Tamai K, Yamazaki T, Chino T, Ishii M, Otsuru S, Kikuchi Y, Iinuma S, Saga K, Nimura K, Shimbo T, Umegaki N, Katayama I, Miyazaki J, Takeda J, McGrath JA, Uitto J, Kaneda Y (2011) Pdgfralpha-positive cells in bone marrow are mobilized by high mobility group box 1 (hmgb1) to regenerate injured epithelia. Proc Natl Acad Sci U S A 108:6609–6614
18. Horwitz EM, Le Blanc K, Dominici M, Mueller I, Slaper-Cortenbach I, Marini FC, Deans RJ, Krause DS, Keating A, International Society for Cellular T (2005) Clarification of the nomenclature for msc: The international society for cellular therapy position statement. Cytotherapy 7:393–395
19. George JC (2010) Stem cell therapy in acute myocardial infarction: a review of clinical trials. Transl Res 155:10–19
20. Fisher SA, Zhang H, Doree C, Mathur A, Martin-Rendon E (2015) Stem cell treatment for acute myocardial infarction. Cochrane Database Syst Rev 9:CD006536
21. Bedini G, Bersano A, Zanier ER, Pischiutta F, Parati EA (2018) Mesenchymal stem cell therapy in intracerebral haemorrhagic stroke. Curr Med Chem 25:2176
22. Castro-Manrreza ME, Montesinos JJ (2015) Immunoregulation by mesenchymal stem cells: biological aspects and clinical applications. J Immunol Res 2015:394917
23. Hofer HR, Tuan RS (2016) Secreted trophic factors of mesenchymal stem cells support neurovascular and musculoskeletal therapies. Stem Cell Res Ther 7:131
24. Dezawa M, Ishikawa H, Itokazu Y, Yoshihara T, Hoshino M, Takeda S, Ide C, Nabeshima Y (2005) Bone marrow stromal cells generate muscle cells and repair muscle degeneration. Science 309:314–317
25. Kamada T, Koda M, Dezawa M, Yoshinaga K, Hashimoto M, Koshizuka S, Nishio Y, Moriya H, Yamazaki M (2005) Transplantation of bone marrow stromal cell-derived schwann cells promotes axonal regeneration and functional recovery after complete transection of adult rat spinal cord. J Neuropathol Exp Neurol 64:37–45
26. Kuroda Y, Kitada M, Wakao S, Nishikawa K, Tanimura Y, Makinoshima H, Goda M, Akashi H, Inutsuka A, Niwa A, Shigemoto T, Nabeshima Y, Nakahata T, Nabeshima Y, Fujiyoshi Y, Dezawa M (2010) Unique multipotent cells in adult human mesenchymal cell populations. Proc Natl Acad Sci U S A 107:8639–8643
27. Amit M, Carpenter MK, Inokuma MS, Chiu CP, Harris CP, Waknitz MA, Itskovitz-Eldor J, Thomson JA (2000) Clonally derived human embryonic stem cell lines maintain pluripotency and proliferative potential for prolonged periods of culture. Dev Biol 227:271–278
28. Prinsloo E, Setati MM, Longshaw VM, Blatch GL (2009) Chaperoning stem cells: a role for heat shock proteins in the modulation of stem cell self-renewal and differentiation? BioEssays 31:370–377
29. Shevinsky LH, Knowles BB, Damjanov I, Solter D (1982) Monoclonal antibody to murine embryos defines a stage-specific embryonic antigen expressed on mouse embryos and human teratocarcinoma cells. Cell 30:697–705
30. Henderson JK, Draper JS, Baillie HS, Fishel S, Thomson JA, Moore H, Andrews PW (2002) Preimplantation human embryos and embryonic stem cells show comparable expression of stage-specific embryonic antigens. Stem Cells 20:329–337
31. Kannagi R, Cochran NA, Ishigami F, Hakomori S, Andrews PW, Knowles BB, Solter D (1983) Stage-specific embryonic antigens (ssea-3 and -4) are epitopes of a unique globo-series ganglioside isolated from human teratocarcinoma cells. EMBO J 2:2355–2361
32. Kuroda Y, Wakao S, Kitada M, Murakami T, Nojima M, Dezawa M (2013) Isolation, culture and evaluation of multilineage-differentiating stress-enduring (muse) cells. Nat Protoc 8:1391–1415
33. Ogura F, Wakao S, Kuroda Y, Tsuchiyama K, Bagheri M, Heneidi S, Chazenbalk G, Aiba S, Dezawa M (2014) Human adipose tissue possesses a unique population of pluripotent stem cells with nontumorigenic and low telomerase activities: potential implications in regenerative medicine. Stem Cells Dev 23:717–728

34. Kinoshita K, Kuno S, Ishimine H, Aoi N, Mineda K, Kato H, Doi K, Kanayama K, Feng J, Mashiko T, Kurisaki A, Yoshimura K (2015) Therapeutic potential of adipose-derived ssea-3-positive muse cells for treating diabetic skin ulcers. Stem Cells Transl Med 4:146–155
35. Liu Q, Zhang RZ, Li D, Cheng S, Yang YH, Tian T, Pan XR (2016) Muse cells, a new type of pluripotent stem cell derived from human fibroblasts. Cell Reprogram 18:67–77
36. Yamauchi T, Yamasaki K, Tsuchiyama K, Koike S, Aiba S (2017) The potential of muse cells for regenerative medicine of skin: procedures to reconstitute skin with muse cell-derived keratinocytes, fibroblasts, and melanocytes. J Invest Dermatol 137:2639–2642
37. Wakao S, Kitada M, Kuroda Y, Shigemoto T, Matsuse D, Akashi H, Tanimura Y, Tsuchiyama K, Kikuchi T, Goda M, Nakahata T, Fujiyoshi Y, Dezawa M (2011) Multilineage-differentiating stress-enduring (muse) cells are a primary source of induced pluripotent stem cells in human fibroblasts. Proc Natl Acad Sci U S A 108:9875–9880

Chapter 2
Basic Characteristics of Muse Cells

Shohei Wakao, Yoshihiro Kushida, and Mari Dezawa

Abstract Multilineage-differentiating stress-enduring (Muse) cells exhibit the core characteristics of pluripotent stem cells, namely, the expression of pluripotency markers and the capacity for trilineage differentiation both in vitro and in vivo and self-renewability. In addition, Muse cells have unique characteristics not observed in other pluripotent stem cells such as embryonic stem cells, control of pluripotency by environmental switch of adherent suspension, symmetric and asymmetric cell division, expression of factors relevant to stress tolerance, and distinctive tissue distribution. Pluripotent stem cells were recently classified into two discrete states, naïve and primed. These two states have multiple functional differences, including their proliferation rate, molecular properties, and growth factor dependency. The properties exhibited by Muse cells are similar to those of primed pluripotent stem cells while with some uniqueness. In this chapter, we provide a comprehensive description of the basic characteristics of Muse cells.

Keywords Trilineage differentiation · Self-renewal · SSEA-3 · Connective tissue · Peripheral blood · Bone marrow · Primed pluripotent · DNA repair · Suspension · Cluster

2.1 The Characteristics of Muse Cells

The basic characteristics of Muse cells are summarized here.

1. The ability to generate cells of all three germ layers and self-renew
2. Express pluripotency markers; such as Nanog, Oct3/4, Sox2 and SSEA-3
3. Non-tumorigenic and consistent with the fact that they reside in the body

---

The original version of this chapter was revised: typographical errors were corrected throughout the chapter. The correction to this chapter can be found at https://doi.org/10.1007/978-4-431-56847-6_19

S. Wakao · Y. Kushida · M. Dezawa (✉)
Department of Stem Cell Biology and Histology, Tohoku University Graduate School of Medicine, Sendai, Japan
e-mail: mdezawa@med.tohoku.ac.jp

M. Dezawa (ed.), *Muse Cells*, Advances in Experimental Medicine and Biology 1103, https://doi.org/10.1007/978-4-431-56847-6_2

4. Stress tolerant and have a robust effective capacity of sensing and repairing DNA damage
5. Distinct from other known somatic stem cells

Pluripotent stem cells have trilineage differentiation and self-renewal abilities [1]. These abilities are especially robust when observed at the single cell level. Muse cells have both abilities at the single cell level, and their potential for differentiating into ectodermal-, endodermal-, and mesodermal-lineage cells was demonstrated in vitro (Fig. 2.1).

Muse cells from the bone marrow (BM), adipose tissue, and dermal fibroblasts form single cell-derived clusters in single cell suspension culture that have a morphology reminiscent of embryonic stem (ES) cell-derived embryoid bodies, as described in Chap. 1 (please see Fig. 1.3a, b in Chap. 1) [2–5]. These clusters are positive for the alkaline phosphatase reaction, a known characteristic of ES cells (please see Fig. 1.3c in Chap. 1 [2, 3]. When the clusters are transferred to a gelatin-coated adherent culture, they adhere to the dish, and cells expand out of the cluster. The expanded cells include cells positive for endodermal (GATA-6, cytokeratin 7,

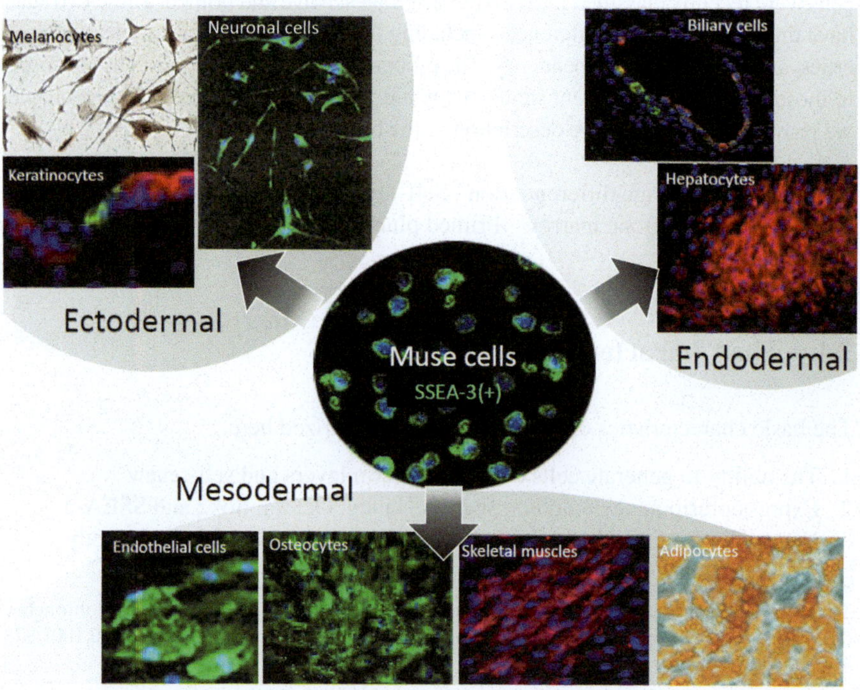

Fig. 2.1 SSEA-3(+) Muse cells differentiate into various kinds of ectodermal-, endodermal- and mesodermal-lineage cells in vitro and in vivo, either spontaneously or by cytokine induction. (Pictures for neuronal cells, hepatocytes, and skeletal muscles were adopted with permission from Ref. [2–4])

alpha-fetoprotein), mesodermal (Nkx2.5, smooth muscle actin, desmin), and ectodermal (MAP-2, neurofilament) markers, suggesting that single Muse cells are able to spontaneously generate trilineage cells (Fig. 1.5 in Chap. 1 and Fig. 2.2 in this chapter) [2, 6]. Because the cells spontaneously differentiate into cells of the three lineages on a gelatin-coated adherent culture and neither based on gene introduction nor cytokine stimulation, the proportion of cells positive for each marker is not very high – several percent are positive for endodermal and ectodermal markers, and 10~15% are positive for mesodermal markers [2].

The trilineage differentiation ability is self-renewable (Fig. 2.2); when first-generation single cell-derived clusters are transferred onto general uncoated adherent culture dishes, the clusters adhere to the dishes, and cells expand out of the cluster. These expanded cells form second-generation single cell-derived clusters that appear similar to ES cell-derived embryoid bodies when transferred to a single cell suspension culture. These second-generation clusters are positive for the alkaline phosphatase reaction, and cells expanded from the clusters differentiate into the three lineages after transfer onto gelatin-coated culture dishes. This cycle can be

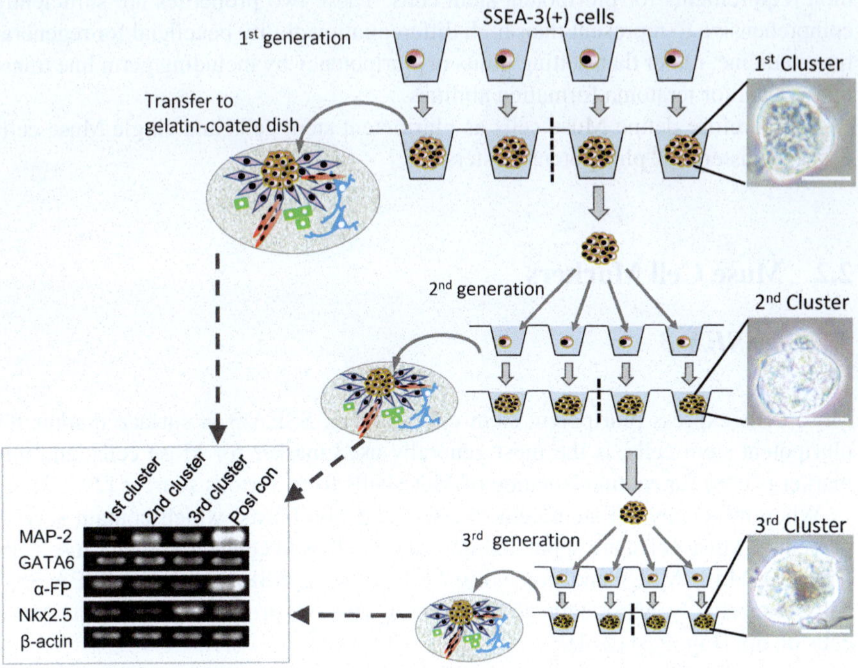

Fig. 2.2 Self-renewal of trilineage differentiation ability in Muse cells. RT-PCR data are from adipose-Muse cells. MAP-2 (ectodermal), GATA-6 (endodermal), α-FP (endodermal), and Nkx2.5 (mesodermal) gene expressions were detected in RT-PCR from cells expanded from each of clusters from first to third generations. Positive controls for MAP-2, α-FP, and Nkx2.5 were human whole embryos and for GATA-6 was a human fetus liver. Scale bar = 25 μm. (Pictures were adopted with permission from Ref. [3] with minor modifications)

repeated for up to five generations, demonstrating that single Muse cell has self-renewability in trilineage differentiation [2, 3].

Muse cells comprise ~0.03% of BM mononucleated cells. When Muse cells are directly collected from adult human BM aspirates as SSEA-3(+) cells, they express Nanog, Oct3/4, and Sox2; form alkaline phosphatase (+) clusters in suspension; and exhibit trilineage differentiation. These characteristics are repeatable in second-generation clusters, indicating that these characteristics are not newly acquired by in vitro manipulation nor are they modified under culture conditions [2, 7].

Pluripotent cells are defined as cells having the ability to generate trilineage cells [8]. In the case of pluripotent stem cells, the "stem cell" concept applies not only to the trilineage differentiation potential but also to the ability to self-renew. Both ES and induced pluripotent stem (iPS) cells show germ line transmission and/or teratoma formation in addition to their ability to generate trilineage cells and self-renew [8, 9]. Epiblast stem cells (EpiSCs), another pluripotent stem cell type, however, do not form teratomas under certain circumstances, nor do they generally participate in germ line transmission [10]. Thus, pluripotent stem cells do not always meet the strict requirements of teratoma formation or germ line transmission, and the self-renewal abilities and differentiation into trilineage cells are more essential and common requirements for pluripotent stem cells. These two properties are sufficiently comprehensive to represent their high differentiation ability beneficial for regenerative medicine, rather than setting limits in pluripotency by including germ line transmission and/or teratoma formation abilities.

We therefore define Muse cells as pluripotent stem cells and single Muse cell-derived clusters as "pluripotent clusters."

2.2 Muse Cell Markers

2.2.1 SSEA-3

Muse cells express pluripotent stem cell markers. SSEA-3, a surface marker for pluripotent stem cells, is the most generally used marker for Muse cells, and this marker is used for routine isolation of Muse cells from various sources [2–4, 7].

When mesenchymal stem cells (MSCs) and fibroblasts, which contain several percent of Muse cells, are separated into Muse cells and cells other than Muse cells (i.e., non-Muse MSCs and non-Muse fibroblasts) as SSEA-3(+) and SSEA-3(−) cells, respectively, Muse cells demonstrate pluripotent properties, while non-Muse cells do not (Fig. 2.3) [2, 4].

Single cell-derived pluripotent cluster formation, trilineage differentiation, self-renewability, and expression of pluripotent markers are all observed in SSEA-3(+)-Muse cells, but not in non-Muse cells. Non-Muse cells basically do not survive or proliferate in single cell suspension (Fig. 2.3). Thus, they are not able to form pluripotent-like clusters. Because single cell-derived pluripotent-like clusters are not formed, trilineage differentiation and self-renewal are not feasible in

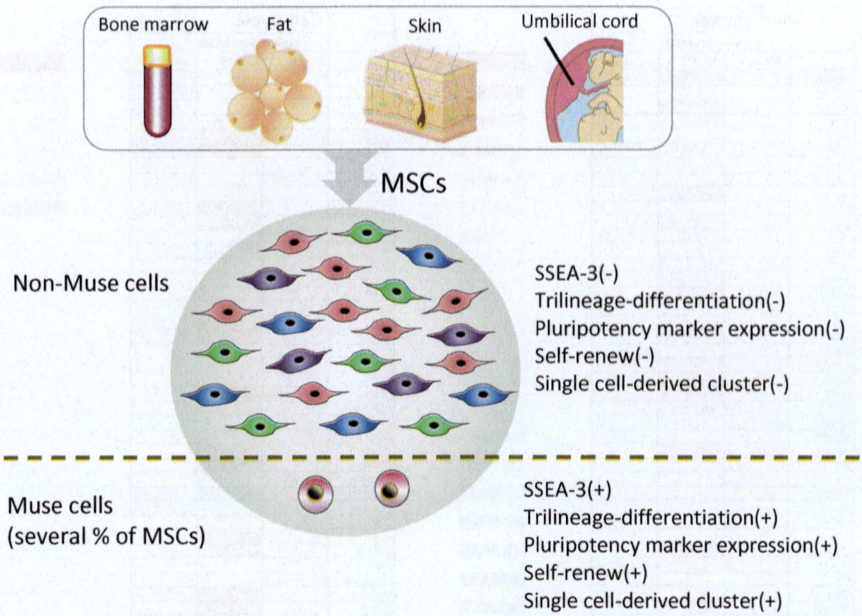

Fig. 2.3 Schematic diagram shows basic differences between Muse and non-Muse cells in MSCs

non-Muse cells. Concomitantly, pluripotency markers are under the detection level or substantially lower in non-Muse cells compared with Muse cells (Fig. 2.4) [2, 4]. Based on the fact that SSEA-3 consistently labels cells at the pluripotent stage in normal development, as well as ES and iPS cells [11, 12], and is clearly the difference in Muse and non-Muse cells separated based on SSEA-3 expression, SSEA-3 is considered a practical marker for isolating Muse cells. The homogeneity of SSEA-3(+) Muse cells, however, is still an open question. The presence of a hierarchy and/or subpopulation of Muse cells is possible because they are naturally existing endogenous cells and not immortalized or clonally expanded tumorigenic cells.

SSEA-3 is an antibody that recognizes a sugar epitope [13, 14], and there is no gene that encodes SSEA-3; thus, SSEA-3 knockout animals are practically difficult to generate. On the other hand, SSEA-3 has an advantage – because SSEA-3 is not encoded by a gene, species differences do not exist, and SSEA-3 can thus be applied to identify Muse cells across species. In fact, Muse cells with pluripotent properties have been identified based on SSEA-3 expression in mouse, rat, rabbit, swine, sheep, and monkey (unpublished data).

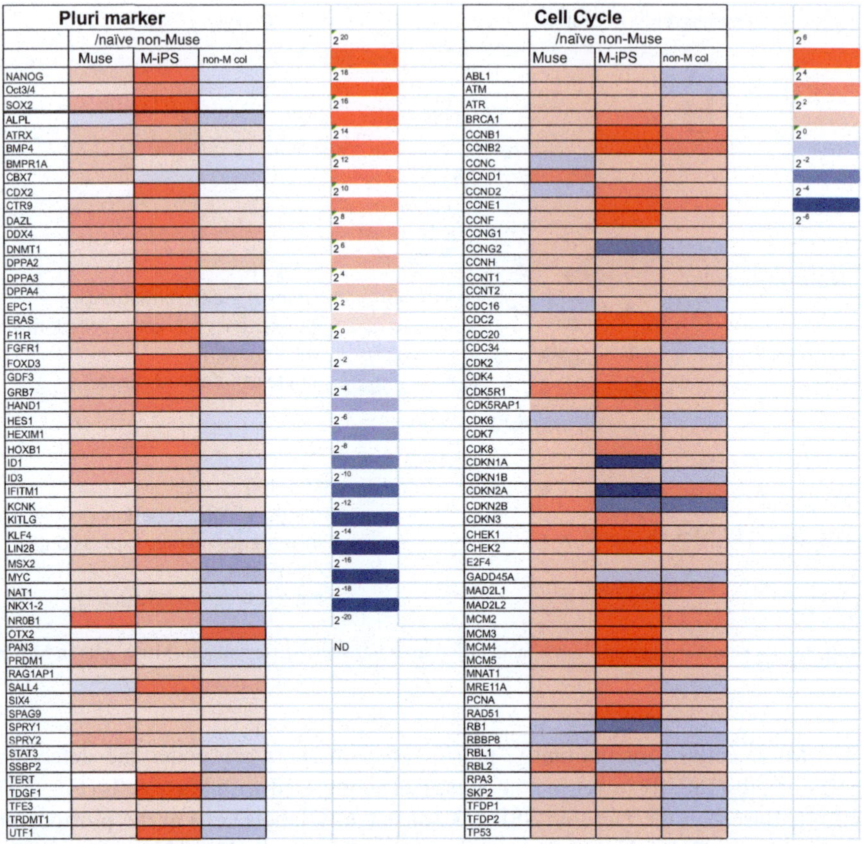

Fig. 2.4 Gene expression pattern in human Muse, iPSCs and non-Muse cells. The expression pattern of pluripotency markers in Muse cells and iPSCs is similar, but the expression level is higher in iPSCs than in Muse cells. However, the expression level of pluripotency genes in Muse cells is generally higher than that in non-Muse cells. Genes related to tumorigenic factors, namely, cell cycle progression in the table, do not largely differ between Muse and non-Muse cells but are highly upregulated in iPSCs. (Pictures adapted and modified with permission from Ref. [4])

2.2.2 *Other Markers*

Other than SSEA-3, Muse cells express higher levels of master pluripotency genes, such as Oct3/4, Sox2, Nanog, Rex1, and PAR4, than non-Muse fibroblasts, one of the representatives of somatic cells [2, 4]. Furthermore, the expression of bone morphogenic protein 4 (BMP4), CBX7, DAZL, DPPA3, DPPA4, FGFR1, GDF3, Hes1, Hoxb1, ID1/3, KLF4, MSX2, Myc, NAT1, Nkx1-2, NR0B1, Prdm1, Six4, SPRY1/2, SSBP2, and UTF1, known to relate to pluripotency/undifferentiation state of stem cells, is recognized in Muse cells [2, 4, 15]. Generally, the expression levels of these markers in Muse cells are not as high as those in ES/iPS cells but are clearly higher than those in non-Muse fibroblasts (Fig. 2.4) [4]. ES and iPS cells are not endogenous, and thus their pluripotency state differs from that of naturally existing endogenous pluripotent cells, such as inner cell mass and epiblast stem cells. Indeed, pluripotent gene expression levels are generally lower in those endogenous pluripotent cells than in ES/iPS cells [16]. For endogenous pluripotent cells, including Muse cells, high expression of pluripotency genes may not be required to maintain their pluripotent state.

Muse cells secrete factors that may play a role in stem cell self-renewal [15]. AKAP13 protein anchors cAMP-dependent protein kinase (PKA) and acts as an adapter protein to selectively couple Gα-13 and Rho. This signaling is reported to sustain the proliferation, growth, and pluripotency of human ES cells [17]. Guanine nucleotide-binding proteins (G proteins) act as transducers in several signaling pathways, including those involved in the activation of adenylyl cyclases. In fact, the guanine nucleotide-binding protein G(s) alpha subunit-PKA pathway has an important role in promoting proper homeostasis for some stem cell types, such as epithelial stem cells in hair follicles, by limiting excessive proliferation that may lead to stem cell exhaustion [18]. The PKA pathway is reported to have a role in the self-renewal and differentiation abilities of mouse ES cells [19]. The retinoid X receptors (RXRs) dimerize with different nuclear receptors, such as the liver X receptor (LXR) or the farnesoid X receptor (FXR) [20]. LXR/RXR and FXR/RXR pathways in Muse cells are suggested to promote self-renewal as this system relates to the maintenance of multipotency in several stem cell types [21–23].

2.3 Adherent Suspension: The Toggle Switch of Pluripotency in Muse Cells

Muse cells survive and proliferate both in suspension and adherent environments. Generally, cell survival and activity largely depends on the environment that surrounds the cell, and whether it is a suspension or adherent system is one of the factors that affects cellular activities and behaviors. For example, hematopoietic cells survive basically in a suspension, namely, peripheral blood, and not in an adherent system except in certain situations, such as rolling of lymphocytes in the blood vessels and migration to inflammatory sites [24]. On the other hand, many other cell types, including those of mesenchymal origin, are generally based on an adherent system and do not successfully survive in suspension, particularly in a single cell suspension. Muse cells are unique because they are able to survive and proliferate in both environments, and more importantly, their pluripotency is regulated by an "adherent-suspension switch."

As mentioned above, Muse cells express Oct3/4, Nanog, and Sox2, higher than those in non-Muse cells in adherent cultures [2–4]. Interestingly, when Muse cells are transferred to a suspension culture, the expression of these genes becomes 50 to several hundred times higher than those in adherent culture (Fig. 2.5) [25].

The interesting point related to this phenomenon is the localization of these transcription factors in the cell according to the adherent-suspension changes. Nanog, Oct3/4, and Sox2 distribute in the cytoplasm when Muse cells are in an adherent

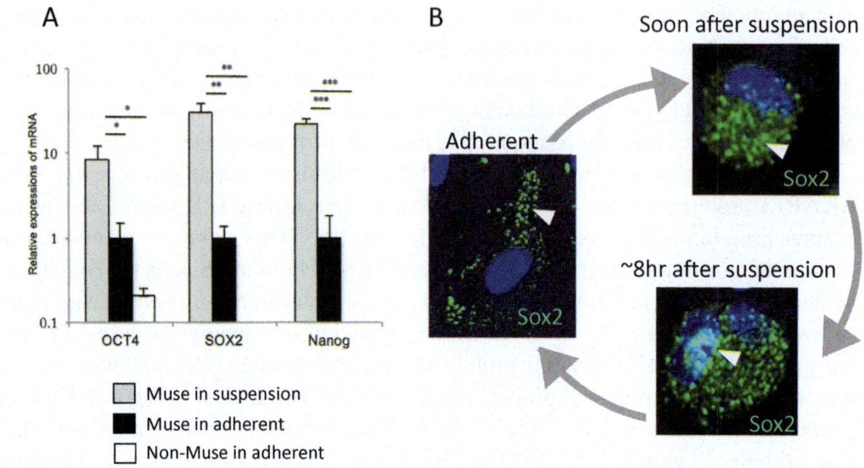

Fig. 2.5 Pluripotency of Muse cells in suspension and adherent. (**A**) Gene expression level of Oct3/4, Sox2, and Nanog in suspension and adherent Muse cells and in adherent non-Muse cells. Muse and non-Muse cells were separated from BM-MSCs (Picture was adopted with permission from Ref. [25]). (**B**) Immunocytochemical localization of Sox2 in Muse cells in adherent, soon after suspension and ~8 h after suspension. White arrowheads are the main location of Sox2 (green)

state. Those factors, however, translocate into the nucleus when the cells are transferred to a suspension (Fig. 2.5) (unpublished data). Considering that these factors are transcription factors, they are not functional when they are located in the cytoplasm. Therefore, those pluripotent factors are on standby and not fully functional in an adherent state and are switched on in a suspension state. Applying this system to the living body, Muse cells are in an adherent state when they are in the BM and the connective tissue of organs, and therefore their pluripotent factors are on standby. When Muse cells are in a suspension state (circulating in the peripheral blood), their pluripotent factors are switched on and highly activated. The molecular mechanism by which adhesion-suspension controls pluripotency gene expression levels requires further investigation.

2.4 Epigenetic Differences Between Muse Cells and Somatic Cells

In an adherent state, the promoter regions of Nanog and Oct3/4 are less methylated in Muse cells than in non-Muse fibroblasts, one of the representatives of somatic cells [4]. This is consistent with the finding that the gene expression levels of Nanog and Oct3/4 are higher in Muse cells than in non-Muse cells. Interestingly, when fibroblasts are separated into Muse and non-Muse cells, and treated with the four Yamanaka factors, only Muse cells generate iPS cells, while non-Muse cells do not

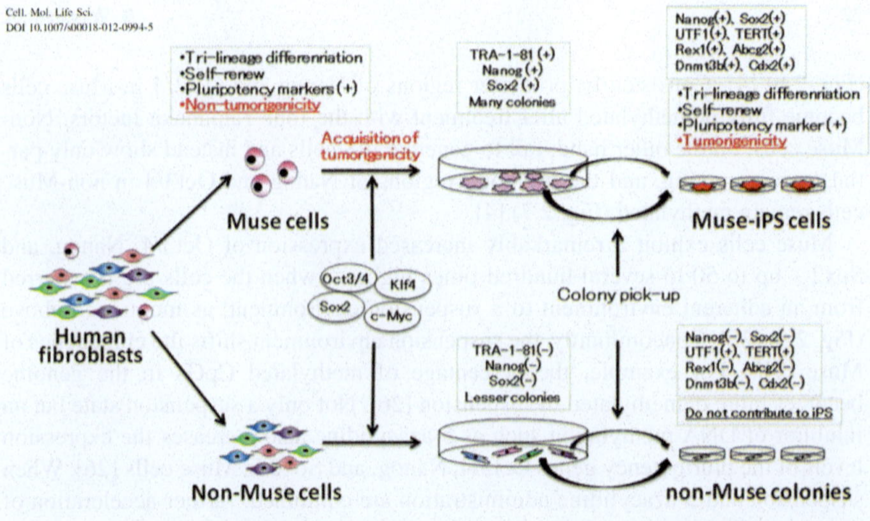

Fig. 2.6 Muse cells are the source of iPS cell generation. Outline of iPS cell generation from Muse and non-Muse cells in human fibroblasts. iPS cells were generated only from Muse cells but not from non-Muse cells. (Picture was adopted with permission from Refs. [75])

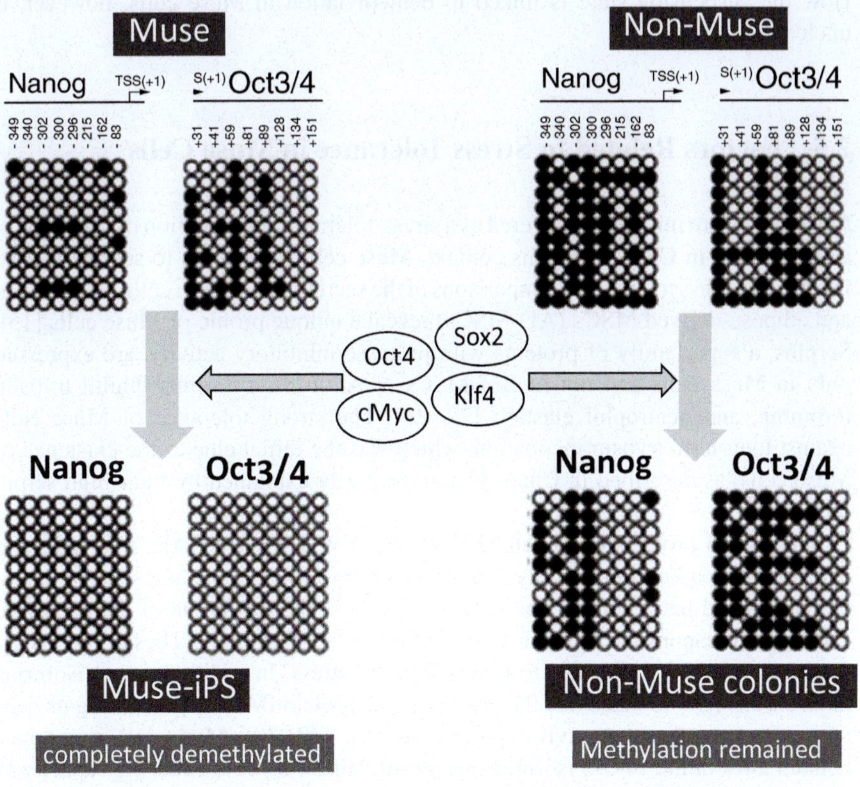

Fig. 2.7 Nanog and Oct3/4 promoters in Muse and non-Muse cells. Methylation of Nanog and Oct3/4 promoters in naïve Muse and non-Muse cells and after receiving Yamanaka four factors. (Pictures adapted and modified with permission from Ref. [4])

(Fig. 2.6) [4]. Consistently, promoter regions of Nanog and Oct3/4 in Muse cells become fully demethylated after treatment with the four Yamanaka factors. Non-Muse cells, on the other hand, fail to generate iPS cells and instead show only partial reprogramming, and the promoter regions of Nanog and Oct3/4 in non-Muse cells remain methylated (Fig. 2.7) [4].

Muse cells exhibit a remarkably increased expression of Oct3/4, Nanog, and Sox2 – up to 50 to several hundred times higher – when the cells are transferred from an adherent environment to a suspension environment, as mentioned above (Fig. 2.5) [25]. Concomitantly, the suspension environment shifts the epigenetics of Muse cells. For example, the percentage of methylated CpGs in the genome becomes more demethylated in suspension [26]. Not only a suspension state but an inhibitor of DNA methylation such as 5-azacytidine also increases the expression levels of the pluripotency genes Oct3/4, Nanog, and Sox2 in Muse cells [26]. When suspension and 5-azacytidine administration are combined, further acceleration of the expression of Nanog, Oct3/4, and Sox2 in Muse cells occurs [26]. Interestingly, when the Muse cells are transferred from a suspension to an adherent state, the higher pluripotency gene levels return to baseline levels. Therefore, the pluripotency gene expression levels are reversible between the adherent-suspension states. How the suspension state is linked to demethylation in Muse cells, however, is unclear currently.

2.5 Factors Related to Stress Tolerance in Muse Cells

Muse cells were initially discovered as a stress-tolerant subpopulation of BM-MSCs, as mentioned in Chap. 1. In this context, Muse cells are known to secrete factors involved in stress tolerance. Comparisons of the secretome of Muse cells, BM-MSCs, and adipose-derived MSCs (AD-MSCs) reveal a unique profile of Muse cells [15]. Serpins, a superfamily of proteins with protease inhibitory activity, are expressed only in Muse cells and not in BM-MSCs or AD-MSCs. Serpins inhibit trypsin, thrombin, and neutrophil elastase [27, 28]. The strong tolerance of Muse cells against long-term trypsin incubation, which was the initial clue to the existence of Muse cells, as described in Chap. 1, may be partly explained by their high serpin production.

The 14-3-3 proteins (also named YWHAQ, YWHAG, YWHAE, YWHAZ, and YWHAB) are a family of highly conserved acidic 30-kDa molecules that form stable homo- and heterodimers and play a key role in the regulation of the cell cycle and the cell response to DNA damage following cellular injury. These proteins act as chaperonin-like molecules to reduce cellular stress. In addition, some isoforms have the capacity to inactivate the pro-apoptotic protein BAD by preventing its negative effect on the pro-survival protein Bcl-XL [29, 30]. Muse cell secretomes contain most of the 14-3-3 isoforms, suggesting that the stress-enduring capacity of Muse cells involves the secretion of pro-survival factors.

2.5.1 *Muse Cells Tolerate Extensive Genotoxic Stimuli and Quickly Repair DNA than Non-Muse Cells*

Muse cells, like other somatic stem cells, may undergo several rounds of intrinsic and extrinsic stresses. Such genotoxic episode generally promotes triggering senescence and/or apoptosis. Due to the long life, Muse cells must have a robust effective DNA damage checkpoint and DNA repair system for the complete recovery of cells to maintain their reparative function.

Muse cells are resistant to chemical and physical genotoxic stresses, namely, peroxide hydrogen (H2O2) treatment and UV irradiation, respectively, better than non-Muse cells [31]. The level of senescence and apoptosis was lower in Muse cells than non-Muse cells in both genotoxic stresses because in Muse cells, the DNA damage repair system (DDR) was properly activated following injury and properly repaired as confirmed in g-H2AX staining while non-Muse cells did not show these responses [31].

Detection of activated H2AX, namely, g-H2AX, evidenced that 1 h following stress, damaged DNA was marked in all the cycle phases in Muse cells, suggesting that repair mechanisms may work throughout the cell cycle. Non-Muse cells, on the other hand, demonstrated activated H2AX staining in G1 cells only 48 h following stress, suggesting that they were not proficient in DNA repair. Notably, single-strand repair systems, the base excision repair (BER) and nucleotide excision repair (NER) pathway, had the same efficiency in Muse and non-Muse cells, whereas the nonhomologous end-joining (NHEJ) enzymatic activity increased promptly in Muse cells compared to other cells. The activation of RAD51 and DNA-PK, involved in the homologous recombination (HR) and NHEJ repair systems, respectively, is activated soon after DNA damage and decline to the basal level 6 h following stress in Muse cells, while in non-Muse cells, these repair mechanisms were lower or absent. NHEJ is the mechanism that can repair double-strand breaks in every phase of the cell cycle, and this fact suggests that Muse cells may have a powerful NHEJ system to survive strong genotoxic stress [31].

Along with g-H2AX staining, data collectively demonstrates the existence of a quick efficient DNA repair process in Muse cells compared to non-Muse cells.

2.6 Differentiation Ability of Muse Cells

2.6.1 *In Vitro*

Not only do Muse cells differentiate spontaneously, they also differentiate in vitro at a high rate (~80%–95%) into various cell types, such as hepatocyte- and neural-lineage cells, as well as into adipocytes, osteocytes, cardiac cells, keratinocytes, and melanocytes, when certain sets of cytokines are supplied (Figs. 2.8 and 2.9) [4, 6, 26, 32, 33].

Lineage	differentiation	Muse cells	Non-Muse cells
Mesodermal			
	Adipocytes	O	O (lower ratio)
	Osteocytes	O	O (lower ratio)
	Chondrocytes	O	O (lower ratio)
	Glomerular cells	O	X
	cardiomyocytes	O	X
	Skeletal muscle	O	X
Endodermal			
	hepatocytes	O	X
	cholangiocytes	O	X
Ectodermal			
	Neuronal cells	O	X
	melanocytes	O	X
	keratinocytes	O	X

Fig. 2.8 Trilineage differentiation ability in Muse and non-Muse cells. Muse cells were able to differentiate into various kinds of mesodermal-, endodermal- and ectodermal-lineage cells. However, the differentiation potential of non-Muse cells was limited to adipocytes, osteocytes and chondrocytes inmesodermal-lineage, and they were unable to differentiate into other cell kinds in mesodermal-lineage nor did they differentiate into endodermal- or ectodermal-lineages

Hepatocyte-lineage cells are differentiated by insulin-transferrin-selenium, dexamethasone, hepatocyte growth factor, and fibroblast growth factor-4 (FGF-4) [4]; neural-lineage cells by neurobasal medium, B27 supplement, 1-methyl-3 isobutylxanthine, dexamethasone, cAMP, valproic acid, and forskolin [34]; and adipocytes and osteocytes by the Human Mesenchymal Stem Cell Functional Identification Kit (R&D Systems), which contain recombinant proteins to induce efficient differentiation of MSCs to adipocytes and osteocytes [3]. Keratinocytes are inducible using defined keratinocyte serum-free medium (Invitrogen) plus BMP-4 and all-trans retinoic acid [32]. Ventricular cardiomyocytes can be induced by multiple induction steps: (1) initial treatment with 5′-azacytidine in suspension; (2) transfer onto adherent culture and supplied with Wnt3a, BMP2/4, and transforming growth factor β1; (3) treated with DKK-1 and Noggin, inhibitors of Wnt3a and BMP-4, respectively; and (4) cytokine cardiotrophin-1 incubation converts Muse cells into cardiomyocytes that express MLC2v, a marker of the mature ventricular subtype, as well as other cardiac markers including alpha-actinin and troponin I with a striated pattern [26]. The administration of ten factors (including Wnt3a, endothelin-3, linoleic acid, stem cell factor, and dexamethasone) is sufficient to convert Muse cells into melanin pigment-producing functional melanocytes [6, 33].

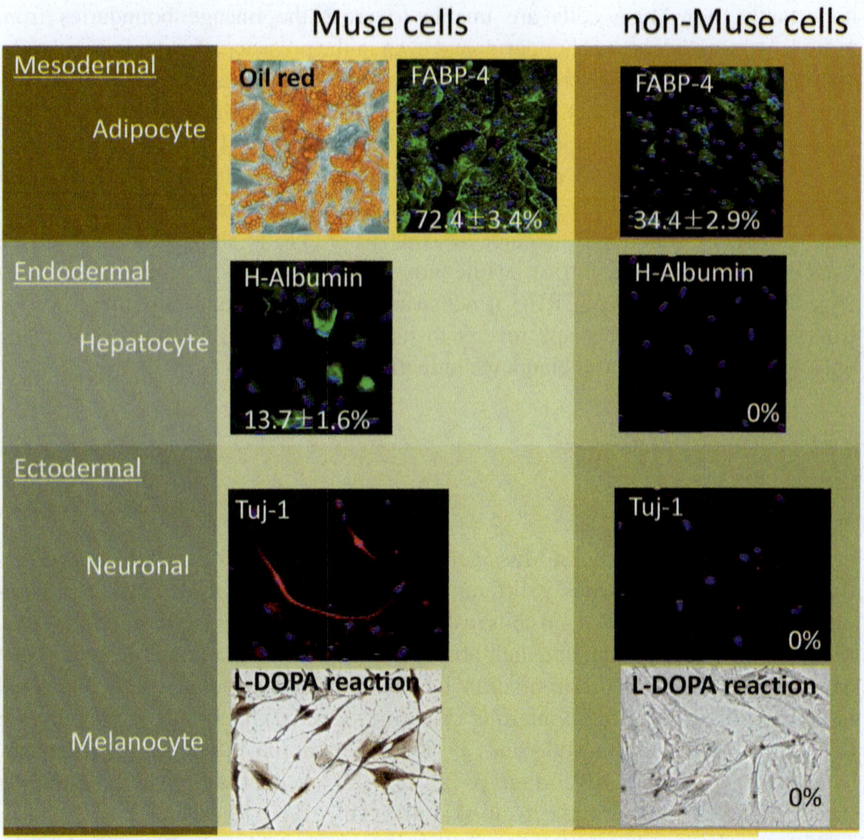

Fig. 2.9 Trilineage differentiation ability in Muse and non-Muse cells (histological analysis). Both Muse and non-Muse cells were able to differentiate into Oil-Red-O- and fatty acid-binding protein 4 (FABP)-positive adipose cells (mesodermal lineage) while the differentiation ratio in non-Muse cells was lower than that in Muse cells. In regard to the differentiation into hepatocytes (endodermal lineage), neuronal cells (ectodermal) and melanocytes (ectodermal), Muse cells could differentiate into cells positive for human albumin (H-Albumin), Tuj-1 and L-DOPA reaction, while none of non-Muse cells demonstrated positivity to these markers. The result suggested that Muse cells are able to differentiate into trilineage cells while non-Muse cells are not able to differentiate into endodermal and ectodermal lineages. (Pictures adapted and modified with permission from Ref. [3])

The differentiation potential of non-Muse cells in MSCs and fibroblasts is in sharp contrast with that of Muse cells. While non-Muse cells can differentiate into osteocytes, cartilage cells, and adipocytes in the presence of certain sets of induction cytokines, their differentiation rate is generally lower than that of Muse cells (Fig. 2.9). In addition, the time required for differentiation is substantially longer than that of Muse cells. For example, adipocyte differentiation efficiency is ~72% in Muse cells within ~7 days and ~34% in non-Muse cells within ~3 weeks [3]. Most importantly, non-Muse cells are unable to cross the lineage boundaries from mesoderm to ectoderm or endoderm, and even in the presence of cytokine cocktails, non-Muse cells do not differentiate into hepatocyte-, neuronal-, or melanocyte-lineages and instead show only partial responses to cytokine induction (Fig. 2.9) [3, 4, 6]. For example, in melanocyte induction, Muse cells newly express tyrosinase-related protein 1 (TRP-1) and gp100 at 3 weeks, dopachrome tautomerase at 5 weeks, and finally tyrosinase at 6 weeks, at which time point they become pigment-producing functional melanocytes that were able to show positivity for the L-DOPA reaction. In contrast, while non-Muse cells newly express TRP-1 at 3 weeks, the expression of TRP-1 is not sustainable and disappears within 5 weeks. Concomitantly, the cell shape reverts to its fibroblast-like morphology, showing only a partial response to melanocyte induction [6].

2.6.2 *In Vivo*

A prominent characteristic of Muse cells that also contrasts with other stem cells is that Muse cells spontaneously differentiate into tissue-compatible cells after homing and replenish lost/damaged cells in damaged tissue to repair the tissue (Fig. 2.10). Muse cells are pluripotent, and their ability to replenish lost/damaged cells has been demonstrated by their differentiation into neuronal cells (ectodermal) in animal models of stroke [35, 36], glomerular cells (mesodermal) in chronic kidney disease [37], cardiomyocytes (mesodermal) in acute myocardial infarction [38], hepatocytes (endodermal) in a liver damage [25, 39, 40], skeletal muscle cells in muscle degeneration, and epidermal cells in skin ulcer (Fig. 2.10) [2, 41].

In mouse/rat stroke models, topically administered Muse cells applied at the ischemic and border areas survived in the tissue and spontaneously differentiated into either neuronal cells or oligodendrocytes. The differentiated neuronal cells not only expressed neuronal markers such as MAP-2, NeuN, and calbindin, but their neurites were incorporated into the pyramidal tract and sensory circuits, leading to long-lasting robust functional recovery [35, 36]. Similar to this, subcutaneously injected Muse cells successfully integrated into skin ulcers in a mouse model of diabetes mellitus and incorporated into the epidermis by spontaneous differentiation into keratinocytes, as well as into the dermis by differentiation into fibroblasts and vascular cells, thereby repairing skin tissue [41].

Similar effects were observed following intravenous injection of Muse cells into animal models. In a mouse model of chronic kidney disease, intravenously administered Muse cells homed into damaged glomeruli and spontaneously differentiated into podocin- and WT-1-positive podocytes, megsin-positive mesangial cells, and von Willebrand factor- and CD31-positive endothelial cells and delivered renal function recovery [37]. In mouse liver damage models, Muse cells spontaneously differentiated into major liver components, namely, hepatocytes, cholangiocytes, sinusoid endothelial cells and Kupffer cells. Regarding hepatocyte differentiation, Muse cells not only express general hepatocyte markers, HepPar-1, antitrypsin and

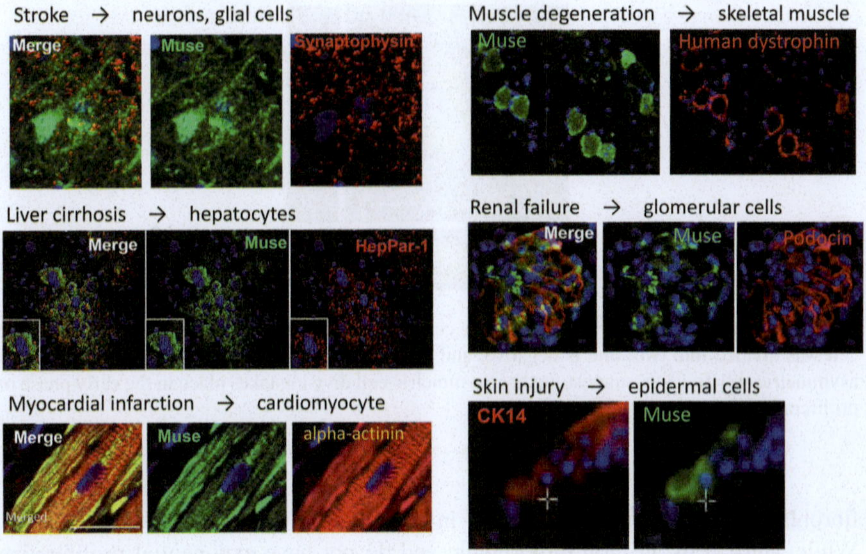

Fig. 2.10 In vivo trilineage differentiation ability of Muse cells. While Muse cells demonstrated trilineage differentiation in vitro either by spontaneous differentiation or by cytokine induction, their differentiation in vivo is shown to be compatible to the tissue they homed. Muse cells spontaneously differentiated into neuronal cells that expressed synaptophysin in rat stroke. (Picture was adopted with permission from Ref. [35]), HepPar-1-positive hepatocytes in mouse liver cirrhosis. (Picture was adopted with permission from Ref. [25]), alpha-actinin-positive cardiomyocytes in rabbit acute myocardial infarction. (Picture was adopted with permission from Ref. [38]), dystrophin-positive skeletal muscle in mouse muscle degeneration. (Picture was adopted with permission from Ref. [2]), glomerular cells including podocin-positive podocytes in chronic kidney disease. (Picture was adopted with permission from Ref. [37]) and cytokeratin 14 (CK14)-positive epidermal cells in mouse skin injury model. (Picture was adopted with permission from Ref. [2])

albumin, but they also express enzymes relevant to detoxification and glycolysis [25, 39, 40]. In a rabbit acute myocardial infarction model, intravenously injected Muse cells spontaneously differentiated into cardiomyocytes positive for troponin I, alpha-actinin, and connexin43. The differentiated cardiomyocytes were electrophysiologically active cardiomyocytes based on the observation of GCaMP fluorescence activity synchronous with electrocardiogram recordings [38]. Individual differentiation of Muse cells will be further discussed in later chapters.

2.7 Proliferation of Muse Cells

Muse cells are non-tumorigenic, consistent with the fact that they reside in the living body. The proliferation speed is ~1.3 days/cell division, both in adherent and suspension states, which is nearly the same as or slightly slower than that of

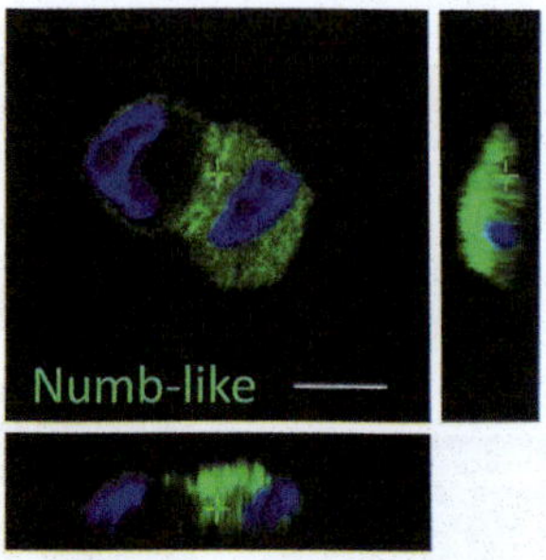

Fig. 2.11 Asymmetric cell division of Muse cells. In single cell suspension culture, single Muse cell was divided into two; one is negative, and the other is positive for Numb-like, a marker for asymmetric cell division, suggesting that asymmetric cell division takes place in the early phase of proliferation in Muse cells. Bar = 10 μm

fibroblasts. Muse cells expand stably in adherent culture until the Hayflick limit. While Muse cells are non-tumorigenic and do not have exponential proliferative activity, they can still grow on a clinically relevant scale [2].

Muse cells are natural somatic stem cells and thus proliferate through symmetric and asymmetric cell division [2]. In fact, Numb-like, related to asymmetric cell division, is expressed in proliferating Muse cells (Fig. 2.11). Therefore, even if Muse cells are purified to 100% by cell sorting, they will gradually generate non-Muse cells during expansion, and thus the proportion of Muse cells will gradually decrease, reaching a plateau at one to several percent of the total cells, corresponding to the proportion of Muse cells seen among MSCs and fibroblasts. Based on this, it is possible that Muse cells, genuine pluripotent stem cells contained in the BM and dermis, are the origin of MSCs and fibroblasts in primary culture, with the production of non-Muse cells through asymmetric cell division of Muse cells during expansion gradually forming the final population of MSCs and fibroblasts we usually see.

The proliferation style of Muse cells in a suspension state is different from that in an adherent state. When single Muse cells are cultured in suspension, slender flat non-Muse cells are generated by asymmetric cell division at a very early stage of proliferation (Figs. 2.12 and 2.13). The non-Muse cells enwrap the Muse cells, and those enwrapped Muse cells continue to proliferate to form embryoid body-like clusters reaching 50~100 μm within 10~14 days (Figs. 2.12 and 2.13) [2–4]. Muse cell morphology is similar to ES cell morphology, but the decisive difference between Muse cell-derived clusters and ES-derived embryoid bodies is that Muse cell-derived clusters are always covered by several slender non-Muse cells, while ES-derived embryoid bodies, resulting from symmetrical clonal proliferation as seen in tumorigenic cells, have no covering cells. Perhaps because of the presence of the covering cells or of intracellular signaling caused by suspension, the proliferation of single Muse cell-derived clusters gradually slows by ~14 days and then ceases. Thus, unlike ES cells, Muse cell clusters do not show unlimited proliferative activity in suspension. When the clusters are transferred to adherent culture, non-Muse cells in

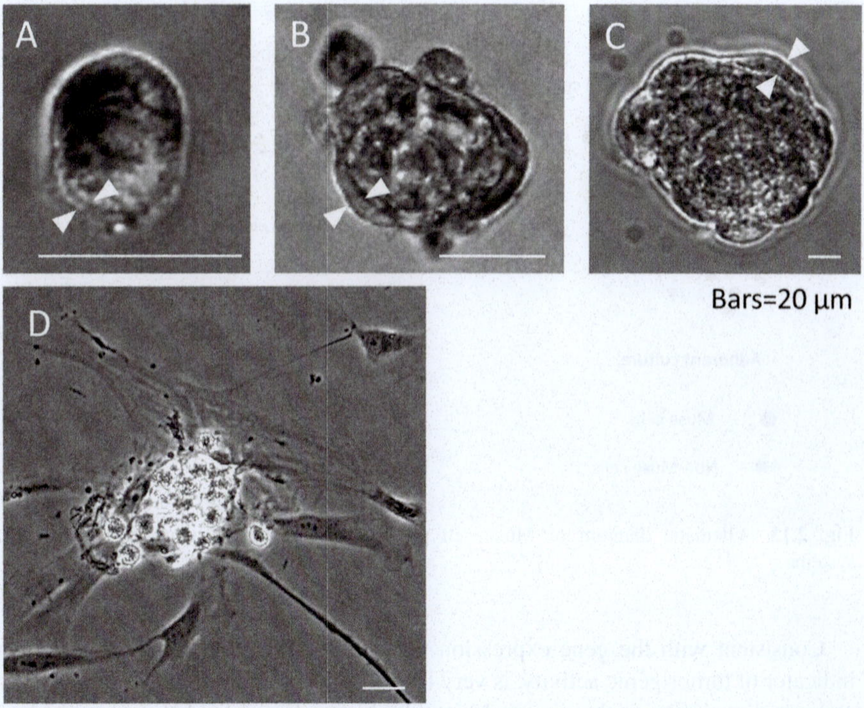

Fig. 2.12 Development of cluster from single Muse cell. (**a**) Small-scale cluster formed from single Muse cell in suspension. White arrowheads display a slender cell sheath enwrapping several cells inside the sheath. (**b**) More advanced stage cluster shows ensheathment (white arrowheads) of multiple cells. (**c**) Mature cluster that contains a number of cells inside the ensheathment (white arrowheads). (**d**) Opening of ensheathing cells and migration out of cluster cells after transferring the cluster to gelatin-coated culture dish. Bar = 20 μm

the outermost layer open the wrap, and the Muse cells again begin to proliferate, migrate out the cluster, and generate expanded cells until reaching the Hayflick limit (Figs. 2.12 and 2.13).

Comparison of the gene expression among three types of cells, ES/iPS cells, Muse, and non-Muse fibroblasts (one of the representatives of somatic cells) reveals that the expression pattern of pluripotency genes is similar between Muse and ES/iPS cells, although the expression level is lower in Muse cells than in ES/iPS cells (Fig. 2.4). Importantly, non-Muse fibroblasts basically do not express pluripotency genes or express at a very low level, if any, which is in sharp contrast to Muse cells [4]. The pattern of gene expression of tumorigenic factors, however, differs between ES/iPS and Muse cells (Fig. 2.4). Generally, those factors are highly expressed in ES/iPS cells, consistent with the fact that they have exponential proliferative activity, while very low expression is observed in Muse cells, and the level and pattern are similar to that observed in non-Muse fibroblasts [4].

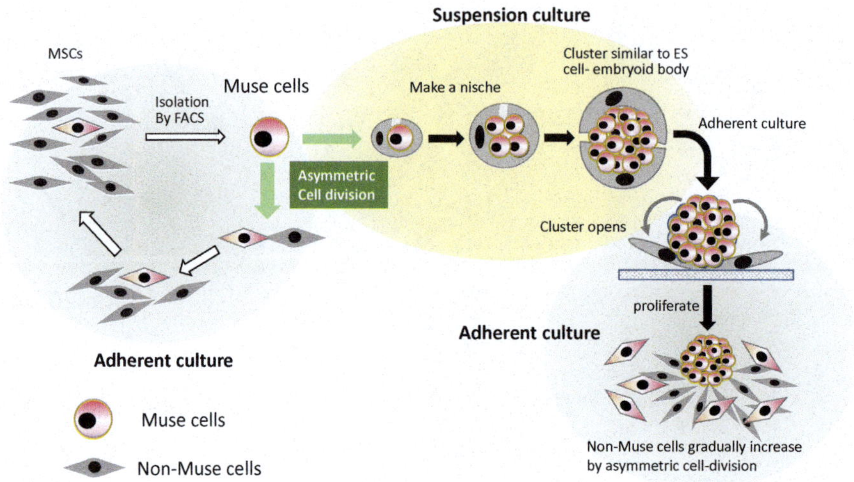

Fig. 2.13 Schematic diagram of Muse cell proliferation style in suspension and adherent systems

Consistent with the gene expression pattern and level, telomerase activity, an indicator of tumorigenic activity, is very low in Muse cells compared with iPS cells and is rather similar to that in non-Muse fibroblasts (Fig. 2.14) [4]. In fact, human BM- and adipose-derived Muse cells transplanted into the testes of immunodeficient mice do not generate any tumors for up to 6 months [2, 3]. Therefore, Muse cells are considered to have a low risk of tumorigenesis (Fig. 2.14).

2.8 Unique Distribution of Muse Cells in the Body

With respect to tissue, Muse cells were first found in the BM, and subsequent studies showed the presence of Muse cells in the adipose tissue, dermis, and the umbilical cord [4, 42, 43]. Because all these tissues are mesenchymal, the home ground of Muse cells was initially considered to be limited to mesenchymal tissues. As research progressed, however, Muse cells were found to be widely distributed in the body.

The current understanding is as follows: Muse cells reside in the BM, forming loose clusters with each other in nearby vessels. The proportion of Muse cells in the BM differs among reports, but in humans, it is assumed to be around 0.1~0.03% of the mononucleated cell fraction, i.e., 1 of 1000~3000 mononucleated cells [2].

Muse cells are also found in the peripheral blood at the proportion of 0.01~0.2% or slightly more of the mononuclear fraction [44]. The major source of peripheral blood Muse cells is assumed to be the BM, as the BM is directly connected to the peripheral blood. Other sources, such as the spleen, cannot be excluded as candidate

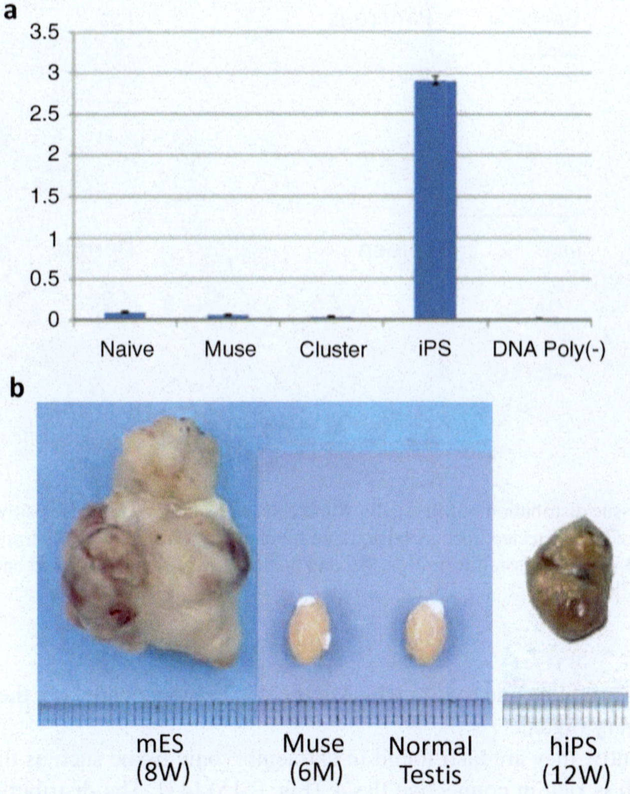

Fig. 2.14 Non-tumorigenicity of Muse cells. (**a**) Telomerase activity of naïve fibroblasts (naïve), Muse cells (Muse in adhesion), Muse cell-derived cluster in suspension (cluster), and iPS cells. DNA poly(−) represents a negative control. (Pictures adapted and modified with permission from Wakao et al. (2011), PNAS [4].) (**b**) Injection into Nog mice testes. Mouse ES cells and human iPS cells formed teratoma by 8 weeks and 12 weeks, respectively, while human Muse cells did not generate teratoma for up to 6 months. (Pictures adapted and modified with permission from Ref. [2] with minor modifications)

suppliers of peripheral blood Muse cells. As described later, the proportion of Muse cells in the peripheral blood drastically changes after injury and in the presence of disease [44].

Other than the BM and peripheral blood, Muse cells sparsely distribute to the connective tissue of every organ. This has been demonstrated in the dermis, spleen, pancreas, trachea, umbilical cord and adipose tissue (Fig. 2.15) [43]. Muse cells are even found in the pia mater and arachnoid of the brain, whose main composition is connective tissue (unpublished data). Vessels are always adjacent to connective tissue, particularly when they enter an organ from the hilus and penetrate into the tissue by branching into small vessels. Therefore, when peripheral blood Muse cells leave the circulation and enter into the tissues, they first locate in the connective

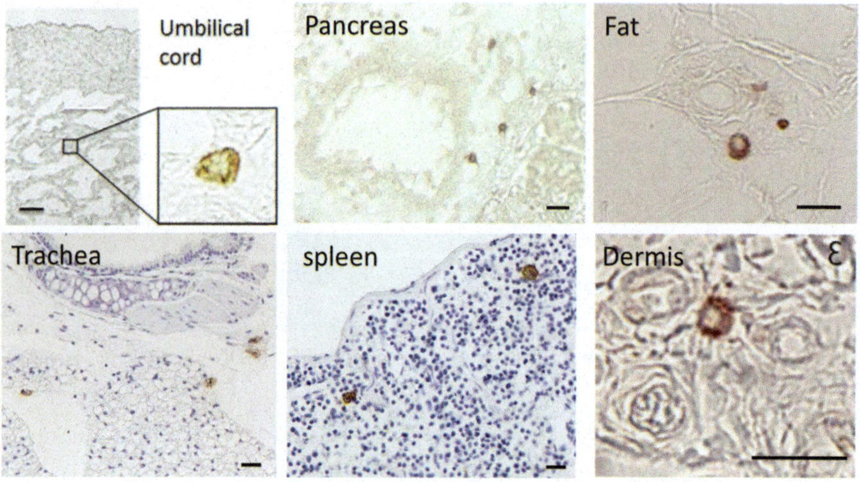

Fig. 2.15 Tissue distribution of Muse cells. Muse cells were detected as cells positive for SSEA-3 (brown). They were detected in connective tissue sporadically. (Picture for the dermis was adapted and modified with permission from Ref. [4], and the trachea, fat, umbilical cord and spleen were from Ref. [42])

tissue. This might be one reason why Muse cells distribute widely in the connective tissue of every organ.

Importantly, they are also found in extraembryonic tissue such as the umbilical cord, which is rich in connective tissue (Fig. 2.15) [43]. The distribution of Muse cells in the living body is thus unique and distinct from other reported somatic/tissue stem cells. Particularly, the significance of the presence of pluripotent Muse cells in extraembryonic tissue should be clarified in future studies.

To date, Muse cells have not been observed in any particular niche-like tissue structures. Rather they are freely and sparsely distributed in the connective tissue.

2.8.1 *Why Do Muse Cells Maintain a Round Shape in Vivo?*

Another unique characteristic of Muse cells located in connective tissue is their different morphology from that of typical spindle-shaped fibroblasts, even though they are surrounded by an extracellular matrix that includes collagen fibers. Interestingly, Muse cells are in most cases observed as round cells, even when surrounded by abundant matrix in vivo (Fig. 2.15) [42, 43]. As mentioned earlier in this chapter, in vitro experiments demonstrated that the pluripotency of Muse cells becomes remarkably higher in a suspension state compared with that in an adherent state. It might be that their shape remains round, the state similar to that in suspension, to maintain their pluripotency in vivo.

2.8.2 *Muse Cells Are Distinct from Other Somatic Stem Cells*

Several kinds of stem cells are reported to exist in various tissues. In the BM and peripheral blood where Muse cells are found, hematopoietic stem cells (HSCs), endothelial progenitors (EPs), and MSCs also exist. Muse cells are also found in connective tissue. For example, the connective tissue of the dermis contains skin-derived precursors (SKPs), neural crest-derived stem cells (NCSCs), melanoblasts, perivascular cells (PCs), EPs, and adipose-derived stem cells (ADSCs) [45–52]. Thus, whether or not Muse cells are distinct from these known stem cells is a critical issue.

Muse cells are negative for C-kit/CD117 (marker for HSCs and melanoblasts), CD34 (HSCs, EPs, and ADSCs), NG2 (PCs), von Willebrand factor (EPs), CD31 (EPs), CD146 (PCs and ADSCs), CD271 (NCSCs), Sox10 (NCSCs), Snail (SKPs), Slug (SKPs), Tyrp1 (melanoblasts), and Dct (melanoblasts), suggesting that Muse cells are distinct from these stem or progenitor cells found in the BM, peripheral blood, and connective tissues [3, 4].

The differences and similarities between Muse and BM-MSCs are other critical issues as Muse cells are a subpopulation of BM-MSCs, corresponding to 1% to several percent of the total cells. Muse cells are positive for nearly all of the major markers known to be expressed by BM-MSCs, such as CD105, CD90, and CD29 (Fig. 2.16) [4]. SSEA-3, however, is the definitive marker that discriminates Muse from cells other than Muse cells, namely, non-Muse cells, in MSCs. Non-Muse MSCs do not express the pluripotent surface marker SSEA-3, nor do they show

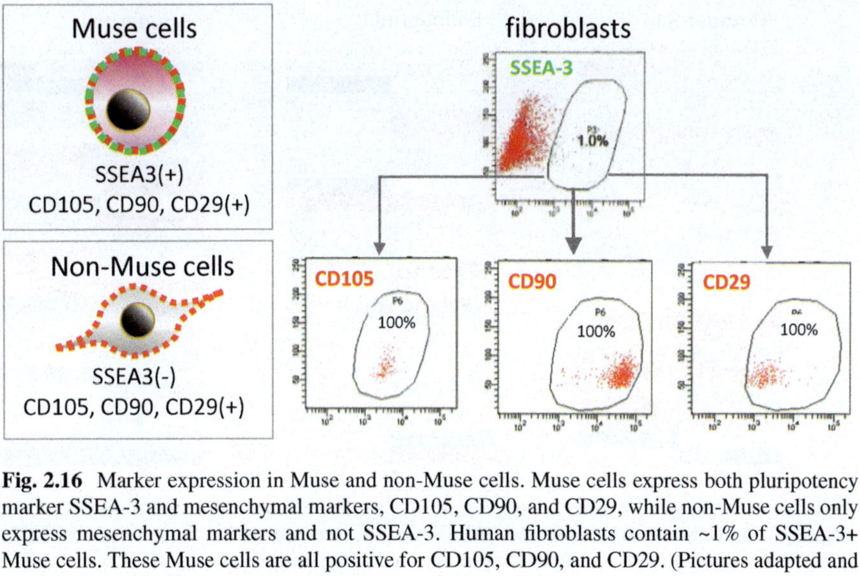

Fig. 2.16 Marker expression in Muse and non-Muse cells. Muse cells express both pluripotency marker SSEA-3 and mesenchymal markers, CD105, CD90, and CD29, while non-Muse cells only express mesenchymal markers and not SSEA-3. Human fibroblasts contain ~1% of SSEA-3+ Muse cells. These Muse cells are all positive for CD105, CD90, and CD29. (Pictures adapted and modified with permission from Ref. [4])

pluripotency, as mentioned before. In other words, Muse cells are double positive for both mesenchymal and pluripotency markers, while non-Muse MSCs are only positive for mesenchymal markers (Fig. 2.16) [4].

2.8.3 Are Muse Cells from Different Sources the Same?

The core properties of Muse cells, pluripotency gene expression, as well as trilineage differentiation ability and self-renewal at a single cell level are consistently exhibited among Muse cells derived from different tissue sources. Muse cells present their own differentiation directivity, however, according to their source [2–4]. Adipose-Muse cells express higher levels of genes related to their differentiation into adipocytes, osteocytes, and skeletal muscle cells, all of which are categorized as mesodermal lineage, than BM- and dermal-Muse cells (Fig. 2.17) [3, 53]. BM-Muse cells, in contrast, contain the highest levels of factors pertinent to endodermal lineage cells, such as hepatocytes and pancreatic cells. Furthermore, both BM- and dermal-Muse cells exhibit higher expression levels of factors related to the ectodermal lineage, such as neuronal-, melanocyte-, and epidermal-related genes, compared with adipose-Muse cells (Fig. 2.17) [3]. Therefore, while pluripotency is demonstrated by all BM-, dermal-, and adipose-Muse cells, their differentiation potential is not the same. This suggests that the appropriate tissue source should be selected according to the target tissue and cell types for the medical use of Muse cells. It is also interesting to know how the microenvironment affects the differentiation potential of Muse cells.

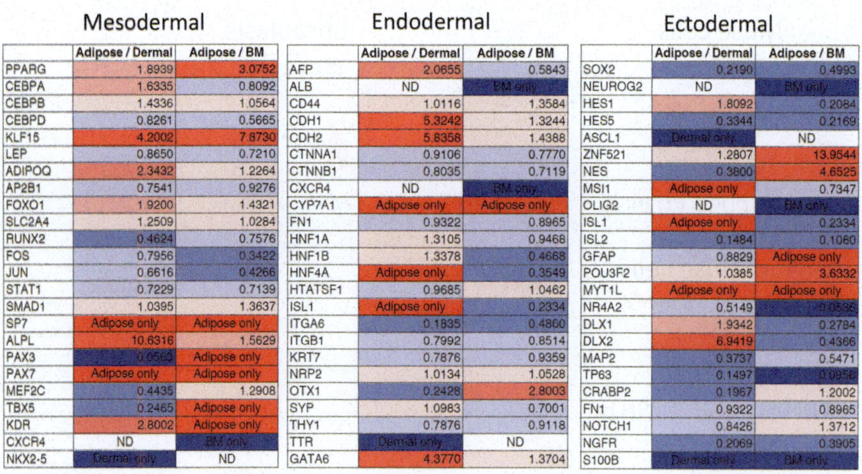

Mesodermal

	Adipose / Dermal	Adipose / BM
PPARG	1.8939	3.0752
CEBPA	1.6335	0.8092
CEBPB	1.4336	1.0564
CEBPD	0.8261	0.5665
KLF15	4.2002	7.8730
LEP	0.8650	0.7210
ADIPOQ	2.3432	1.2264
AP2B1	0.7541	0.9276
FOXO1	1.9200	1.4321
SLC2A4	1.2509	1.0284
RUNX2	0.4624	0.7576
FOS	0.7956	0.3422
JUN	0.6616	0.4266
STAT1	0.7229	0.7139
SMAD1	1.0395	1.3637
SP7	Adipose only	Adipose only
ALPL	10.6316	1.5629
PAX3	0.0563	Adipose only
PAX7	Adipose only	Adipose only
MEF2C	0.4435	1.2908
TBX5	0.2465	Adipose only
KDR	2.8002	Adipose only
CXCR4	ND	BM only
NKX2-5	Dermal only	ND

Endodermal

	Adipose / Dermal	Adipose / BM
AFP	2.0655	0.5843
ALB	ND	BM only
CD44	1.0116	1.3584
CDH1	5.3242	1.3244
CDH2	5.8358	1.4388
CTNNA1	0.9106	0.7770
CTNNB1	0.8035	0.7119
CXCR4	ND	BM only
CYP7A1	Adipose only	Adipose only
FN1	0.9322	0.8965
HNF1A	1.3105	0.9468
HNF1B	1.3378	0.4668
HNF4A	Adipose only	0.3549
HTATSF1	0.9685	1.0462
ISL1	Adipose only	0.2334
ITGA6	0.1835	0.4860
ITGB1	0.7992	0.8514
KRT7	0.7876	0.9359
NRP2	1.0134	1.0528
OTX1	0.2428	2.8003
SYP	1.0983	0.7001
THY1	0.7876	0.9118
TTR	Dermal only	ND
GATA6	4.3770	1.3704

Ectodermal

	Adipose / Dermal	Adipose / BM
SOX2	0.2190	0.4993
NEUROG2	ND	BM only
HES1	1.8092	0.2084
HES5	0.3344	0.2169
ASCL1	Dermal only	ND
ZNF521	1.2807	13.9544
NES	0.3800	4.6525
MSI1	Adipose only	0.7347
OLIG2	ND	BM only
ISL1	Adipose only	0.2334
ISL2	0.1484	0.1060
GFAP	0.8829	Adipose only
POU3F2	1.0385	3.6332
MYT1L	Adipose only	Adipose only
NR4A2	0.5149	0.0536
DLX1	1.9342	0.2784
DLX2	6.9419	0.4366
MAP2	0.3737	0.5471
TP63	0.1497	0.0956
CRABP2	0.1967	1.2002
FN1	0.9322	0.8965
NOTCH1	0.8426	1.3712
NGFR	0.2069	0.3905
S100B	Dermal only	BM only

Fig. 2.17 Comparison of gene expression related to the differentiation of trilineages among adipose-Muse cells versus dermal- and BM-Muse cells. Red is higher in the numerator than the dominator, and blue is lower. (Pictures were adopted with permission from Ref. [3] with minor modifications)

2.9 Muse Cells Are More Like Primed Pluripotent Stem Cells than Naïve Pluripotent Stem Cells

After identification of EpiSCs, a new stem cell type isolated from post implantation epiblasts in mice that are distinct from mouse embryonic stem cells (mES cells) [54–56], two pluripotent stem cell states with slightly different characteristics, namely naïve (for mES cells) and primed (for EpiSCs) [57], were proposed. Both naïve and primed states are pluripotent states, in which the pluripotency factors Oct3/4 and Sox2 are expressed and self-renewability is observed. The two pluripotent stem cell types, however, exhibit a distinct developmental potential, as evidenced by the fact that naïve pluripotent stem cells are able to contribute to blastocyst chimeras, and exhibit a very stable self-renewal capacity and survival, allowing for efficient expansion under the inhospitable selection conditions required for genome editing and clonal drug selection. Primed pluripotent stem cells, on the other hand, do not have these properties. Moreover, naïve pluripotent stem cells, unlike primed pluripotent stem cells, are capable of generating functional hematopoietic progenitors with in vivo repopulation competence [58, 59].

The growth factor dependency of these two stem cell types is also different [57, 60]. Naïve pluripotent stem cells derived from the inner cell mass of the preimplantation blastocyst are dependent on leukemia inhibitory factor (LIF)/STAT3 pathway or the combination of LIF and small molecule inhibitors (usually called 2i) to sustain their self-renewal [61–63]. In contrast, primed pluripotent stem cells require active FGF and activin/transforming growth factor β, but not LIF and BMP4 signaling, for stable promotion of self-renewal [55, 56]. Primed pluripotent stem cells also have slower proliferation rates and are unable to propagate as single cells, leading to less efficient genetic manipulation and a lower ability to generate germ line chimeras [64–66].

The naïve and primed pluripotent stem cells also use different modes of respiration for generating energy. While metabolism in naïve stem cells utilizes both oxidative phosphorylation, namely, mitochondrial respiration, and glycolysis, primed pluripotent stem cells preferentially utilize the glycolytic pathway [67, 68]. An interesting epigenetic difference between these stem cell types is that both X chromosomes are active in naïve female mES cells and undergo random X chromosome inactivation upon differentiation in vitro, while X chromosome inactivation is already established in primed female mEpiSCs, and this feature can thus be used as a reliable marker to distinguish between the two pluripotent states in female stem cells [69]. Other epigenetic differences are global DNA hypomethylation [70, 71], reduced prevalence of the repressive histone mark H3K27me3 at promoters, and fewer bivalent domains in naïve than in pluripotent stem cells [72]. Human ES cells derived from preimplantation blastocysts exhibit characteristics reminiscent of primed state EpiSCs [73, 74], except for in vivo chimerism, which cannot be tested in humans.

While further detailed investigation is required, human Muse cells appear to be more similar to primed prepotent stem cells than naïve pluripotent stem cells in several aspects. First, Muse cells depend strongly on FGF to maintain their proliferation and self-renewal abilities. The withdrawal of FGF substantially slows the speed of Muse cell proliferation and accelerates asymmetric cell division rather than symmetric cell division, leading to a higher proportion of non-Muse cells and a lower proportion of Muse cells. On the other hand, unlike naïve pluripotent stem cells, Muse cells do not respond to LIF or BMP4 (unpublished data).

When mouse Muse cells derived from embryoid body-like clusters were simply introduced into mouse blastocysts, they did not show commitment to germ line chimeras (unpublished data). Regarding DNA methylation, Muse cell DNA is more demethylated than somatic cell DNA but more methylated than the DNA of naïve pluripotent stem cells, such as iPS cells. Because Muse cells can be obtained from adult tissues, it might be reasonable that they are more similar to primed pluripotent stem cells than naïve pluripotent stem cells. How Muse cells and primed pluripotent stem cells, represented by EpiSCs, are similar and how they differ is an important subject for future studies. Muse cells have several unique characteristics; they are found in extraembryonic tissues such as the umbilical cord, while EpiSCs are confined to embryonic tissue. The metabolism and epigenetics of these two types of cells are another interesting topic to be clarified in future studies.

Muse cells and ES/iPS cells share common properties such as pluripotency gene expression and trilineage differentiation and self-renewal abilities [2–4]. Other properties, however, are different, such as chimera formation, growth factor dependency, and proliferation style. The expression level of pluripotency gene markers in Muse cells is not as high as that in ES/iPS cells [4]. ES and iPS cells are not endogenous and are rather established/generated by manipulations. Thus, their pluripotency state differs from that of naturally existing endogenous pluripotent cells. Indeed, inner cell masses in normal development express lower levels of pluripotency genes than ES/iPS cells [17]. For endogenous pluripotent cells, including Muse cells, pluripotency gene levels do not necessarily need to be high to maintain pluripotency in vivo. Their main purpose is to maintain minute daily repair of tissues and to maintain tissue homeostasis. In this regard, the moderate pluripotent state of Muse cells seems to be rational and adaptable to the living body.

References

1. Efthymiou AG, Chen G, Rao M, Chen G, Boehm M (2014) Self-renewal and cell lineage differentiation strategies in human embryonic stem cells and induced pluripotent stem cells. Expert Opin Biol Ther 14(9):1333–1344. https://doi.org/10.1517/14712598.2014.922533
2. Kuroda Y, Kitada M, Wakao S, Nishikawa K, Tanimura Y, Makinoshima H, Goda M, Akashi H, Inutsuka A, Niwa A, Shigemoto T, Nabeshima Y, Nakahata T, Nabeshima Y, Fujiyoshi Y, Dezawa M (2010) Unique multipotent cells in adult human mesenchymal cell populations. Proc Natl Acad Sci U S A 107(19):8639–8643. https://doi.org/10.1073/pnas.0911647107
3. Ogura F, Wakao S, Kuroda Y, Tsuchiyama K, Bagheri M, Heneidi S, Chazenbalk G, Aiba S, Dezawa M (2014) Human adipose tissue possesses a unique population of pluripotent stem cells with nontumorigenic and low telomerase activities: potential implications in regenerative medicine. Stem Cells Dev 23(7):717–728. https://doi.org/10.1089/scd.2013.0473

4. Wakao S, Kitada M, Kuroda Y, Shigemoto T, Matsuse D, Akashi H, Tanimura Y, Tsuchiyama K, Kikuchi T, Goda M, Nakahata T, Fujiyoshi Y, Dezawa M (2011) Multilineage-differentiating stress-enduring (Muse) cells are a primary source of induced pluripotent stem cells in human fibroblasts. Proc Natl Acad Sci U S A 108(24):9875–9880. https://doi.org/10.1073/pnas.1100816108
5. Liu Q, Zhang RZ, Li D, Cheng S, Yang YH, Tian T, Pan XR (2016) Muse cells, a new type of pluripotent stem cells derived from human fibroblasts. Cell Reprogram 18(2):67–77. https://doi.org/10.1089/cell.2015.0085
6. Tsuchiyama K, Wakao S, Kuroda Y, Ogura F, Nojima M, Sawaya N, Yamasaki K, Aiba S, Dezawa M (2013) Functional melanocytes are readily reprogrammable from multilineage-differentiating stress-enduring (Muse) cells, distinct stem cells in human fibroblasts. J Invest Dermatol 133(10):2425–2435. https://doi.org/10.1038/jid.2013.172
7. Kuroda Y, Wakao S, Kitada M, Murakami T, Nojima M, Dezawa M (2013) Isolation, culture and evaluation of multilineage-differentiating stress-enduring (Muse) cells. Nat Protoc 8(7):1391–1415. https://doi.org/10.1038/nprot.2013.076
8. Wesselschmidt RL (2011) The teratoma assay: an in vivo assessment of pluripotency. Methods Mol Biol 767:231–241. https://doi.org/10.1007/978-1-61779-201-4_17
9. Chou YF, Yabuuchi A Murine embryonic stem cell derivation, in vitro pluripotency characterization, and in vivo teratoma formation. Curr Protoc Toxicol 2011 Chapter 2:Unit2.22. https://doi.org/10.1002/0471140856.tx0222s50
10. Chou YF, Chen HH, Eijpe M, Yabuuchi A, Chenoweth JG, Tesar P, Lu J, McKay RD, Geijsen N (2008) The growth factor environment defines distinct pluripotent ground states in novel blastocyst-derived stem cells. Cell 135(3):449–461. https://doi.org/10.1016/j.cell.2008.08.035
11. Draper JS, Pigott C, Thomson JA, Andrews PW (2002) Surface antigens of human embryonic stem cells: changes upon differentiation in culture. J Anat 200(Pt 3):249–258
12. Takahashi K, Tanabe K, Ohnuki M, Narita M, Ichisaka T, Tomoda K, Yamanaka S (2007) Induction of pluripotent stem cells from adult human fibroblasts by defined factors. Cell 131(5):861–872
13. Shevinsky LH, Knowles BB, Damjanov I, Solter D (1982) Monoclonal antibody to murine embryos defines a stage-specific embryonic antigen expressed on mouse embryos and human teratocarcinoma cells. Cell 30(3):697–705
14. Kannagi R, Cochran NA, Ishigami F, Hakomori S, Andrews PW, Knowles BB, Solter D (1983) Stage-specific embryonic antigens (SSEA-3 and SSEA-4) are epitopes of a unique globo-series ganglioside isolation from human teratocarcinoma cells. EMBO J 2(12):2355–2361
15. Alessio N, Özcan S, Tatsumi K, Murat A, Peluso G, Dezawa M, Galderisi U (2017) The secretome of MUSE cells contains factors that may play a role in regulation of stemness, apoptosis and immunomodulation. Cell Cycle 16(1):33–44. https://doi.org/10.1080/15384101.2016.1211215
16. O'Leary T, Heindryckx B, Lierman S, van Bruggen D, Goeman JJ, Vandewoestyne M, Deforce D, de Sousa Lopes SM, De Sutter P (2012) Tracking the progression of the human inner cell mass during embryonic stem cell derivation. Nat Biotechnol 30(3):278–282. https://doi.org/10.1038/nbt.2135
17. Ohgushi M, Minaguchi M, Sasai Y (2015) Rho-signaling-directed YAP/TAZ activity underlies the long-term survival and expansion of human embryonic stem cells. Cell Stem Cell 17:448–461. PMID: 26321201; https://doi.org/10.1016/j.stem.2015.07.009
18. Iglesias-Bartolome R, Torres D, Marone R, Feng X, Martin D, Simaan M, Chen M, Weinstein LS, Taylor SS, Molinolo AA, Gutkind JS (2015) Inactivation of a Galpha(s)-PKA tumour suppressor pathway in skin stem cells initiates basal-cell carcinogenesis. Nat Cell Biol 17:793–803. PMID:25961504; https://doi.org/10.1038/ncb3164
19. Faherty S, Fitzgerald A, Keohan M, Quinlan LR (2007) Self-renewal and differentiation of mouse embryonic stem cells as measured by Oct4 expression: the role of the cAMP/PKA pathway. In Vitro Cell Dev Biol Anim 43:37–47. PMID:17570033; https://doi.org/10.1007/s11626-006-9001-5

20. Lefebvre P, Benomar Y, Staels B (2010) Retinoid X receptors: common heterodimerization partners with distinct functions. Trends Endocrinol Metab 21:676–683. PMID:20674387; https://doi.org/10.1016/j.tem.2010.06.009
21. Elahy M, Baindur-Hudson S, Dass CR (2012) The emerging role of PEDF in stem cell biology. J Biomed Biotechnol 2012:239091. PMID:22675247; https://doi.org/10.1155/2012/239091
22. Ho TC, Chen SL, Wu JY, Ho MY, Chen LJ, Hsieh JW, Cheng HC, Tsao YP (2013) PEDF promotes self-renewal of limbal stem cell and accelerates corneal epithelial wound healing. Stem Cells 31:1775–1784. PMID:23553951; https://doi.org/10.1002/stem.1393
23. Zille M, Riabinska A, Terzi MY, Balkaya M, Prinz V, Schmerl B, Nieminen-Kelh€a M, Endres M, Vajkoczy P, Pina AL (2014) Influence of pigment epithelium-derived factor on outcome after striatal cerebral ischemia in the mouse. PLoS One 9:e114595. PMID:25470280; https://doi.org/10.1371/journal.pone.0114595
24. Mazo IB, Massberg S, vo Andrian UH (2011) Hematopoietic stem and progenitor cell trafficking. Trends Immunol 32(10):493–503. https://doi.org/10.1016/j.it.2011.06.011
25. Iseki M, Kushida Y, Wakao S, Akimoto T, Mizuma M, Motoi F, Asada R, Shimizu S, Unno M, Chazenbalk G, Dezawa M (2017) Muse cells, nontumorigenic pluripotent-like stem cells, have liver regeneration capacity through specific homing and cell replacement in a mouse model of liver fibrosis. Cell Transplant 26(5):821–840. https://doi.org/10.3727/096368916X693662
26. Amin M, Kushida Y, Wakao S, Kitada M, Tatsumi K, Dezawa M (2017) Cardiotrophic growth factors-driven induction of human Muse cells into cardiomyocyte-like phenotype. Cell Transplant [Epub ahead of print]. https://doi.org/10.1177/0963689717721514
27. Silverman GA, Whisstock JC, Bottomley SP, Huntington JA, Kaiserman D, Luke CJ, Pak SC, Reichhart JM, Bird PI (2015) Serpins flex their muscle: I. Putting the clamps on proteolysis in diverse biological systems. J Biol Chem 285(32):24299–24305. https://doi.org/10.1074/jbc.R110.112771
28. Whisstock JC, Silverman GA, Bird PI, Bottomley SP, Kaiserman D, Luke CJ, Pak SC, Reichhart JM, Huntington JA (2010) Serpins flex their muscle: II. Structural insights into target peptidase recognition, polymerization, and transport functions. J Biol Chem 285(32):24307–24312. https://doi.org/10.1074/jbc.R110.141408
29. Clapp C, Portt L, Khoury C, Sheibani S, Norman G, Ebner P, Eid R, Vali H, Mandato CA, Madeo F, Greenwood MT (2012) 14-3-3 protects against stress-induced apoptosis. Cell Death Dis 3:e348. PMID:22785534; https://doi.org/10.1038/cddis.2012.90
30. Gardino AK, Yaffe MB (2011) 14-3-3 proteins as signaling integration points for cell cycle control and apoptosis. Semin Cell Dev Biol 22:688–695. PMID:21945648; https://doi.org/10.1016/j.semcdb.2011.09.008
31. Alessio N, Squillaro T, Özcan S, Di Bernardo G, Venditti M, Melone M, Peluso G, Galderisi U (2018) Stress and stem cells: adult Muse cells tolerate extensive genotoxic stimuli better than mesenchymal stromal cells. Oncotarget 9(27):19328–19341. https://doi.org/10.18632/oncotarget.25039
32. Yamauchi T, Yamasaki K, Tsuchiyama K, Koike S, Aiba S (2017) The potential of Muse cells for regenerative medicine of skin: procedures to reconstitute skin with Muse cell-derived keratinocytes, fibroblasts, and melanocytes. J Invest Dermatol 137(12):2639–2642. https://doi.org/10.1016/j.jid.2017.06.021
33. Yamauchi T, Yamasaki K, Tsuchiyama K, Koike S, Aiba S (2017 Jun) A quantitative analysis of multilineage-differentiating stress-enduring (Muse) cells in human adipose tissue and efficacy of melanocytes induction. J Dermatol Sci 86(3):198–205. https://doi.org/10.1016/j.jdermsci.2017.03.001
34. Boulland JL, Mastrangelopoulou M, Boquest AC, Jakobsen R, Noer A, Glover JC, Collas P (2013) Epigenetic regulation of nestin expression during neurogenic differentiation of adipose tissue stem cells. Stem Cells Dev 22(7):1042–1052. https://doi.org/10.1089/scd.2012.0560
35. Uchida H, Morita T, Niizuma K, Kushida Y, Kuroda Y, Wakao S, Sakata H, Matsuzaka Y, Mushiake H, Tominaga T, Borlongan CV, Dezawa M (2016) Transplantation of unique subpopulation of fibroblasts, Muse cells, ameliorates experimental stroke possibly via robust neuronal differentiation. Stem Cells 34(1):160–173. https://doi.org/10.1002/stem.2206

36. Uchida H, Niizuma K, Kushida Y, Wakao S, Tominaga T, Borlongan CV, Dezawa M (2017) Human Muse cells reconstruct neuronal circuitry in subacute lacunar stroke model. Stroke 48(2):428–435. https://doi.org/10.1161/STROKEAHA.116.014950
37. Uchida N, Kushida Y, Kitada M, Wakao S, Kumagai N, Kuroda Y, Kondo Y, Hirohara Y, Kure S, Chazenbalk G, Dezawa M (2017) Beneficial effects of systemically administered human Muse cells in adriamycin nephropathy. J Am Soc Nephrol 28(10):2946–2960. https://doi.org/10.1681/ASN.2016070775
38. Yamada Y, Wakao S, Kushida Y, Minatoguchi S, Mikami A, Higashi K, Baba S, Shigemoto T, Kuroda Y, Kanamori H, Amin M, Kawasaki M, Nishigaki K, Hosoyama K, Taoka M, Isobe T, Muramatsu C, Dezawa M, Minatoguchi S (2018) S1P-S1PR2 axis mediates homing of Muse cells into damaged heart for long lasting tissue repair and functional recovery after acute myocardial infarction. Circ Res 2018 13;122(8):1069–1083. https://doi.org/10.1161/CIRCRESAHA.117.311648
39. Wakao S, Kitada M, Kuroda Y, Dezawa M (2012) Isolation of adult human pluripotent stem cells from mesenchymal cell populations and their application to liver damages. Methods Mol Biol 826:89–102. https://doi.org/10.1007/978-1-61779-468-1_8
40. Katagiri H, Kushida Y, Nojima M, Kuroda Y, Wakao S, Ishida K, Endo F, Kume K, Takahara T, Nitta H, Tsuda H, Dezawa M, Nishizuka SS (2016) A distinct subpopulation of bone marrow mesenchymal stem cells, Muse cells, directly commit to the replacement of liver components. Am J Transplant 16(2):468–483. https://doi.org/10.1111/ajt.13537
41. Kinoshita K, Kuno S, Ishimine H, Aoi N, Mineda K, Kato H, Doi K, Kanayama K, Feng J, Mashiko T, Kurisaki A, Yoshimura K (2015) Therapeutic potential of adipose-derived SSEA-3-positive Muse cells for treating diabetic skin ulcers. Stem Cells Transl Med 4(2):146–155. https://doi.org/10.5966/sctm.2014-0181
42. Wakao S, Akashi H, Kushida Y, Dezawa M (2014) Muse cells, newly found non-tumorigenic pluripotent stem cells, reside in human mesenchymal tissues. Pathol Int 64(1):1–9. https://doi.org/10.1111/pin.12129 Review
43. Dezawa M (2016) Muse cells provide the pluripotency of mesenchymal stem cells: direct contribution of Muse cells to tissue regeneration. Cell Transplant 25(5):849–861. https://doi.org/10.3727/096368916X690881
44. Tanaka T, Nishigaki K, Minatoguchi S, Nawa T, Yamada Y, Kanamori H, Mikami A, Ushikoshi H, Kawasaki M, Dezawa M, Minatoguchi S (2018) Mobilized Muse cells after acute myocardial infarction predict cardiac function and remodeling in the chronic phase. Circ J 82(2):561–571. https://doi.org/10.1253/circj.CJ-17-0552
45. Nishimura EK, Jordan SA, Oshima H, Yoshida H, Osawa M, Moriyama M, Jackson IJ, Barrandon Y, Miyachi Y, Nishikawa S (2002) Dominant role of the niche in melanocyte stem-cell fate determination. Nature 416(6883):854–860
46. Murga M, Yao L, Tosato G (2004) Derivation of endothelial cells from CD34- umbilical cord blood. Stem Cells 22(3):385–395
47. Middleton J, Americh L, Gayon R, Julien D, Mansat M, Mansat P, Anract P, Cantagrel A, Cattan P, Reimund JM, Aguilar L, Amalric F, Girard JP (2005) A comparative study of endothelial cell markers expressed in chronically inflamed human tissues: MECA-79, Duffy antigen receptor for chemokines, von Willebrand factor, CD31, CD34, CD105 and CD146. J Pathol 206:260–268
48. Crisan M, Yap S, Casteilla L, Chen CW, Corselli M, Park TS, Andriolo G, Sun B, Zheng B, Zhang L, Norotte C, Teng PN, Traas J, Schugar R, Deasy BM, Badylak S, Buhring HJ, Giacobino JP, Lazzari L, Huard J, Péault B (2008) A perivascular origin for mesenchymal stem cells in multiple human organs. Cell Stem Cell 3(3):301–313. https://doi.org/10.1016/j.stem.2008.07.003
49. Nagoshi N, Shibata S, Kubota Y, Nakamura M, Nagai Y, Satoh E, Morikawa S, Okada Y, Mabuchi Y, Katoh H, Okada S, Fukuda K, Suda T, Matsuzaki Y, Toyama Y, Okano H (2008) Ontogeny and multipotency of neural crest-derived stem cells in mouse bone marrow, dorsal root ganglia, and whisker pad. Cell Stem Cell 2(4):392–403. https://doi.org/10.1016/j.stem.2008.03.005

50. Fernandes KJ, McKenzie IA, Mill P, Smith KM, Akhavan M, Barnabé-Heider F, Biernaskie J, Junek A, Kobayashi NR, Toma JG, Kaplan DR, Labosky PA, Rafuse V, Hui CC, Miller FD (2004) A dermal niche for multipotent adult skin-derived precursor cells. Nat Cell Biol 6(11):1082–1093
51. Biernaskie J, Paris M, Morozova O, Fagan BM, Marra M, Pevny L, Miller FD (2009) SKPs derive from hair follicle precursors and exhibit properties of adult dermal stem cells. Cell Stem Cell 5(6):610–623. https://doi.org/10.1016/j.stem.2009.10.019
52. Gimble JM, Katz AJ, Bunnell BA (2007) Adipose-derived stem cells for regenerative medicine. Circ Res 100(9):1249–1260 Review
53. Heneidi S, Simerman AA, Keller E, Singh P, Li X, Dumesic DA, Chazenbalk G (2013) Awakened by cellular stress: isolation and characterization of a novel population of pluripotent stem cells derived from human adipose tissue. PLoS One 8(6):e64752. https://doi.org/10.1371/journal.pone.0064752
54. Pera MF, Trounson AO (2004) Human embryonic stem cells: prospects for development. Development 131(22):5515–5525
55. Brons IG, Smithers LE, Trotter MW, Rugg-Gunn P, Sun B, de Sousa C, Lopes SM, Howlett SK, Clarkson A, Ahrlund-Richter L, Pedersen RA, Vallier L (2007) Derivation of pluripotent epiblast stem cells from mammalian embryos. Nature 448(7150):191–195
56. Tesar PJ, Chenoweth JG, Brook FA, Davies TJ, Evans EP, Mack DL, Gardner RL, McKay RD (2007) New cell lines from mouse epiblast share defining features with human embryonic stem cells. Nature 448(7150):196–199
57. Nichols J, Smith A (2009) Naive and primed pluripotent states. Cell Stem Cell 4(6):487–492. https://doi.org/10.1016/j.stem.2009.05.015 Review
58. Kyba M, Perlingeiro RC, Daley GQ (2002) HoxB4 confers definitive lymphoid-myeloid engraftment potential on embryonic stem cell and yolk sac hematopoietic progenitors. Cell 109(1):29–37
59. Wang L, Menendez P, Shojaei F, Li L, Mazurier F, Dick JE, Cerdan C, Levac K, Bhatia M (2005) Generation of hematopoietic repopulating cells from human embryonic stem cells independent of ectopic HOXB4 expression. J Exp Med 201(10):1603–1614
60. Buecker C, Geijsen N (2010) Different flavors of pluripotency, molecular mechanisms, and practical implications. Cell Stem Cell 7(5):559–564. https://doi.org/10.1016/j.stem.2010.10.007
61. Niwa H, Ogawa K, Shimosato D, Adachi K (2009) A parallel circuit of LIF signalling pathways maintains pluripotency of mouse ES cells. Nature 460(7251):118–122. https://doi.org/10.1038/nature08113
62. Ying QL, Nichols J, Chambers I, Smith A (2003) BMP induction of id proteins suppresses differentiation and sustains embryonic stem cell self- renewal in collaboration with STAT3. Cell 115(3):281–292
63. Ying QL, Wray J, Nichols J, Batlle-Morera L, Doble B, Woodgett J, Cohen P, Smith A (2008) The ground state of embryonic stem cell self-renewal. Nature 453(7194):519–523. https://doi.org/10.1038/nature06968
64. Zwaka TP, Thomson JA (2003) Homologous recombination in human embryonic stem cells. Nat Biotechnol 21(3):319–321
65. Ware CB, Nelson AM, Mecham B, Hesson J, Zhou W, Jonlin EC, Jimenez-Caliani AJ, Deng X, Cavanaugh C, Cook S, Tesar PJ, Okada J, Margaretha L, Sperber H, Choi M, Blau CA, Treuting PM, Hawkins RD, Cirulli V, Ruohola-Baker H (2014) Derivation of naive human embryonic stem cells. Proc Natl Acad Sci U S A 111(12):4484–4489. https://doi.org/10.1073/pnas.1319738111
66. Sancho-Martinez I, Li M, Izpisua Belmonte JC (2011) Disease correction the iPSC way: advances in iPSC-based therapy. Clin Pharmacol Ther 89(5):746–749. https://doi.org/10.1038/clpt.2010.341
67. Zhou W, Choi M, Margineantu D, Margaretha L, Hesson J, Cavanaugh C, Blau CA, Horwitz MS, Hockenbery D, Ware C, Ruohola-Baker H (2012) HIF1alpha induced switch from bivalent to exclusively glycolytic metabolism during ESC-to-EpiSC/hESC transition. EMBO J 31(9):2103–2116. https://doi.org/10.1038/emboj.2012.71

68. Sperber H, Mathieu J, Wang Y, Ferreccio A, Hesson J, Xu Z, Fischer KA, Devi A, Detraux D, Gu H, Battle SL, Showalter M, Valensisi C, Bielas JH, Ericson NG, Margaretha L, Robitaille AM, Margineantu D, Fiehn O, Hockenbery D, Blau CA, Raftery D, Margolin AA, Hawkins RD, Moon RT, Ware CB, Ruohola-Baker H (2015) The metabolome regulates the epigenetic landscape during naive-to-primed human embryonic stem cell transition. Nat Cell Biol 17(12):1523–1535. https://doi.org/10.1038/ncb3264
69. Nichols J, Smith A (2011) The origin and identity of embryonic stem cells. Development 138(1):3–8. https://doi.org/10.1242/dev.050831
70. Ficz G, Hore TA, Santos F, Lee HJ, Dean W, Arand J, Krueger F, Oxley D, Paul YL, Walter J, Cook SJ, Andrews S, Branco MR, Reik W (2013) FGF signaling inhibition in ESCs drives rapid genome-wide demethylation to the epigenetic ground state of pluripotency. Cell Stem Cell 13(3):351–359. https://doi.org/10.1016/j.stem.2013.06.004
71. Habibi E, Brinkman AB, Arand J, Kroeze LI, Kerstens HH, Matarese F, Lepikhov K, Gut M, Brun-Heath I, Hubner NC, Benedetti R, Altucci L, Jansen JH, Walter J, Gut IG, Marks H, Stunnenberg HG (2013) Whole-genome bisulfite sequencing of two distinct interconvertible DNA methylomes of mouse embryonic stem cells. Cell Stem Cell 13(3):360–369. https://doi.org/10.1016/j.stem.2013.06.002
72. Marks H, Kalkan T, Menafra R, Denissov S, Jones K, Hofemeister H, Nichols J, Kranz A, Stewart AF, Smith A, Stunnenberg HG (2012) The transcriptional and epigenomic foundations of ground state pluripotency. Cell 149(3):590–604. https://doi.org/10.1016/j.cell.2012.03.026
73. Thomson JA, Itskovitz-Eldor J, Shapiro SS, Waknitz MA, Swiergiel JJ, Marshall VS, Jones JM (1998) Embryonic stem cell lines derived from human blastocysts. Science 282(5391):1145–1147
74. Greber B, Wu G, Bernemann C, Joo JY, Han DW, Ko K, Tapia N, Sabour D, Sterneckert J, Tesar P, Schöler HR (2010) Conserved and divergent roles of FGF signaling in mouse epiblast stem cells and human embryonic stem cells. Cell Stem Cell 6(3):215–226. https://doi.org/10.1016/j.stem.2010.01.003
75. Kitada M, Wakao S, Dezawa M (2012) Muse cells and induced pluripotent stem cell: implication of the elite model. Cell Mol Life Sci 69(22):3739–3750. https://doi.org/10.1007/s00018-012-0994-5

Chapter 3
Muse Cells Are Endogenous Reparative Stem Cells

Yoshihiro Kushida, Shohei Wakao, and Mari Dezawa

Abstract The dynamics and actions of Muse cells at a time of physical crisis are unique and highly remarkable compared with other stem cell types. When the living body is in a steady state, low levels of Muse cells are mobilized to the peripheral blood, possibly from the bone marrow, and supplied to the connective tissue of nearly every organ. Under conditions of serious tissue damage, such as acute myocardial infarction and stroke, Muse cells are highly mobilized to the peripheral blood, drastically increasing their numbers in the peripheral blood within 24 h after the onset of tissue injury. The alerting signal, sphingosine-1-phosphate, attracts Muse cells to the damaged site mainly via the sphingosine-1-phosphate receptor 2, enabling them to preferentially home to site of injury. After homing, Muse cells spontaneously differentiate into tissue-compatible cells and replenish new functional cells for tissue repair. Because Muse cells have pleiotropic effects, including paracrine, anti-inflammatory, anti-fibrotic, and anti-apoptotic effects, these cells synergistically deliver long-lasting functional and structural recovery. This chapter describes how Muse cells exert their reparative effects in vivo.

Keywords Sphingosine-1-phosphate (S1P) · Migration · Homing · Repair · Anti-inflammation · Anti-fibrosis · Immunosuppression · Intravenous injection · Allograft · Paracrine effect

3.1 Muse Cell Function as Reparative Stem Cells

Endogenous Muse cells are considered to mobilize from the bone marrow (BM) into the peripheral blood and circulate throughout the body, widely supplying organs through vessels and migration into the connective tissues of the organs (Fig. 3.1) [1, 2]. A series of studies in which exogenous Muse cells were administered unveiled many mechanisms underlying the tissue reparative effects of

Y. Kushida · S. Wakao · M. Dezawa (✉)
Department of Stem Cell Biology and Histology, Tohoku University Graduate School of Medicine, Sendai, Japan
e-mail: mdezawa@med.tohoku.ac.jp

M. Dezawa (ed.), *Muse Cells*, Advances in Experimental Medicine and Biology 1103, https://doi.org/10.1007/978-4-431-56847-6_3

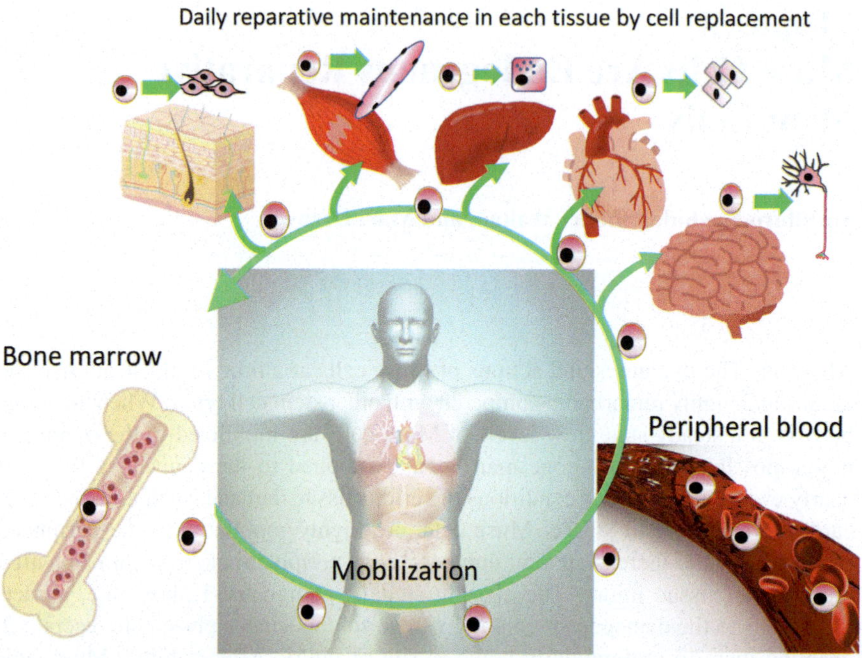

Fig. 3.1 Schematic diagram for Muse cell dynamics in the steady state

endogenous/exogenous Muse cells as well as the contribution of endogenous Muse cells to daily maintenance of the living body. The innate reparative functions of Muse cells substantiate the feasibility of Muse cells for clinical application.

The reparative functions of Muse cells are summarized as follows:

1. Both endogenous Muse cells and exogenously administered Muse cells sense the location of the damaged site and migrate to and home to the site [3–9] mainly via the sphingosine-1-phosphate (S1P)-S1P receptor 2 (S1PR2) system (Fig. 3.2) [10].
2. Because Muse cells are stress tolerant, they are able to survive in the hostile microenvironment of damaged tissue. The immunomodulatory effect of Muse cells allows allograft and xenograft cells to escape from the host immunologic attack, enabling them to efficiently home to the damaged site after administration [9, 10].
3. After homing, Muse cells replenish new functional cells by spontaneous differentiation into tissue-compatible cells (Fig. 3.2) [3–10].
4. Muse cells survive and remain integrated in the host tissue for an extended period. Therefore, their anti-fibrotic, anti-inflammatory, anti-apoptotic, and paracrine effects are long-lasting and effective [3, 7, 9, 10].
5. Together, the above actions lead to the functional and structural repair of the damaged tissue.
6. Muse cells can be supplied to patients in three simple steps: (1) collection and (2) expansion from tissue sources and (3) intravenous administration (Fig. 3.3).

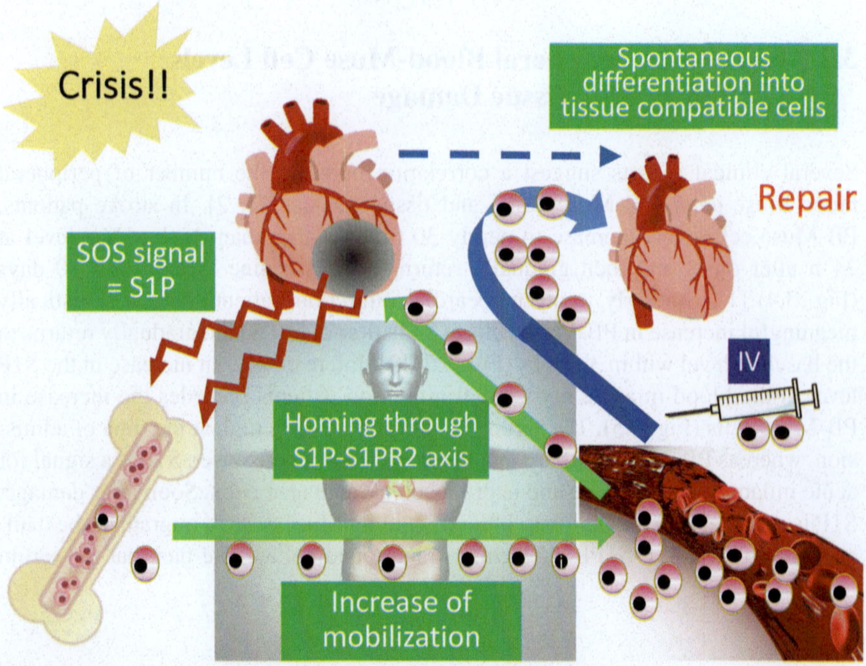

Fig. 3.2 Schematic diagram for Muse cell dynamics and its mechanism under critical situation such as acute myocardial infarction

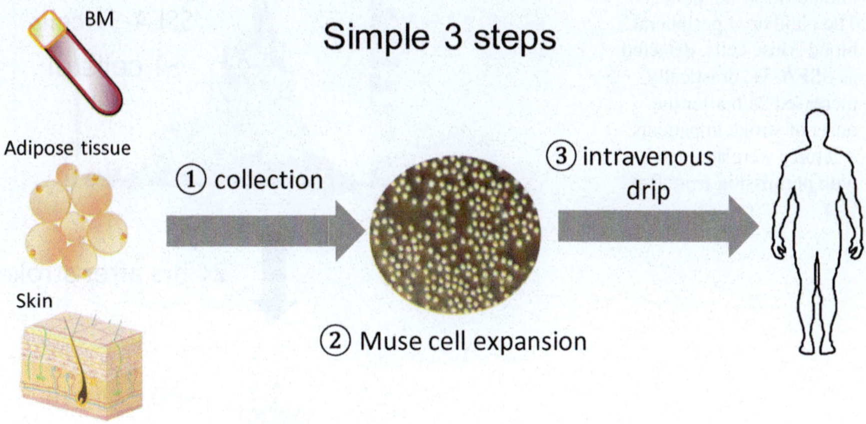

Fig. 3.3 Schematic diagram for Muse cell therapy

7. In this process, Muse cells require no gene introduction or cytokine induction prior to treatment, providing feasible regenerative therapy through an intravenous drip, which is an expedient approach for patients.

Based on these characteristics, Life Science Institute Inc., a group company of Mitsubishi Chemical Holdings, started clinical trial for acute myocardial infarction in 2018.

3.2 Increase in Peripheral Blood-Muse Cell Levels Correlates with Tissue Damage

Several clinical reports suggest a correlation between the number of peripheral blood-Muse cells (PB-Muse cells) and tissue damage [1, 2]. In stroke patients, PB-Muse cell levels increase to nearly 30 times higher than the baseline level at 24 h after onset and then gradually return to the baseline level within 30 days (Fig. 3.4) [1]. Similarly, acute myocardial infarction patients have a statistically meaningful increase in PB-Muse cells at 24 h after onset, which gradually returns to the baseline level within 3 weeks (Fig. 3.5) [2]. Interestingly, an increase in the S1P level in the blood in acute myocardial infarction patients precedes the increase in PB-Muse cells (Fig. 3.5). The serum S1P level is already high at the time of admission, whereas PB-Muse cells are increased at 24 h after the onset. S1P is a signal for acute inflammation/damage and is produced by damaged cells. Soon after damage, S1P is released into the peripheral blood, guiding Muse cells to migrate to the damaged site by the S1P-S1PR2 system, as mentioned below. The fact that the serum

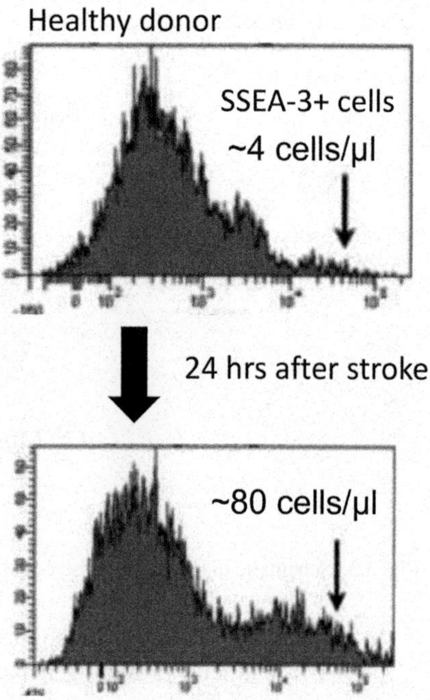

Fig. 3.4 The increase of peripheral blood-Muse cell number in stroke patients. The number of peripheral blood-Muse cells, detected as SSEA-3+, drastically increased 24 h after the onset of stroke in patients. (Pictures were adopted with permission from Ref. [1])

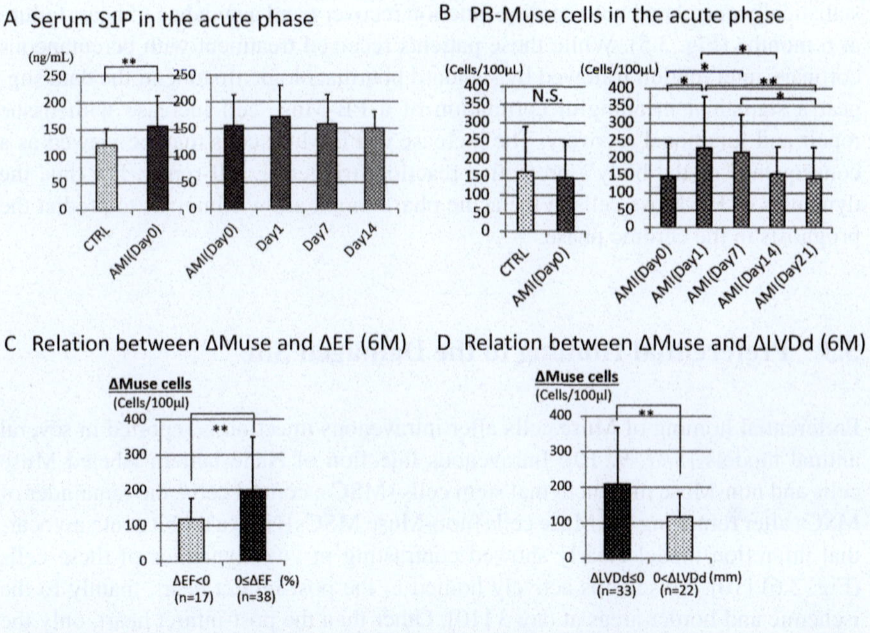

Fig. 3.5 Correlation between the increase of peripheral blood-Muse cells and cardiac function recovery. (**A**) Serum S1P is already elevated at day0, the onset of acute myocardial infarction, with statistical difference to control (CTRL). The maximum S1P level is at day1. (**B**) The number of peripheral blood-Muse cells, on the other hand, did not increase at day0 but elevated at day 1 with statistical significance, and gradually returned to the baseline level by several weeks. (**C**) Patients who showed the substantial increase of peripheral blood-Muse cells in the acute phase (Δ Muse) tended to have better recovery in ejection fraction (EF), an indicator of cardiac function, at 6 months compared to that in the acute phase (0<Δ EF). (**D**) Patients who showed the substantial increase of peripheral blood-Muse cells in the acute phase (Δ Muse) tended to have the lower left ventricular LV end-diastolic dimension (LVDd), an indicator of heart failure, at 6 months compared to the acute phase (Δ LVDd<0). Thus, patients who showed higher peripheral blood-Muse cell mobilization after the onset in the early phase showed the tendency of better recovery in cardiac function. *; $p<0.05$, **; $p<0.01$. (Pictures were adopted with permission from Ref. [10])

S1P increase precedes the increase in PB-Muse cells supports the notion that Muse cell migration is controlled by the S1P-S1PR2 system.

Findings from patients with acute myocardial infarction also demonstrate that Muse cells work as endogenous reparative stem cells in vivo [2]. Some patients did not show an increase in PB-Muse cells in the acute phase (within 7 days). In these patients, the ejection fraction, a parameter of cardiac function, tended to deteriorate, and the left ventricular end-diastolic dimension, an indicator of heart failure, increased in the chronic phase at 6 months after onset (Fig. 3.5). Conversely, an increase in PB-Muse cells from the baseline level in patients in the acute phase statistically correlated with cardiac function recovery and avoidance of heart failure at 6 months (Fig. 3.5). While those patients received treatment with percutaneous coronary intervention followed by standard pharmacologic treatment, the data suggest a statistical meaningful correlation of a PB-Muse cell increase with tissue repair and functional recovery. The increase in PB-Muse cells may be viewed as a consequence of the body's protective reaction, triggering self-repair. For this, the dynamics of PB-Muse cells in the acute phase might allow clinicians to predict the prognosis in the chronic phase.

3.3 Preferential Homing to the Damaged Site

Preferential homing of Muse cells after intravenous injection is reported in several animal models [5, 7, 9, 10]. Intravenous injection of Nano-lantern-labeled Muse cells and non-Muse mesenchymal stem cells (MSCs; control cells, the remainder of MSCs after removing the Muse cells [non-Muse MSCs]) into a rabbit acute myocardial infarction model clearly showed contrasting in vivo dynamics of these cells (Fig. 3.6) [10]. Muse cells actively homed to the post-infarct heart, mainly to the ischemic and border areas at day 3 [10]. Other than the post-infarct heart, only the lung contained detectable levels of the exogenous Muse cells. On the other hand, non-Muse MSCs mostly accumulated in the lung and basically did not home to the post-infarct heart (Fig. 3.6). The cellular distribution became more impressive and contrasting at 2 weeks; homing of Muse cells became more apparent in the

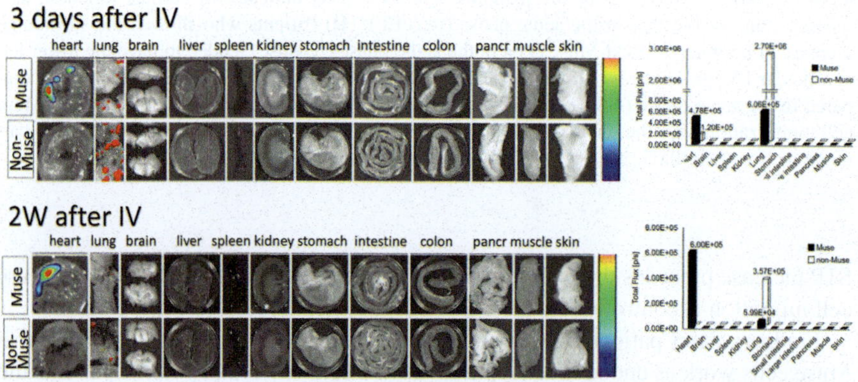

Fig. 3.6 In vivo dynamics of intravenously injected Muse and non-Muse cells in rabbit acute myocardial infarction at 3 days and 2 W after administration. In vivo dynamics of intravenously injected Nano-lantern-introduced human Muse and non-Muse cells in acute myocardial infarction model rabbit. Muse cells preferentially homed into the post-infarct heart at 3 days and lesser extent into the lung, while non-Muse cells did not home into the heart. Majority of non-Muse cells were detected in the lung. Muse cells remained in the post-infarct heart at 2 weeks. The signal of non-Muse cells was still negative in the heart and that in the lung decreased compared to 3 days. (Pictures were adopted with permission from Ref. [10])

post-infarct heart compared with other organs while that in the lung declined to a negligible level (Fig. 3.6). The amount of Muse in cells in other organs remained under the detection level. The homing of non-Muse MSCs to the lung was reduced compared to that at day 3, and the low number of the non-Muse MSCs recognized at day 3 in the post-infarct heart became under the detection level at 2 weeks.

Severe combined immunodeficiency (SCID) mice do not reject human cells. When BM-derived human Muse cells were administered to SCID mice models of partial hepatectomy, acute and chronic liver damage (Fig. 3.7), and chronic kidney disease (Fig. 3.8), the human Muse cells preferentially homed to the damaged tissue rather than the lung and spleen in the early phase [5, 7, 9]. Human cells in mouse tissue are distinguishable by human-specific genes, such as the Alu sequence [7–9] and human specific-prostaglandin E receptor 2 gene [5], as well as antibodies that are specific for human Golgi complex and mitochondria [3–9]. Green fluorescent protein (GFP) can be introduced into Muse cells by lentivirus prior to their administration and is another strong tool to label and trace Muse cells [4–9]. In these liver and kidney disease models, the majority of Muse cells detected in the body accumulated in the damaged organ in the early phase of until 1 week [7, 9]. The lung and spleen had a very small number of homing cells and the amount of Muse cells in the other organs was under the detection limit at this phase. In the later phase, such as at 4 weeks and 2 months after inducing damage, the damaged organ was the only organ in which Muse cells were detectable while the number of Muse cells in the

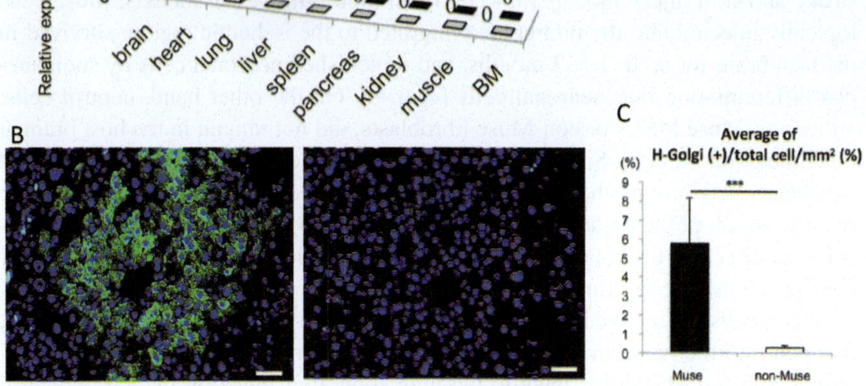

Fig. 3.7 In vivo dynamics of intravenously injected human Muse and non-Muse cells in SCID mouse acute liver damage model. In vivo dynamics at 2 W after administration (**A**). (**B**) GFP-Muse cell and non-Muse cell distribution in the liver at 2 W, detected by anti-GFP immunohistochemistry. (**C**) The average of integrated Muse and non-Muse cell number per area in the mouse liver. Bars = 50 μm. (Pictures were adopted with permission from Ref. [7])

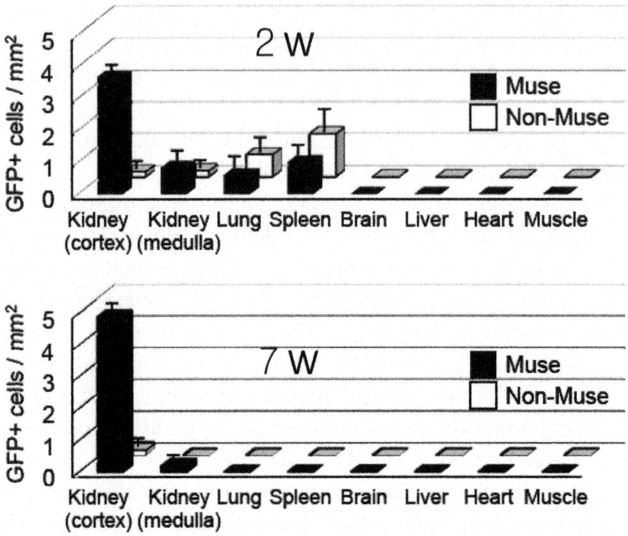

Fig. 3.8 In vivo dynamics of intravenously injected human Muse and non-Muse cells in SCID mouse chronic kidney disease model. In vivo dynamics at 2 W and 7 W after administration. (Pictures were adopted with permission from Ref. [9])

other organs, including the lung and spleen, was under the detection limit. In contrast to Muse cells, control cells such as non-Muse MSCs exhibited different dynamics; non-Muse MSCs did not home to the damaged site in the early phase, but instead, the majority distributed to the lung and spleen at high levels. Non-Muse MSCs in the lung and spleen eventually disappeared after ~2 weeks and did not remain in the body.

Locally injected Muse cells also migrate to and integrate into the damaged site in stroke and skin injury models [3, 4, 6, 8]. In mouse/rat stroke models, Muse cells topically injected into the infarct area migrated to the ischemic region, survived in the host brain for up to 1.5~3 months, and replenished neuronal cells by spontaneous differentiation into neuronal cells [4, 6, 8]. On the other hand, control cells, either non-Muse MSCs or non-Muse fibroblasts, did not remain in the host brain at this time point [4, 6, 8]. Similarly, in a mouse skin ulcer model, Muse cells injected subcutaneously around the ulcer migrated to the defect site and participated in the reconstruction of the epidermis and dermis by in vivo differentiation into keratinocytes, vessels, and fibroblasts, while non-Muse cells did not remain in the skin – nor did they contribute to skin tissue reconstruction [3].

When cells are injected into a vein, they are first delivered to the right ventricle, pumped into the lung, and flow into the lung capillaries. The majority of intravenously injected MSCs distribute to the lung soon after injection by entrapment in the lung capillaries [11–19]. Cells that escape from the entrapment join the arterial flow and circulate around the body to be distributed elsewhere. Other than the lung, MSCs preferentially home to the spleen [11, 14–19]. After more than 2 weeks,

MSCs usually disappear from the lung and spleen due to their removal by macrophages and vanish from the whole body [13, 15–17].

Compared with MSCs, the dynamics of Muse cells are unique and different. The most important difference is that the majority of Muse cells escape from lung capillary entrapment. This is why Muse cells can efficiently home to the site of damage. Indeed, in a rabbit model of acute myocardial infarction, autologous Muse cells were shown to home to the post-infarct heart at a high rate, ~14.5% of the total number of injected cells, at 3 days after intravenous injection [10]. The homing rate is remarkably high compared with that MSCs; intravenously injected MSCs basically do not home to the post-infarct heart, as mentioned above, and even if they are detected in the heart, the homing ratio is generally less than 1% if it occurs at all.

When separating Muse and non-Muse cells from MSCs by cell sorting, the Muse cells and non-Muse cells do not largely differ in cell size. Both fractions contain a range of cell sizes, and there is no obvious tendency toward a size difference. Thus, the mechanism by which Muse cells travel through the lung capillaries may not depend on their smaller size but rather on some unknown mechanism.

3.4 Muse Cell Migration Is Mainly Controlled by the S1P-S1PR2 System

In an in vitro Boyden chamber experiment, Muse cells exhibited potent migration toward tissue slices of damaged organs or the serum from animal models of disease. Using liver damage and chronic kidney disease models, slice cultures of damaged tissue and the serum from disease model animals placed in the lower chamber elicited strong migration of Muse cells from the upper chamber through the slit (Fig. 3.9) [7, 9]. The extent of the migration of Muse cells was consistently higher than that of non-Muse MSCs with statistical significance. This is consistent with the fact that Muse cells actively migrate and home to damaged tissue, while non-Muse MSCs do not in vivo.

Several factors and ligands participate in stem cell migration and homing. For example, stromal-derived factor 1 (SDF-1), which binds the CXCR4 receptor, directs the migration and homing of CXCR4+ hematopoietic stem cells to the BM [20–24]. Similarly, migration of MSCs is controlled by CXCR4 [25–29] and c-Met, ligands for SDF-1and hepatocyte growth factor (HGF) [25, 30, 31], respectively. Muse cells express both CXCR4 and c-Met to the same extent as non-Muse MSCs [7]. The presence of AMD3100, a CXCR4 antagonist, however, partially suppressed the migration of Muse cells toward the serum from a fulminant hepatitis model in the Boyden chamber experiment, but failed to completely abrogate the migration. Furthermore, the extent of suppression was nearly the same as that observed for non-Muse MSCs, and there was no Muse cell-specific phenomenon observed in this experiment. Therefore, specific migration of Muse cells to the damaged tissue/serum that sharply contrasts with non-Muse MSCs cannot be explained by the SDF-1-CXCR4 system.

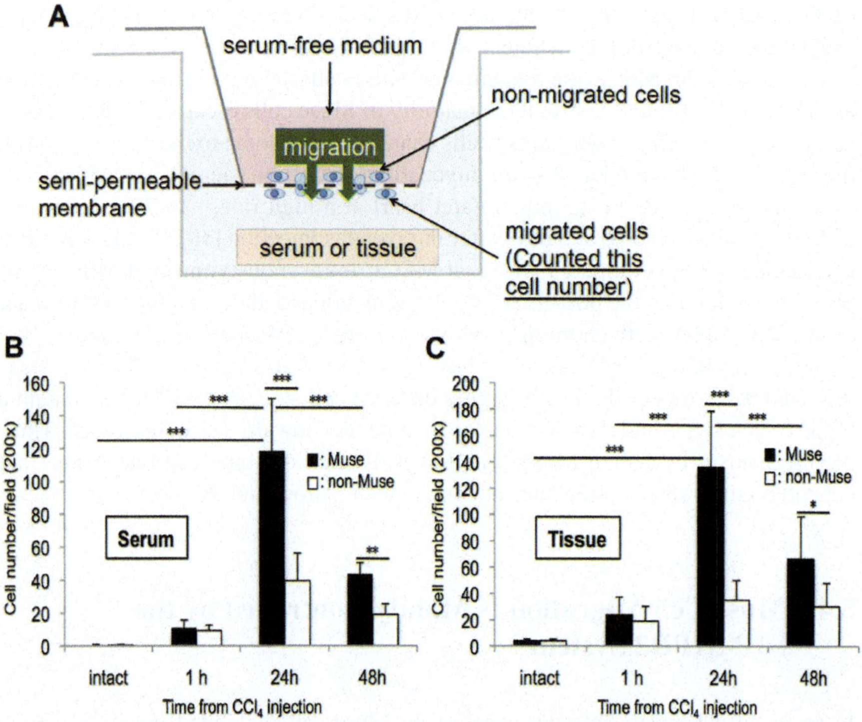

Fig. 3.9 In vitro migration of Muse cells toward the serum and tissue of damaged animal in vitro. The serum and liver tissue were obtained from carbon tetrachloride (CCl_4)-treated mouse. (**A**) Outline of the experiment. Human Muse or non-Muse cells were placed in the upper chamber of the semipermeable membrane, and either the serum or the liver tissue collected at 1, 24, and 28 h after intraperitoneal (IP) injection of CCl_4 or, from intact animal, was placed in the lower chamber. After 22 h of incubation, the number of migrated cells through the semipermeable membrane was counted. Muse cells migrate to the (**B**) serum and (**C**) liver tissue with higher efficiency than that observed in non-Muse cells. $^{**}p < 0.01$, $^{***}p < 0.001$. (Pictures were adopted with permission from Ref. [7])

S1P is a signaling sphingolipid and is also referred to as a bioactive lipid mediator [32–36]. S1P is less abundant in tissue fluids, creating an S1P gradient [37, 38]. This seems to have biologic significance in immune cell trafficking [38–40]. S1P was originally considered to be an intracellular second messenger, but the discovery of an extracellular ligand for G protein-coupled receptor S1PR1 indicates that most of the biologic effects of S1P are mediated by signaling through cell surface receptors [23, 38–42]. Five S1P receptors, S1PR1~S1PR5, have been described to date [38–42].

S1P is produced from sphingomyelin by the enzymatic reaction cascade converting sphingomyelin to ceramide, sphingosine, and finally S1P [35, 38, 39, 43]. While sphingomyelin locates widely in the plasma membrane, nucleus, lysosomes, and mitochondria, S1P is produced in the cytosol and plasma membrane at baseline levels, and production is activated when cells are damaged. In the cell membrane, sphingomyelin locates mainly in the outer leaflet [39, 44, 45]. When the cell is intact, converting enzymes in the cytoplasm are not accessible to the outer leaflet of the cell membrane because the inner leaflet of the cell membrane lies between them [39, 44, 46]. However, once the cell membrane is damaged, the active production of S1P begins (Fig. 3.10).

The active production of S1P in damaged tissue was demonstrated in a rabbit acute myocardial infarction model: 24 h after the onset, S1P levels in the border area (adjacent to infarcted area where living cardiomyocytes remained) were significantly higher than that in the infarct area (the area where the most of cardiomyocytes had died by severe ischemia) and intact areas [10]. Compared with normal heart tissue, the S1P level in the border area at 6 and 24 h after acute myocardial infarction was substantially higher compared to the baseline level with statistical significance (Fig. 3.11A). Consistently, intravenously administered Muse cells mainly homed to the border area rather than the infarct or intact area in the post-infarct heart (Fig. 3.11B). Why Muse cells homed to the infarct area to a lower extent than to the border area might be because cells in the infarct area were already killed by the severe ischemia and would not be able to produce S1P efficiently.

A series of experiments demonstrated that S1P and its receptor, S1PR2, comprise the most relevant system that controls the preferential migration of Muse cells to the damaged site [10]. Among five S1PR1~S1PR5 subtypes, S1PR2 was the receptor highly expressed in Muse cells than in non-Muse cells (Fig. 3.11C, D). In this context, Muse cells, but not non-Muse MSCs, migrated toward an S1PR2-specific agonist (SID46371153) in a dose-dependent manner in a Boyden chamber in vitro (Fig. 3.11E).

In vitro, JTE-013, an antagonist particularly selective for S1PR2 [47], inhibited the migration of Muse cells, but not non-Muse cells, toward a post-infarct heart slice culture in a dose-dependent manner (Fig. 3.11F) [10]. In vivo, when JTE-013 was co-injected, integration of Nano-lantern Muse cells was substantially attenuated, and the signal of Muse cell homing became under the detection limit at 2 weeks [10]. Consistently, the total estimated number of Muse cells that engrafted into the whole left ventricle at 2 weeks decreased to less than 7%~8% of the number integrated without JTE-013 (Fig. 3.12) [10]. Compared to Muse cell injection without JTE-013, the reduction of the infarct size by Muse cells co-injected with JTE-013 was lower and recovery of the EF was substantially attenuated (Fig. 3.12) [10]. The results suggested that S1PR2 is the most relevant receptor for Muse cell migration and homing and that efficient homing of Muse cells to the damaged site is a key point for delivery of the beneficial effects because tissue repair and functional recovery were significantly attenuated following S1PR2 blockade. Furthermore, knockdown of S1PR2 in Muse cells by siRNA gene silencing substantially attenuated migration toward the S1PR2-specific agonist SID46371153 as well as toward tissue slices of the post-infarct heart [10].

These findings together confirm the strong central role of the S1P-S1PR2 axis in the specific homing of systemically administered Muse cells to the site of damage.

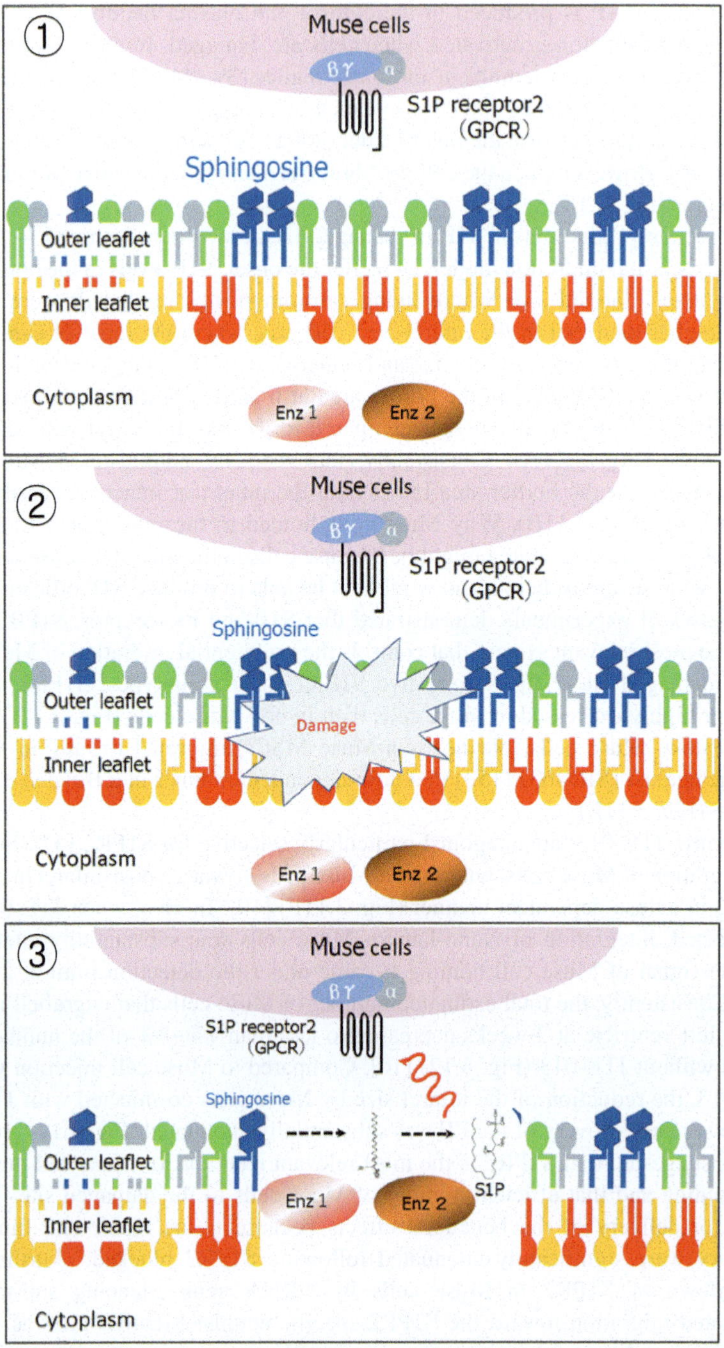

Fig. 3.10 Schematic diagram for production of S1P under cell damage and S1PR2 in Muse cells

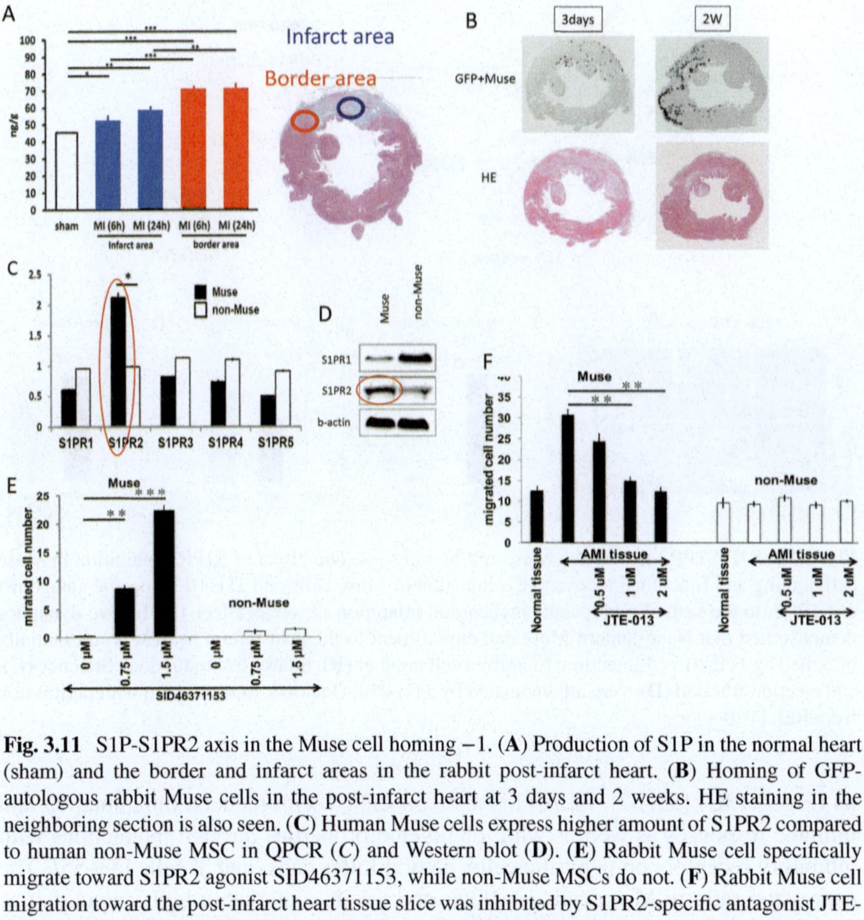

Fig. 3.11 S1P-S1PR2 axis in the Muse cell homing –1. (**A**) Production of S1P in the normal heart (sham) and the border and infarct areas in the rabbit post-infarct heart. (**B**) Homing of GFP-autologous rabbit Muse cells in the post-infarct heart at 3 days and 2 weeks. HE staining in the neighboring section is also seen. (**C**) Human Muse cells express higher amount of S1PR2 compared to human non-Muse MSC in QPCR (*C*) and Western blot (**D**). (**E**) Rabbit Muse cell specifically migrate toward S1PR2 agonist SID46371153, while non-Muse MSCs do not. (**F**) Rabbit Muse cell migration toward the post-infarct heart tissue slice was inhibited by S1PR2-specific antagonist JTE-013 in Boyden chamber experiment. (Pictures were adopted with permission from Ref. [10])

3.5 Immunomodulatory Effects of Muse Cells

The immunomodulatory effects of Muse cells are displayed in multiple aspects. Muse cells activate regulatory T cells and suppress dendritic cell differentiation. The regulation of T cells was demonstrated by co-culture of Muse cells and naïve human T cells, leading to upregulation of the regulatory T cell factors interleukin-10 and CD25 (Fig. 3.13A) [10, 48, 49]. The suppressed dendritic cell differentiation was displayed by co-culture of Muse cell-monocytes and Muse cell-monocyte-dendritic cell progenitors, which led to significant suppression of the differentiation of monocytes into monocyte-dendritic cell progenitors and into mature monocyte-dendritic cells, respectively (Fig. 3.13B) [10].

One potential explanation for the immunomodulatory effects of Muse cells is the expression of human leukocyte antigen (HLA)-G, an immunotolerance factor expressed in immune-privileged organs such as the placenta, thymus, ovary, and tes-

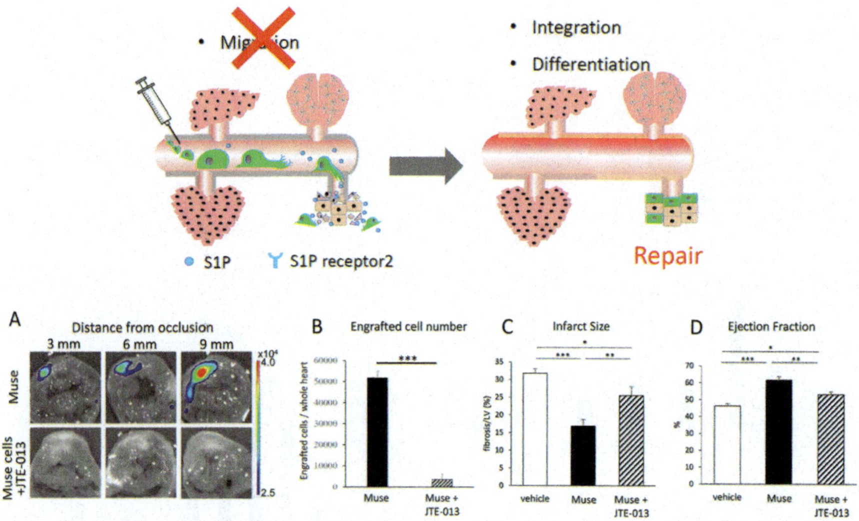

Fig. 3.12 S1P-S1PR2 axis in the Muse cell homing –2. The effect of S1PR2 inhibition in Muse cell homing and functional recovery. Co-injection of Muse cells and JTE-013, specific antagonist of S1PR2, to the vein of rabbit acute myocardial infarction (2 weeks after). (**A**) In vivo dynamics demonstrated that Nano-lantern Muse cell engraftment to the post-infarct heart was substantially inhibited by JTE-013 co-injection. Engrafted cell number (**B**), infarct size reduction efficiency (**C**), and ejection fraction (**D**) were all attenuated by JTE-013. (Pictures were adopted with permission from Ref. [10])

tis (Fig. 3.13C) [50–52]. HLA-G is also associated with reduced inflammation and immune responses as well as with tolerogenic properties through interactions with inhibitory receptors on dendritic cells, natural killer cells, and T cells [53–55]. The expression ratio in Muse cells is ~90%, and this is remarkably higher than that of other stem cells (Fig. 3.13) [10]. For example, human embryonic stem (ES) and induced pluripotent stem (iPS) cells do not express HLA-G [56–58], and less than 20% of adult BM-MSCs express HLA-G [59, 60]. Because HLA-G is suggested to promote graft tolerance in heart transplantation [61], the high expression of HLA-G together with the immunomodulatory effects of Muse cells may contribute to their escape from immunologic attack during the early phase of integration into the tissue. In fact, in a rabbit acute myocardial infarction model, allograft Muse cells that engrafted to the infarct border area expressed HLA-G on day 3 after intravenous injection (Fig. 3.13D) [10].

Another factor relevant to the immunomodulation of Muse cells is interferon gamma-induced expression of indoleamine-2,3 dioxygenase (IDO), a mediator of immunosuppression in MSCs [9]. Mixed lymphocyte proliferation assay displayed that the presence of Muse cells inhibited the proliferation of T lymphocytes and interferon-gamma-stimulated expression of IDO in Muse cells (FIg. 3.13E, F) [9].

To assess the effectiveness of the immunomodulatory effects of Muse cells, tissue repair and functional recovery were compared between autologous and alloge-

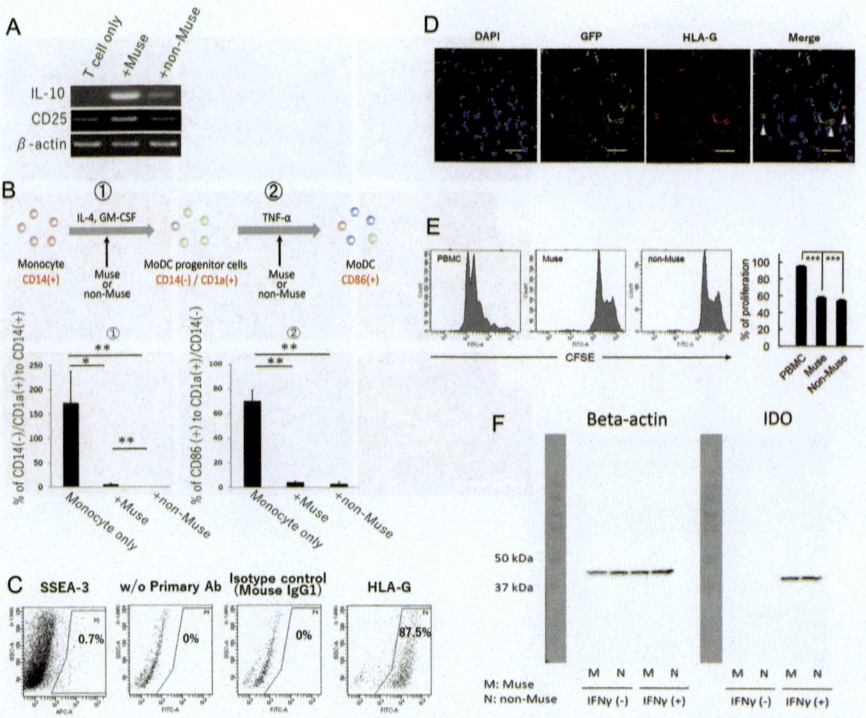

Fig. 3.13 Immunomodulatory effect of Muse cells. Muse cells stimulated conversion of naïve T cells to Treg (**A**). They also inhibited development of monocytes to dendritic progenitor and to mature dendritic cells (**B**). (**C**) Expression of HLA-G in Muse cells. (**D**) Muse cells after engraftment into the post-infarct heart expressed HLA-G in vivo at 3 days. (**E**) Mixed lymphocyte assay showed that Muse cells suppressed activation of lymphocytes. (**F**) By the stimulation of interferon-gamma, Muse cells produce IDO as the same extent as that in non-Muse MSCs. Bars = 50 μm. (Pictures were adopted with permission from Ref. [9] and [10])

neic Muse cells in the rabbit acute myocardial infarction model [10]. The number of engrafted allogeneic Muse cells in the post-infarct heart after intravenous injection was lower than that of autologous Muse cells, corresponding to nearly 85% of that of autologous Muse cells. The infarct size and ejection fraction, however, were not significantly different between the two groups. Overall, the reparative effects are expected to be achieved by allogeneic Muse cells (Fig. 3.14).

3.6 Spontaneous Differentiation into Tissue-Compatible Cells After Homing

The most prominent feature of Muse cells relevant to their reparative functions is that they spontaneously differentiate into cells compatible with the tissue to which they homed, leading to robust tissue repair by the replenishment of functional cells.

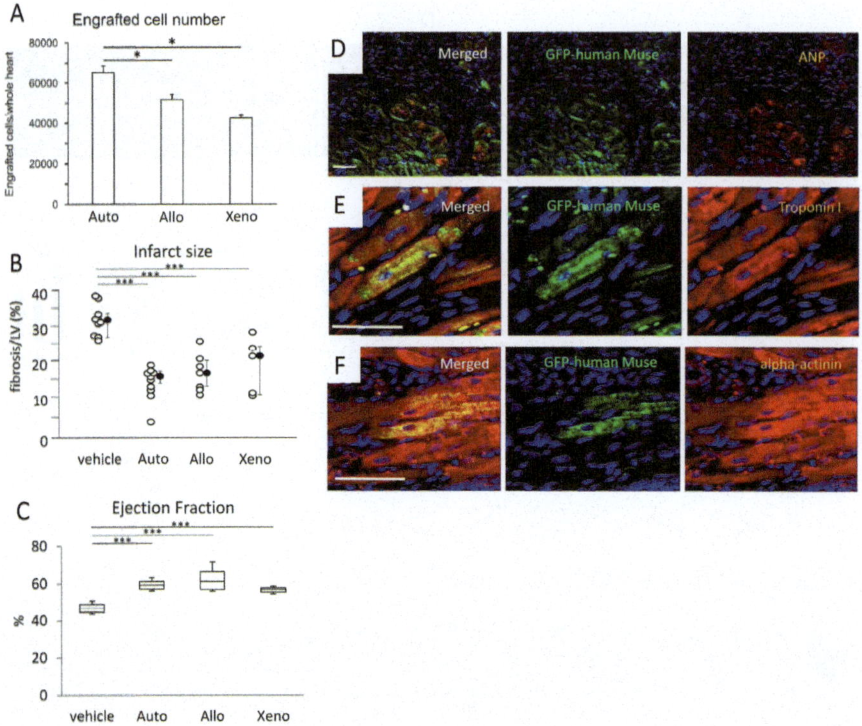

Fig. 3.14 Comparison of effectiveness among autograft, allograft, and xenograft Muse cells in rabbit acute myocardial infarction (2 weeks). Engrafted Muse cell number (**A**), infarct size (**B**), and ejection fraction (**C**) in acute myocardial infarction rabbit model (2 weeks) treated with intravenous injection of vehicle, autograft, allograft, and xenograft (human) Muse cells (3×10^5 cells each). While the engrafted cell number marked statistical difference between autograft and allograft, infarct size and ejection fraction approximated in these two groups. (**D–F**) Expression of atrium nitric peptide (ANP), cardiac troponin-I, and alpha-actinin in integrated GFP-allograft Muse cells at 2 weeks. Bars = 50 μm. (Pictures were adopted with permission from Ref. [10])

In a rabbit acute myocardial infarction model, Muse cells homed to the post-infarct heart and spontaneously differentiated into cardiomyocytes and vascular cells without fusing with host cells, thereby contributing to tissue repair [10]. The functionality of the Muse cell-derived cardiomyocytes was demonstrated not only by the expression of typical cardiac markers, troponin-I and sarcomeric α-actinin, but also by their connections to neighboring cardiomyocytes by connexin 43. Most importantly, GCaMP3-labeled Muse cells that engrafted into the ischemic region exhibited increased GCaMP3 fluorescence during systole and decreased fluorescence during diastole, synchronous with cardiac electrical excitation, suggesting the functionality of spontaneously differentiated Muse cells as cardiomyocytes (see Chap. 8).

In a mouse liver cirrhosis model, intravenously injected Muse cells differentiated into albumin-, HepPar-1-, and anti-trypsin-positive hepatocytes without fusing with the host hepatocytes and also expressed Cytochrome P450 1A2 (CYP1A2) and

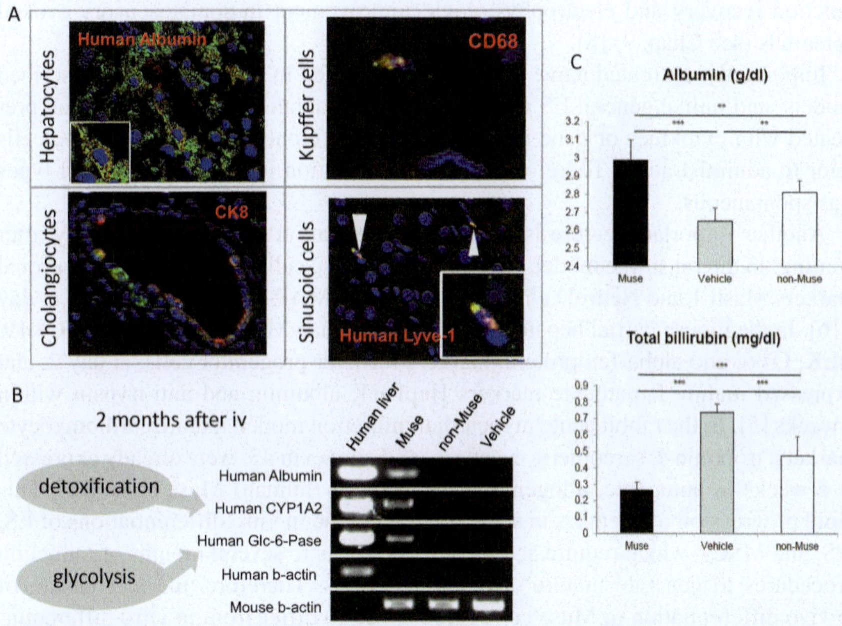

Fig. 3.15 Spontaneous differentiation of Muse cells into functional hepatocyte. (**A**) Human Muse cells integrated into the injured site of the liver of SCID mice partial hepatectomy model (4 weeks). Muse cells spontaneously differentiated into human albumin+ hepatocytes, cytokerain8 (CK8)+ cholangiocytes, CD68+ Kupffer cells, and human Lyve-1+ sinusoid endothelial cells after integration. (Pictures were adopted with permission from Ref. [5].) (**B–C**) Human Muse cells integrated into the SCID mouse liver cirrhosis model expressed not only general hepatocyte marker, human albumin, but also CYP1A2, detoxification enzyme, and Glc-6-Pase, glycolysis enzyme, suggesting the functionality of Muse cell-derived hepatocytes (*B*). In these animals, increase of serum albumin and decrease of total bilirubin were recognized compared to the vehicle and non-Muse groups with statistical differences. **, $p < 0.01$; ***, $p < 0.001$. (Pictures were adopted with permission from Ref. [7])

glucose-6-phosphatase, (Glc-6-ase), enzymes related to drug metabolism and glycolysis, respectively (Fig. 3.15) [7]. Indeed, animals injected with Muse cells had increased serum albumin and decreased total bilirubin levels. These results suggest that the integrated Muse cells functioned as hepatocytes in the liver cirrhosis model.

In a mouse chronic kidney disease model, intravenously injected Muse cells homed to the damaged glomeruli and spontaneously differentiated into podocytes (positive for WT-1 and podocin), mesangial cells (megsin), and endothelial cells (CD31 and von Willebrand factor), components of the glomerulus, without fusion, and delivered improvement in creatinine clearance, urine protein, and plasma creatinine (see Chap. 11) [9].

In the mouse stroke model, as mentioned above, Muse cells topically administered to the ischemic region spontaneously differentiated into neuronal cells (~65% of engrafted cells) and oligodendrocytes (~25%) and incorporated into the pyramidal tract, including the pyramidal decussation, and sensory tracts. This led to motor function recovery and electrophysiologic improvement in somatosensory evoked potentials (see Chap. 9) [8].

Importantly, untreated naïve Muse cells were used in all of the above-described models, and unlike general ES and iPS cell transplantation, they had not been pre-treated with cytokines or gene introduction for differentiation into purposive cells prior to administration. Therefore, their differentiation into the relevant cell types was spontaneous.

Another important feature is that the differentiation was initiated swiftly after homing. In the rat stroke model, homed human Muse cells expressed early neuronal markers Mash 1 and NeuroD already at day 3 and MAP-2, DCX, and NeuN at day 7 [6]. In the mouse partial hepatectomy model, human Muse cells expressed CK19, DLK, OV6, and alpha-fetoprotein, markers for liver progenitor cells, at day 2, and expressed mature hepatocyte markers HepPar1, albumin, and anti-trypsin within 2 weeks [5]. In the rabbit acute myocardial infarction model, mature cardiomyocyte markers, troponin-I, sarcomeric α-actinin, and connexin 43, were already expressed at 2 weeks in autogenic, allogenic, and xenograft (human) Muse cells [10]. This rapid progression of events is in sharp contrast to the in vitro differentiations of ES, iPS, and MSCs, which require at least several weeks to several months of induction procedures to generate mature differentiated cells. Therefore, the mechanism of in vivo differentiation in Muse cells is presumed to differ from in vitro differentiation of ES and iPS cells and MSCs.

The mechanism of Muse cell differentiation has not yet been clarified in detail, but the results of one experiment suggested a unique differentiation system in Muse cells. The conditioned medium of the "mouse" hepatoblast cell line Hepa-1-6 cells, either (1) intact or (2) treated with etoposide, which induces cell death, was supplied to "human" naïve Muse cells (Fig. 3.16). The naïve human Muse cells originally did not express human-specific hepatocyte lineage markers, Sox17, CK18, Prox1, or alpha-fetoprotein. Even after culturing the human Muse cells with conditioned medium from conditions (1) and (2) for up to 3 weeks, none of the human-specific hepatocyte lineage markers were detected in the Muse cells. When Muse cells were mixed with "intact" mouse Hepa-1-6 and co-cultured, the human-specific hepatocyte markers were still all negative. Interestingly, however, mixed co-culture of human Muse cells with "apoptotic" mouse Hepa-1-6 newly induced human CK18 expression in Muse cells at 3 days, human Sox17 and alpha-fetoprotein at 1 week, and human Prox1 at 2 weeks, suggesting the importance of "direct cell-cell interactions" between Muse cells and the damaged host cells for the differentiation commitment of Muse cells and not humoral factors provided by the host cells (Fig. 3.16) [7]. The precise mechanisms underlying how Muse cells interact with damaged host cells and how they receive the instruction from damaged cells, however, require further investigation.

The importance of the in vivo differentiation of Muse cells for tissue repair and functional recovery was evaluated in a rabbit acute myocardial infarction model [10]. GATA4 is a critical factor for cardiomyocyte differentiation. When GATA4-gene-silenced Muse cells were intravenously administered, the reparative functions of Muse cells were substantially impeded. The infarct size was significantly larger

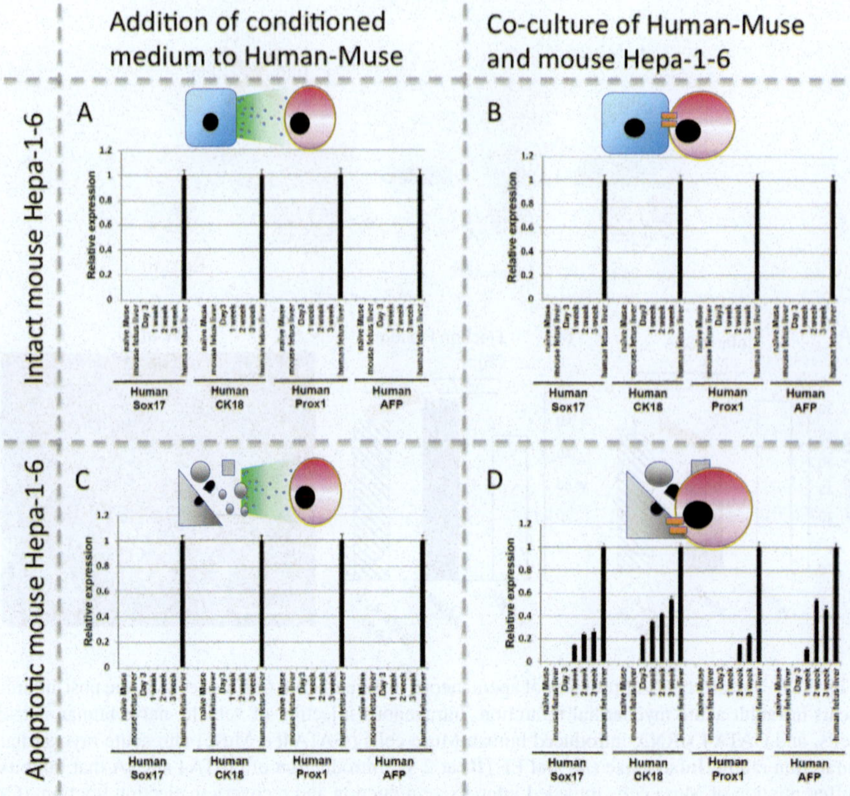

Fig. 3.16 The mechanism of spontaneous differentiation of Muse cells into tissue-compatible cells. Differentiation of human Muse cells into hepatocyte lineage cells by co-culture with apoptotic mouse hepatocytes Hepa-1-6. Expression of human-specific sex-determining region Y-box 17 (Sox17), human cytokeratin 18 (CK18), human prospero homeobox protein 1 (Prox1), and human a-fetoprotein (AFP) in quantitative polymerase chain reaction (qPCR). Since primers were specific for human, mouse fetus liver was consistently negative for all markers, while human fetus liver was positive. Human Muse cells, originally negative for all the markers, became positive for CK18 at day 3, Sox17 and AFP at 1 week, and Prox1 at 2 weeks only when co-cultured with etoposide-treated apoptotic mouse Hepa-1-6. On the other hand, supply of conditioned medium from intact or apoptotic mouse Hepa-1-6 and co-culture with intact mouse Hepa-1-6 did not induce human-specific marker expressions. (Pictures were adopted with permission from Ref. [7])

than that in animals that received naïve Muse cells but significantly smaller than that of the vehicle-injected animals. The ejection fraction was significantly lower in the GATA4-gene-silenced Muse group compared with that in the naïve Muse group but still greater than that in the vehicle group (Fig. 3.17). Because cardiac differentiation was impeded, the homed Muse cells remained undifferentiated. These data suggest the major role of Muse cell differentiation in structural and functional recovery.

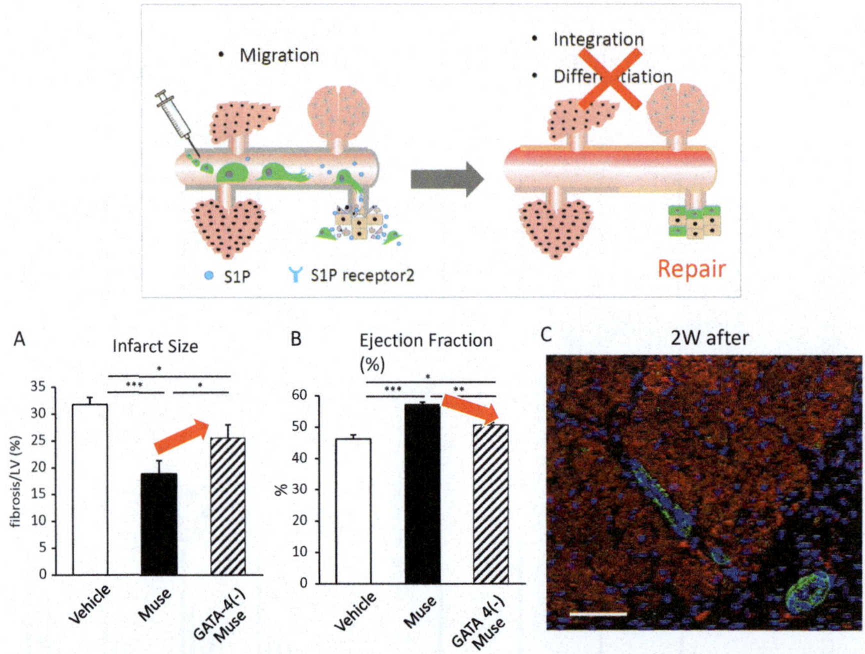

Fig. 3.17 The effect of suppression of spontaneous differentiation of Muse cells in the post-infarct heart in rabbit acute myocardial infarction. Intravenous injection of vehicle, naïve human Muse cells, and GATA4 siRNA-introduced human Muse cells (GATA4(-) Muse) into acute myocardial infarction rabbit. Infarct size (**A**) and EF (**B**) at 2 W. Introduction of GATA4 siRNA that inhibits differentiation of Muse cells impeded infarct size reduction and recovery in ejection fraction. (C) GFP-introduced GATA4 siRNA-introduced human Muse cells integrated into the post-infarct heart did not differentiate into cardiac cells but could only differentiate into vascular cells. Bar = 50 μm, $*p < 0.05$, $**p < 0.01$, $***p < 0.001$. (Pictures were adopted with permission from Ref. [10])

Other than differentiation, Muse cells have pleiotropic effects including paracrine effects, as discussed below. Therefore, just the presence of Muse cells, even if not differentiated, is assumed to have some beneficial effects. To deduce the significance of the presence of Muse cells, suicide gene-introduced Muse cells were intravenously injected into a rabbit acute myocardial infarction model [10]. HSVtk-introduced human Muse cells were intravenously injected into animals, and ganciclovir was administered 5 days after the injection to induce Muse cell suicide after homing. Two weeks later, reduction of the infarct size and recovery of ejection fraction were substantially attenuated compared with naïve human Muse cell injection and became more similar to those in the vehicle group [10]. Thus, most of the beneficial effects of Muse cells were offset when their presence in the tissue was impeded.

The adherent-suspension switch is involved in the control of Muse cell pluripotency, as mentioned in Chap. 2. This system may also apply to in vivo differentiation of Muse cells. Muse cells in the tissue, including those in the BM, are considered to be less pluripotent because they are in an adherent state. Mobilization of the Muse cells into the peripheral blood, however, puts them in a suspension state. Because the pluripotency of Muse cells is prominently upregulated in suspension, PB-Muse cells are considered to have very high pluripotency. This is rational from the viewpoint of Muse cell preconditioning, which prepares them for the upcoming mission of their differentiation into tissue-compatible cells after homing to the damaged tissue. If the pluripotency and differentiation potential are not potentiated before homing, Muse cells may not be able to initiate differentiation commitment into purposive cells, and their reparative effects would be incomplete. Conversely, if Muse cells in the tissue initiated differentiation randomly, this would be harmful from the viewpoint of tissue homeostasis. Therefore, it is entirely rational for tissue-Muse cells to be less pluripotent and PB-Muse cells to be more pluripotent in terms of tissue homeostasis and reparative maintenance.

3.7 Pleiotropic Effects of Muse Cells

In addition to their differentiation and immunomodulatory capacities, Muse cells have other beneficial effects, namely, paracrine, anti-apoptotic, and fibrinolysis/anti-fibrosis effects [3, 7, 9, 10]. Several studies have demonstrated the involvement of paracrine effects in tissue repair [3, 9, 10]. Muse cells produce cytokines and trophic factors that are produced by MSCs, such as tissue growth factor-alpha, tissue growth factor-beta, insulin-like growth factor-1, HGF, vascular endothelial growth factor (VEGF), and epidermal growth factor [3, 9, 10]. In addition, a Muse cell secretome analysis revealed that leukemia inhibitory factor and corticotropin-releasing hormone are uniquely produced by Muse cells and not BM-MSCs or adipose-MSCs [62]. These cytokines have multiple functions, including anti-apoptotic effects and activation of endogenous tissue progenitors. In fact, less apoptosis and increased proliferative activity of endogenous tissue progenitor cells is observed in animal models of chronic kidney disease and acute myocardial infarction treated with Muse cells [9, 10].

Importantly, VEGF and HGF are involved in neovascularization and are indispensable for tissue nutrition and maintenance [63–66]. The production capacity for VEGF and HGF in Muse cells is similar to or higher than that in non-Muse MSCs at both the protein and messenger RNA levels [9]. In addition to these factors, Muse cells directly contribute to neovascularization through their incorporation into vessels and subsequent differentiation into vascular cells. These two mechanisms may act synergistically on neovascularization. In fact, in a rabbit model of acute myocardial infarction, the Muse group showed significantly higher neovascularization in the border area of the infarcted region compared with the non-Muse MSC or MSC groups, with statistical significance [10].

In animal models of liver cirrhosis, chronic kidney disease, and acute myocardial infarction, Muse cell-injected animals consistently exhibited significantly lower levels of fibrosis compared with animals injected with MSCs or non-Muse MSCs (Fig. 3.18) [7, 9, 10]. Muse cells produce matrix metalloprotease-1(MMP-1),

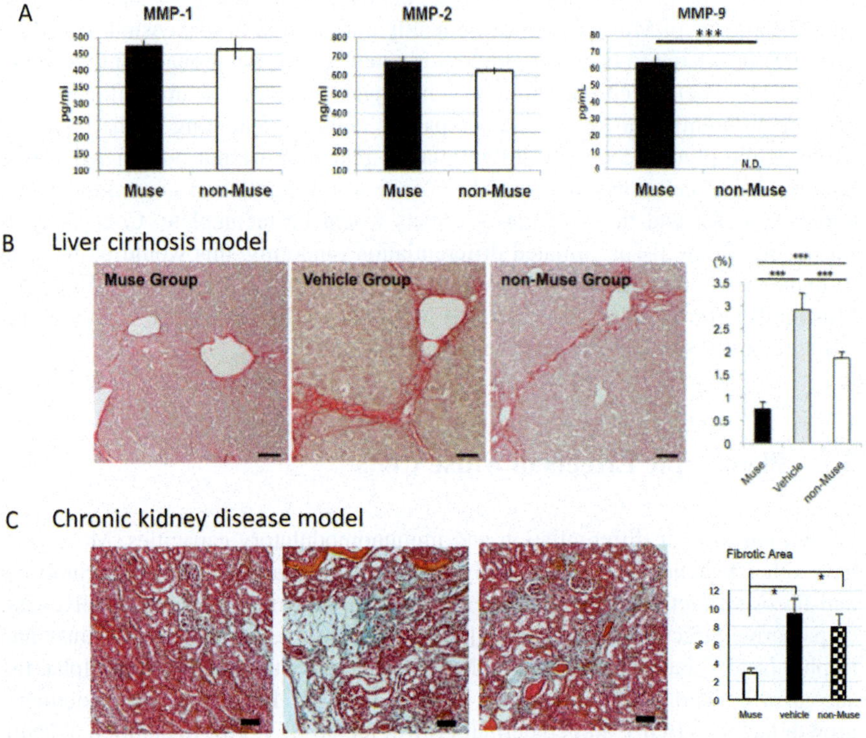

Fig. 3.18 Anti-fibrotic effect of Muse cells. (**A**) Production of MMP-1, MMP-2, and MMP-9 in human Muse and non-Muse cells. (**B**) Fibrotic area evaluated by Sirius red in SCID mouse liver cirrhosis model which received human Muse and non-Muse cell intravenous injection (8 weeks). The Muse group showed substantial reduction of fibrosis. (Pictures were adopted with permission from Ref. [7]). (**C**) Fibrotic area evaluated by Masson trichrome staining in BALB/c mouse chronic kidney disease model which received human Muse and non-Muse cell intravenous injection (7 weeks). Similar to the liver cirrhosis model, the Muse group showed the smallest fibrotic area compared to the vehicle and non-Muse groups. (Pictures were adopted with permission from Ref. [9])

MMP-2 and MMP-9 [7, 10], which act on fibrinolysis and suppression of fibrosis [67–70], at levels similar to those in MSCs and non-Muse MSCs (Fig. 3.18).

There is a discrepancy between the production levels of these factors in vitro and the outcomes in vivo after Muse cell administration [7, 9, 10]. The production capacity of these factors in Muse cells does not largely differ between MSCs and non-Muse MSCs at the mRNA and protein levels in vitro. When animal models receive a cell infusion, however, the Muse group consistently shows less apoptosis, higher endogenous progenitor cell activation, and less fibrosis than animals in the MSC and non-Muse MSC groups. This discrepancy might be explained by the ability of the cells to home to and survive in the target tissue. In animal models of stroke [4, 6, 8], acute myocardial infarction [10], liver cirrhosis [7], partial hepatectomy [5], and chronic kidney disease [9], survival of Muse cells in the damaged tissue is consistently observed, while non-Muse MSCs do not home to the target tissue dur-

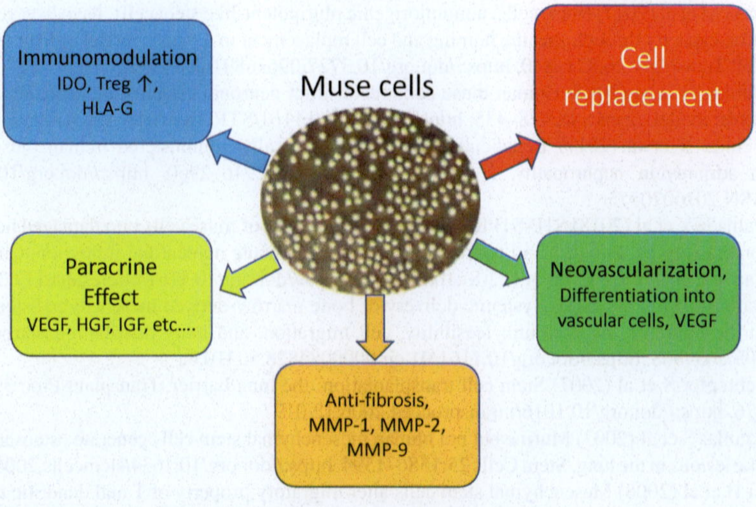

Fig. 3.19 Schematic diagram of pleiotropic effect of Muse cells

ing the earlier period (i.e., 3 days or 1 week after administration) and also do not remain in the tissue in later stages (i.e., several months after). In a rabbit acute myocardial infarction model, allogeneic Muse cells were observed in the post-infarct heart as cardiomyocytes 6 months after their infusion [10]. Because Muse cells remain and survive in the target tissue for a longer period, they might exert continuous pleiotropic effects (Fig. 3.19). This could partially explain why the production levels of beneficial factors are similar among Muse cells, non-Muse MSCs, and MSCs, but the outcomes in vivo are different.

References

1. Hori E et al (2016) Mobilization of pluripotent multilineage-differentiating stress-enduring cells in ischemic stroke. J Stroke Cerebrovas Dis 25:1473–1481. https://doi.org/10.1016/j.jstrokecerebrovasdis.2015.12.033
2. Tanaka T et al (2018) Mobilized muse cells after acute myocardial infarction predict cardiac function and remodeling in the chronic phase. Circ J. Tanaka et al., 2018 Cir J 82(2):561–571. https://doi.org/10.1253/circj.CJ-17-0552
3. Kinoshita K et al (2015) Therapeutic potential of adipose-derived SSEA-3-positive muse cells for treating diabetic skin ulcers. Stem Cells Transl Med 4:146–155. https://doi.org/10.5966/sctm.2014-0181
4. Yamauchi T et al (2015) Therapeutic effects of human multilineage-differentiating stress enduring (MUSE) cell transplantation into infarct brain of mice. PLoS One 10:e0116009. https://doi.org/10.1371/journal.pone.0116009
5. Katagiri H et al (2016) A distinct subpopulation of bone marrow mesenchymal stem cells, muse cells, directly commit to the replacement of liver components. Am J Transplant 16:468–483. https://doi.org/10.1111/ajt.13537
6. Uchida H et al (2016) Transplantation of unique subpopulation of fibroblasts, muse cells, ameliorates experimental stroke possibly via robust neuronal differentiation. Stem Cells 34:160–173. https://doi.org/10.1002/stem.2206

7. Iseki M et al (2017) Muse cells, nontumorigenic pluripotent-like stem cells, have liver regeneration capacity through specific homing and cell replacement in a mouse model of liver fibrosis. Cell Transplant 26:821–840. https://doi.org/10.3727/096368916X693662
8. Uchida H et al (2017) Human muse cells reconstruct neuronal circuitry in subacute lacunar stroke model. Stroke 48:428–435. https://doi.org/10.1161/STROKEAHA.116.014950
9. Uchida N et al (2017) Beneficial effects of systemically administered human muse cells in adriamycin nephropathy. J Am Soc Nephrol 28:2946–2960. https://doi.org/10.1681/ASN.2016070775
10. Yamada Y et al (2018) S1P-S1PR2 axis mediates homing of muse cells into damaged heart for long lasting tissue repair and functional recovery after acute myocardial infarction. Circ Res. Yamada et al, 2018 Cir Res 122(8):1069-1083. https://doi.org/10.1161/circresaha.117.311648
11. Barbash IM et al (2003) Systemic delivery of bone marrow-derived mesenchymal stem cells to the infarcted myocardium: feasibility, cell migration, and body distribution. Circulation 108:863–868. https://doi.org/10.1161/01.cir.0000084828.50310.6a
12. Schrepfer S et al (2007) Stem cell transplantation: the lung barrier. Transplant Proc 39:573–576. https://doi.org/10.1016/j.transproceed.2006.12.019
13. Aguilar S et al (2007) Murine but not human mesenchymal stem cells generate osteosarcoma-like lesions in the lung. Stem Cells 25:1586–1594. https://doi.org/10.1634/stemcells.2006-0762
14. Li H et al (2008) Mesenchymal stem cells alter migratory property of T and dendritic cells to delay the development of murine lethal acute graft-versus-host disease. Stem Cells 26:2531–2541. https://doi.org/10.1634/stemcells.2008-0146
15. Togel F, Yang Y, Zhang P, Hu Z, Westenfelder C (2008) Bioluminescence imaging to monitor the in vivo distribution of administered mesenchymal stem cells in acute kidney injury. Am J Physiol Renal Physiol 295:F315–F321. https://doi.org/10.1152/ajprenal.00098.2008
16. Kurtz A (2008) Mesenchymal stem cell delivery routes and fate. Int J Stem Cells 1:1–7
17. Assis AC et al (2010) Time-dependent migration of systemically delivered bone marrow mesenchymal stem cells to the infarcted heart. Cell Transplant 19:219–230. https://doi.org/10.3727/096368909x479677
18. Leibacher J, Henschler R (2016) Biodistribution, migration and homing of systemically applied mesenchymal stem/stromal cells. Stem Cell Res Ther 7:7. https://doi.org/10.1186/s13287-015-0271-2
19. Brooks A et al (2018) Concise review: quantitative detection and modeling the in vivo kinetics of therapeutic mesenchymal stem/stromal cells. Stem Cells Transl Med 7:78–86. https://doi.org/10.1002/sctm.17-0209
20. Peled A et al (1999) Dependence of human stem cell engraftment and repopulation of NOD/SCID mice on CXCR4. Science 283:845–848
21. Bonig H, Priestley GV, Papayannopoulou T (2006) Hierarchy of molecular-pathway usage in bone marrow homing and its shift by cytokines. Blood 107:79–86. https://doi.org/10.1182/blood-2005-05-2023
22. Vessey DA, Li L, Honbo N, Karliner JS (2009) Sphingosine 1-phosphate is an important endogenous cardioprotectant released by ischemic pre- and postconditioning. Am J Physiol Heart Circ Physiol 297:H1429–H1435. https://doi.org/10.1152/ajpheart.00358.2009
23. Ratajczak MZ, Suszynska M, Borkowska S, Ratajczak J, Schneider G (2014) The role of sphingosine-1 phosphate and ceramide-1 phosphate in trafficking of normal stem cells and cancer cells. Expert Opin Ther Targets 18:95–107. https://doi.org/10.1517/14728222.2014.851671
24. Karpova D, Bonig H (2015) Concise review: CXCR4/CXCL12 signaling in immature hematopoiesis–lessons from pharmacological and genetic models. Stem Cells 33:2391–2399. https://doi.org/10.1002/stem.2054
25. Son BR et al (2006) Migration of bone marrow and cord blood mesenchymal stem cells in vitro is regulated by stromal-derived factor-1-CXCR4 and hepatocyte growth factor-c-met axes and involves matrix metalloproteinases. Stem Cells 24:1254–1264. https://doi.org/10.1634/stemcells.2005-0271
26. Petit I, Jin D, Rafii S (2007) The SDF-1-CXCR4 signaling pathway: a molecular hub modulating neo-angiogenesis. Trends Immunol 28:299–307. https://doi.org/10.1016/j.it.2007.05.007

27. Liu X et al (2011) SDF-1/CXCR4 axis modulates bone marrow mesenchymal stem cell apoptosis, migration and cytokine secretion. Protein Cell 2:845–854. https://doi.org/10.1007/s13238-011-1097-z
28. Cencioni C, Capogrossi MC, Napolitano M (2012) The SDF-1/CXCR4 axis in stem cell preconditioning. Cardiovasc Res 94:400–407. https://doi.org/10.1093/cvr/cvs132
29. Wu Y, Zhao RC (2012) The role of chemokines in mesenchymal stem cell homing to myocardium. Stem Cell Rev 8:243–250. https://doi.org/10.1007/s12015-011-9293-z
30. van de Kamp J, Jahnen-Dechent W, Rath B, Knuechel R, Neuss S (2013) Hepatocyte growth factor-loaded biomaterials for mesenchymal stem cell recruitment. Stem Cells Int 2013:892065. https://doi.org/10.1155/2013/892065
31. Vogel S et al (2013) Migration of mesenchymal stem cells towards glioblastoma cells depends on hepatocyte-growth factor and is enhanced by aminolaevulinic acid-mediated photodynamic treatment. Biochem Biophys Res Commun 431:428–432. https://doi.org/10.1016/j.bbrc.2012.12.153
32. Mathias S, Pena LA, Kolesnick RN (1998) Signal transduction of stress via ceramide. Biochem J 335(Pt 3):465–480
33. Hannun YA, Luberto C, Argraves KM (2001) Enzymes of sphingolipid metabolism: from modular to integrative signaling. Biochemistry 40:4893–4903
34. Levade T et al (2002) Ceramide in apoptosis: a revisited role. Neurochem Res 27:601–607
35. Spiegel S, Milstien S (2003) Sphingosine-1-phosphate: an enigmatic signalling lipid. Nat Rev Mol Cell Biol 4:397–407. https://doi.org/10.1038/nrm1103
36. Ogretmen B, Hannun YA (2004) Biologically active sphingolipids in cancer pathogenesis and treatment. Nat Rev Cancer 4:604–616. https://doi.org/10.1038/nrc1411
37. Schwab SR et al (2005) Lymphocyte sequestration through s1p lyase inhibition and disruption of S1P gradients. Science 309:1735–1739. https://doi.org/10.1126/science.1113640
38. Rivera J, Proia RL, Olivera A (2008) The alliance of sphingosine-1-phosphate and its receptors in immunity. Nat Rev Immunol 8:753–763. https://doi.org/10.1038/nri2400
39. Obinata H, Hla T (2012) Sphingosine 1-phosphate in coagulation and inflammation. Semin Immunopathol 34:73–91. https://doi.org/10.1007/s00281-011-0287-3
40. Blaho VA, Hla T (2014) An update on the biology of sphingosine 1-phosphate receptors. J Lipid Res 55:1596–1608. https://doi.org/10.1194/jlr.R046300
41. Sanchez T, Hla T (2004) Structural and functional characteristics of S1P receptors. J Cell Biochem 92:913–922. https://doi.org/10.1002/jcb.20127
42. Patmanathan SN, Wang W, Yap LF, Herr DR, Paterson IC (2017) Mechanisms of sphingosine 1-phosphate receptor signalling in cancer. Cell Signal 34:66–75. https://doi.org/10.1016/j.cellsig.2017.03.002
43. Hannun YA, Obeid LM (2008) Principles of bioactive lipid signalling: lessons from sphingolipids. Nat Rev Mol Cell Biol 9:139. https://doi.org/10.1038/nrm2329
44. Proia RL, Hla T (2015) Emerging biology of sphingosine-1-phosphate: its role in pathogenesis and therapy. J Clin Invest 125:1379–1387. https://doi.org/10.1172/jci76369
45. Ghasemi R, Dargahi L, Ahmadiani A (2016) Integrated sphingosine-1 phosphate signaling in the central nervous system: from physiological equilibrium to pathological damage. Pharmacol Res 104:156–164. https://doi.org/10.1016/j.phrs.2015.11.006
46. Spiegel S, Milstien S (2007) Functions of the multifaceted family of sphingosine kinases and some close relatives. J Biol Chem 282:2125–2129. https://doi.org/10.1074/jbc.R600028200
47. Osada M, Yatomi Y, Ohmori T, Ikeda H, Ozaki Y (2002) Enhancement of sphingosine 1-phosphate-induced migration of vascular endothelial cells and smooth muscle cells by an EDG-5 antagonist. Biochem Biophys Res Commun 299:483–487
48. Gimeno ML et al (2017) Pluripotent nontumorigenic adipose tissue-derived muse cells have immunomodulatory capacity mediated by transforming growth factor-beta1. Stem Cells Transl Med 6:161–173. https://doi.org/10.5966/sctm.2016-0014
49. Fisch SC et al (2017) Pluripotent nontumorigenic multilineage differentiating stress enduring cells (muse cells): a seven-year retrospective. Stem Cell Res Ther 8:227. https://doi.org/10.1186/s13287-017-0674-3

50. Kovats S et al (1990) A class I antigen, HLA-G, expressed in human trophoblasts. Science 248:220–223
51. Onno M et al (1994) The HLA-G gene is expressed at a low mRNA level in different human cells and tissues. Human Immunol 41:79–86. doi:https://doi.org/10.1016/0198-8859(94)90089-2
52. Ferreira LM, Meissner TB, Tilburgs T, Strominger JL (2017) HLA-G: at the interface of maternal-fetal tolerance. Trends Immunol 38:272–286. https://doi.org/10.1016/j.it.2017.01.009
53. Rizzo R, Bortolotti D, Bolzani S, Fainardi E (2014) HLA-G molecules in autoimmune diseases and infections. Front Immunol 5:592. https://doi.org/10.3389/fimmu.2014.00592
54. Shiroishi M et al (2003) Human inhibitory receptors Ig-like transcript 2 (ILT2) and ILT4 compete with CD8 for MHC class I binding and bind preferentially to HLA-G. Proc Natl Acad Sci U S A 100:8856–8861. https://doi.org/10.1073/pnas.1431057100
55. Endo S, Sakamoto Y, Kobayashi E, Nakamura A, Takai T (2008) Regulation of cytotoxic T lymphocyte triggering by PIR-B on dendritic cells. Proc Natl Acad Sci U S A 105:14515–14520. https://doi.org/10.1073/pnas.0804571105
56. Drukker M et al (2002) Characterization of the expression of MHC proteins in human embryonic stem cells. Proc Natl Acad Sci U S A 99:9864–9869. https://doi.org/10.1073/pnas.142298299
57. Kim EM, Manzar G, Zavazava N (2013) Human iPS cell-derived hematopoietic progenitor cells induce T-cell anergy in in vitro-generated alloreactive CD8(+) T cells. Blood 121:5167–5175. https://doi.org/10.1182/blood-2012-11-467753
58. Kyurkchiev S et al (2012) Stem cells in the reproductive system. Am J of Reprod Immunol (NEW YORK, NY: 1989) 67:445–462. https://doi.org/10.1111/j.1600-0897.2012.01140.x
59. Nasef A et al (2007) Immunosuppressive effects of mesenchymal stem cells: involvement of HLA-G. Transplantation 84:231–237. https://doi.org/10.1097/01.tp.0000267918.07906.08
60. Ivanova-Todorova E et al (2009) HLA-G expression is up-regulated by progesterone in mesenchymal stem cells. Am J Reprod Immunol 62:25–33. https://doi.org/10.1111/j.1600-0897.2009.00707.x
61. Lila N et al (2002) Human leukocyte antigen-G expression after heart transplantation is associated with a reduced incidence of rejection. Circulation 105:1949–1954
62. Alessio N et al (2017) The secretome of MUSE cells contains factors that may play a role in regulation of stemness, apoptosis and immunomodulation. Cell cycle (Georgetown, Tex) 16:33–44. https://doi.org/10.1080/15384101.2016.1211215
63. Chen XH et al (2007) In vivo hepatocyte growth factor gene transfer reduces myocardial ischemia-reperfusion injury through its multiple actions. J Card Fail 13:874–883. https://doi.org/10.1016/j.cardfail.2007.07.004
64. Isner JM, Asahara T (1999) Angiogenesis and vasculogenesis as therapeutic strategies for postnatal neovascularization. J Clin Invest 103:1231–1236. https://doi.org/10.1172/jci6889
65. Mias C et al (2009) Mesenchymal stem cells promote matrix metalloproteinase secretion by cardiac fibroblasts and reduce cardiac ventricular fibrosis after myocardial infarction. Stem Cells 27:2734–2743. https://doi.org/10.1002/stem.169
66. Loffredo FS, Steinhauser ML, Gannon J, Lee RT (2011) Bone marrow-derived cell therapy stimulates endogenous cardiomyocyte progenitors and promotes cardiac repair. Cell Stem Cell 8:389–398. https://doi.org/10.1016/j.stem.2011.02.002
67. Duarte S, Baber J, Fujii T, Coito AJ (2015) Matrix metalloproteinases in liver injury, repair and fibrosis. Matrix Biol J Int Soc Matrix Biol 44-46:147–156. https://doi.org/10.1016/j.matbio.2015.01.004
68. Somerville RPT, Oblander SA, Apte SS (2003) Matrix metalloproteinases: old dogs with new tricks. Genome Biol 4:216. https://doi.org/10.1186/gb-2003-4-6-216
69. Hemmann S, Graf J, Roderfeld M, Roeb E (2007) Expression of MMPs and TIMPs in liver fibrosis – a systematic review with special emphasis on anti-fibrotic strategies. J Hepatol 46:955–975. doi:https://doi.org/10.1016/j.jhep.2007.02.003
70. Roeb E (2017) Matrix metalloproteinases and liver fibrosis (translational aspects). Matrix Biol J Int Soc Matrix Biol. https://doi.org/10.1016/j.matbio.2017.12.012

Chapter 4
Protocols for Isolation and Evaluation of Muse Cells

Kazuki Tatsumi, Yoshihiro Kushida, Shohei Wakao, Yasumasa Kuroda, and Mari Dezawa

Contents

Be sure to read the following comments because these are important issues influencing the Muse cell yield.

K. Tatsumi
Department of Stem Cell Biology and Histology, Tohoku University Graduate School of Medicine, Sendai, Japan

Analytical Research Department, Technology Development Unit, Regenerative Medicine Division, Life Science Institute, Inc., Tokyo, Japan

Y. Kushida · S. Wakao · Y. Kuroda · M. Dezawa (✉)
Department of Stem Cell Biology and Histology, Tohoku University Graduate School of Medicine, Sendai, Japan
e-mail: mdezawa@med.tohoku.ac.jp

M. Dezawa (ed.), *Muse Cells*, Advances in Experimental Medicine and Biology 1103, https://doi.org/10.1007/978-4-431-56847-6_4

Abstract This chapter provides the detailed method for isolation of Muse cells and evaluation of their pluripotency. The basic population of Muse cells is cultured mesenchymal stem cells such as bone marrow-mesenchymal stem cells, fibroblasts, and adipose-derived stem cells. The detailed method for handling mesenchymal stem cells is also provided in this protocol.

Keywords MSCs · Fibroblasts · Adipose-derived stem cells · Serum lot check · Anti-SSEA-3 antibody · Cell sorting · FACS · Clusters

The outline of Muse cell collection, corresponding to 4.3.3 ~ 4.4.2, is shown in Fig. 4.1.

4.1 Materials

4.1.1 Commercial Culture Cells as a Source for Muse Cells (Cells Used in Our Laboratory)

Cultured mesenchymal stem cells are the source of Muse cells. Mesenchymal stem cells can be obtained commercially or by primary culture of tissue sources. The below four mesenchymal stem cells are commercially obtainable cells confirmed in Dezawa's lab to yield a reasonable proportion of Muse cells.

- Human bone marrow (BM)-mesenchymal stem cells (MSCs), BM-MSCs (Cat#PT-2501, Lonza)
- Normal human dermal fibroblasts-adult skin, NHDF (Cat#CC-2511, Lonza)
- Human dermal fibroblasts-adult, HDFa (Cat#2320, ScienCell Research Laboratories)
- Human adipose-derived stem cells, ADSCs (Cat#PT-5006, Lonza)

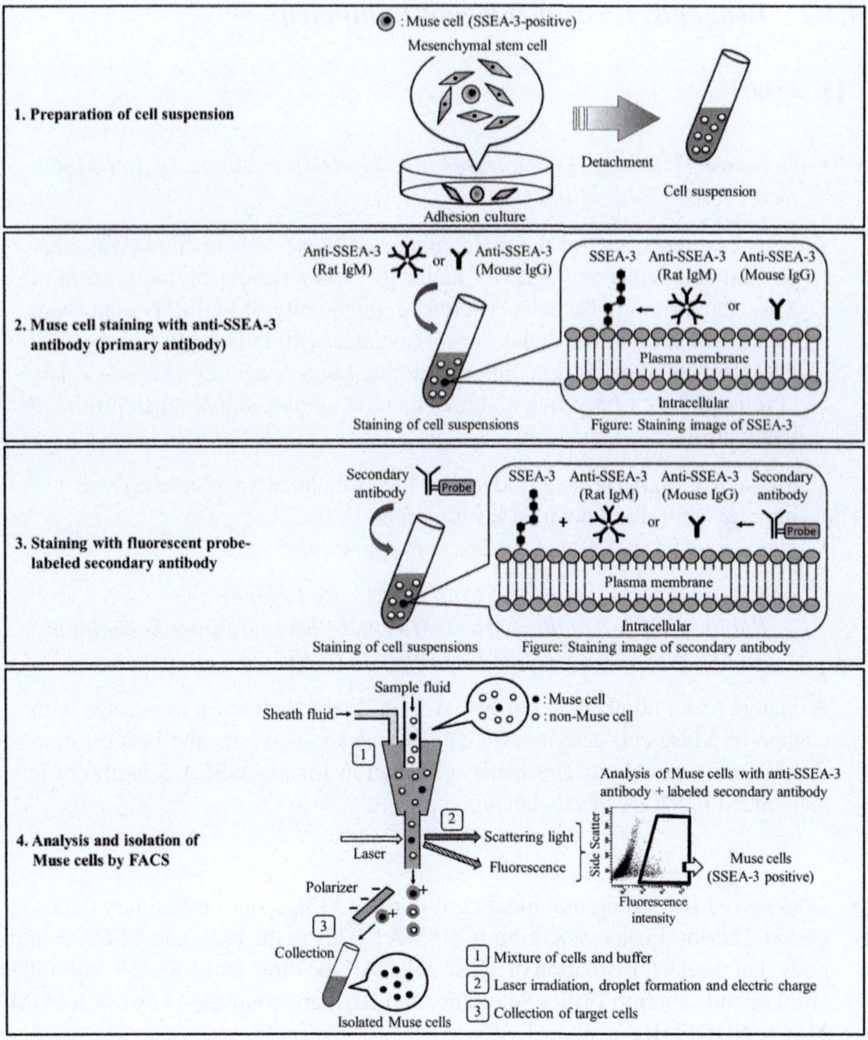

Fig. 4.1 Flowchart of Muse cell collection
The outline of Muse cell collection is comprised of 1) preparation of cell suspension, 2) staining the cells with anti-SSEA-3 antibody (primary antibody), 3) staining the cells with fluorescent probe-labeled secondary antibody, and 4) analysis and isolation of Muse cells by FACS

4.1.2 Reagents, Instruments and Equipment

【Reagents】

- *[Important] Human-FGF-2, premium grade (for culture of BM-MSCs, Cat#130-093-840, Miltenyi)*
- FGF-2 is necessary for the maintenance of Muse cells in cultured mesenchymal stem cells and is a key factor to yield a reasonable proportion of Muse cells from these cells. Therefore, the quality of FGF-2 is important. Change of lot number sometimes associates with change of activity. The latest information is announced in the home page of Dezawa's lab, "Protocol." http://www.stemcells.med.tohoku.ac.jp/english/protocol/index.html
- Please ask Mari Dezawa (mdezawa*med.tohoku.ac.jp, please convert "*" into "@") for more detailed information.
- *[Important] Anti-SSEA-3 antibody: Prepare either of:*
  - *Rat anti-SSEA-3 antibody (Cat#330302, BioLegend)*
  - *Rat anti-SSEA-3 antibody (Cat#MA1-020, Thermo Fisher Scientific)*
  - *Mouse anti-SSEA-3 antibody (Cat#10431, IBL)*

A change of lot number of anti-SSEA-3 antibody sometimes associates with change of Muse cell activity. We ask the readers to verify the best dilution condition for new lots. The latest information for anti-SSEA-3 antibody is announced in the Dezawa's lab home page.

- *[Reference]* Regarding umbilical cord-derived MSCs, our preliminary data suggested Thermo Fisher Scientific (Cat#MA1-020) is the only anti-SSEA-3 antibody that works for isolation of Muse cells. All the three anti-SSEA-3 antibodies (BioLegend, Thermo Fisher Scientific, and IBL) are confirmed to work for BM-MSCs, NHDF, HDFa, and ADSCs.
- Isotype control: Prepare either of:
  - Rat IgM Isotype Control (Cat#400801, BioLegend)
  - Mouse IgG2b Isotype Control (Cat#401201, BioLegend)
- Secondary antibody: Please prepare either of FITC- or APC-conjugated secondary antibodies depending on the situation.
  - Goat anti-rat IgM antibody (FITC-labeled) (Cat#112-095-075, Jackson ImmunoResearch)
  - Goat anti-rat IgM antibody (APC-labeled) (Cat#112-136-075, Jackson ImmunoResearch)
  - Goat anti-mouse IgG antibody (FITC-labeled) (Cat#115-096-146, Jackson ImmunoResearch)

– Goat anti-mouse IgG antibody (APC labeled) (Cat#115-136-146, Jackson ImmunoResearch)

Summary of sets of primary and secondary antibodies and isotype control:

Anti-SSEA-3 antibody	Isotype control	Secondary antibody
BioLegend (Cat#330302)	Rat IgM isotype control (Cat#400801, BioLegend)	FITC-Goat anti-Rat IgM Ab (Cat#112-095-075, Jackson ImmunoResearch)
		APC-Goat anti-Rat IgM Ab (Cat#112-136-075, Jackson ImmunoResearch)
Thermo Fisher Scientific (Cat#MA1-020)	Rat IgM isotype control (Cat#A400801, BioLegend)	FITC-Goat anti-Rat IgM Ab (Cat#112-095-075, Jackson ImmunoResearch)
		APC-Goat anti-Rat IgM Ab (Cat#112-136-075, Jackson ImmunoResearch)
IBL (Cat#10431)	Mouse IgG2b isotype control (Cat#401201, BioLegend)	FITC-Goat anti-Mouse IgG Ab (Cat#115-096-146, Jackson ImmunoResearch)
		APC-Goat anti-Mouse IgG Ab (Cat#115-136-146, Jackson ImmunoResearch)

- Fetal bovine serum (FBS)

> - *[Important]* Selection of FBS is the key point because Muse cell activity and proportion are largely dependent on the quality of FBS for culturing mesenchymal stem cells. *Please be sure to read "Lot check for FBS" in the next section for selection of FBS.*

- FBS (for inactivation of trypsin, no manufacturer specified)

> - *[Important] Low-glucose DMEM + GlutaMAX (Cat#10567, Thermo Fisher Scientific) → for culturing BM-MSCs, NHDF, and HDFa*
> - *[Important] High-glucose DMEM + GlutaMAX (Cat#10569, Thermo Fisher Scientific) → for culturing ADSCs*

- Kanamycin Sulfate (100 x) (Use at 1x dilution in media, Cat#15160-054, Thermo Fisher Scientific)
- PBS (10 x) (Cat#27575-31, Nacalai Tesque)
- Sterile water (for PBS preparation) (Cat#06442-95, Nacalai Tesque)
- Trypsin (0.25%)/EDTA (Cat#25200-072, Thermo Fisher Scientific)
- FluoroBrite DMEM (Cat#A18967-01, Thermo Fisher Scientific)

- BSA (bovine serum albumin) (Cat#01860-65, Nacalai Tesque)
- EDTA (Cat#15111-45, Nacalai Tesque)
- Gelatin (Cat#G-1890, Sigma)
- Poly-HEMA [poly(2-hydroxyethyl methacrylate)] (Cat#P3932, Sigma)
- MethoCult H4100 (Cat#04100, STEMCELL Technologies)
- 【Instruments and Equipment】
- Cell counter plate (disposable) (Cat# WC2-100S, Waken BTech Co., Ltd)
- Cellbanker 1 plus, cryopreservation solution for culture cells (Cat#CB021, ZENOAQ)
- CryoTube vials (Cat#377267, Thermo Fisher Scientific)
- BICELL, cryotube container (for preserving culture cells in deep freezer) BICELL/SANO910/ask office@tech-jam.com
- [Recommended] 10-cm dish(Cat#150464, Thermo Fisher Scientific)
- 1.5-mL tube (Cat#BM-15, BMBio)
- 15-mL tube (Cat#352096, Corning)
- 50-mL tube (Cat#352070, Corning)
- Cell strainer (40 μm) (Cat#352340, Falcon)
- 0.22 μm filter (Cat#SLGV033RS, Merck Millipore)
- Centrifuge with swing rotor (15 mL, 50 mL) (no manufacturer specified)
- Centrifuge with swing rotor and cooler (1.5 mL) (no manufacturer specified)
- Cell sorter (BD FACS Aria II) (used in Dezawa's lab)
- FACS analysis software (BD FACSDiva) (used in Dezawa's lab)

4.2 Cell Culture

4.2.1 Culture Medium

- BM-MSCs: Low-glucose DMEM, 10% FBS, 1 ng/mL FGF-2, 0.1 mg/mL kanamycin
- NHDF and HDFa: Low-glucose DMEM, 10% FBS, 0.1 mg/mL kanamycin
- ADSCs: High-glucose DMEM, 15% FBS, 0.1 mg/mL kanamycin

Attention! Use low-glucose DMEM or high-glucose DMEM depending on the cell type. For cultures of BM-MSCs, NHDF, and HDFa, always use 10% FBS (HyClone)/low-glucose DMEM (Thermo Fisher Scientific) and not high-glucose DMEM. Use of high-glucose DMEM induces hypoproliferative capacity and decreases Muse cell yield. On the other hand, for culture of ADSCs, use 15% FBS (HyClone)/high-glucose DMEM (Thermo Fisher Scientific).

Attention! Use FGF-2 (bFGF) for culture of BM-MSCs. For culture of BM-MSCs, add FGF-2 (concentration: 1 ng/mL, Cat#130-093-840, Miltenyi). Be aware that Muse cell yields change markedly by using products from several manufacturers (Fig. 4.2). For the latest information, please go to the home page of Dezawa's lab, "Protocol."

Attention! (The cells listed in 1-1) Commercial culture cells as a source for Muse cells are examples of sources that can be used to reproduce our data. Other mesenchymal stem cells may also work as a source for Muse cells. However, the outcome cannot be guaranteed unless the above cell types are used.

Attention! When purchasing cells, culture media specialized for the cells should also be bought from the same company to maintain the cells according to the manufacturer's instructions.

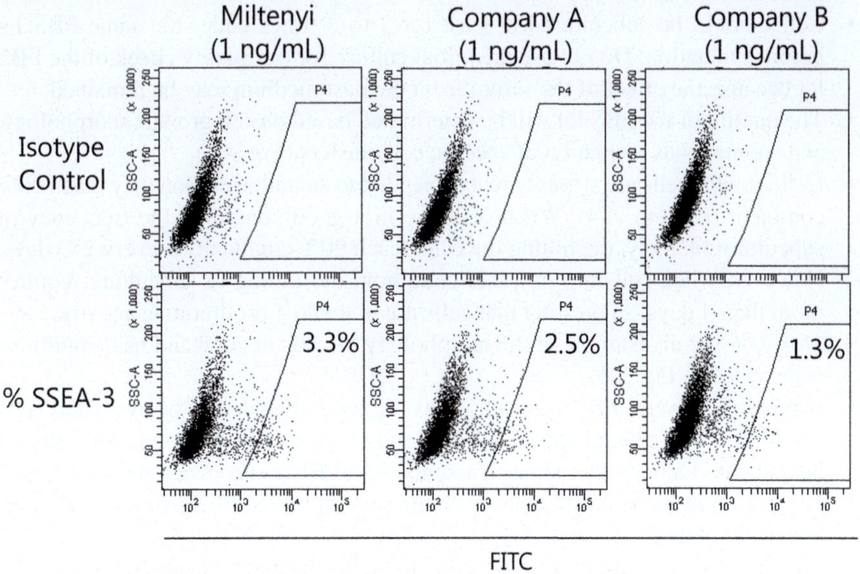

Fig. 4.2 Quality of FGF-2 produces a difference in SSEA-3-positive cell ratio
Comparison of Muse cell percent in human BM-MSCs by different companies FGF-2. The experimental condition is as follows:
Rat anti-SSEA-3 monoclonal antibody (MC-631) (BioLegend, Cat#330302, 0.5 μg/100 μL)
Isotype control: Purified Rat IgM, κ Isotype Control (BioLegend, Cat#400801, 0.5 μg/100 μL)
Human BM-MSCs (Passage 7)

Attention! (The cells listed in 1-1) Commercial culture cells as a source for Muse cells have a number of different lots. Because the cell growth rate and Muse cell ratio may differ among lots, we recommend purchasing a couple of lots and then selecting the best lot for the experiment.

4.2.2 Serum Lot Check

- Mesenchymal stem cells (BM-MSCs are recommended rather than NHDF/HDFa and ADSCs) are used for the lot check. Expanded mesenchymal stem cells are collected by trypsin incubation.
- Count the number of cells, and suspend cells at a concentration of 1×10^4 cells/500 μL in 10% FBS (for serum lot check) in DMEM (+kanamycin), and plate them individually in a 24-well plate.
- Incubate cells at 37 °C, 5% CO_2.
- Change medium the next day. Culture medium is exchanged every 2–3 days.
- Perform subculture when cells reach at 90% confluence. Cells should be expanded to 1:2 (one 90% confluent plate is subcultured to two plates). *Never exceed 1:3.*
- Incubate cells at 37 °C, 5% CO_2.
- Cells should be subcultured at least for 2 to 3 times under the same FBS lot before evaluation. Do not use the initial culture for the quality check of the FBS lot because the effect of the serum from the past medium may be remained.
- The quality of the FBS lot will be determined based on cell growth, morphology, and spontaneous formation of specific cell clusters (Fig. 4.3).
- *Cell growth;* cells are strongly recommended to subculture when they reach 90% confluent (see Fig. 4.4). When cells are in a good condition, the frequency of subculture (namely, the timing the cells reach 90% confluent) is every 2–3 days. However, when cells are in a bad condition, the timing of subculture requires more than 4 days, suggesting that cells are in the low proliferative activity.
- *Morphology:* an example of the morphology of cells in good and bad conditions is provided in Fig. 4.5.
- If possible, several FBS concentrations are recommended to be evaluated (i.e., 5% ~ 20%, the general use is at 10%), and cell growth and morphology should be compared among lots to determine the best FBS and concentration for use.
- For the lot check, it is desirable to obtain several lots of FBS from several companies (a total of ~20 lots).
- The FBS for cell culture does *not* need to be inactivated by heating.

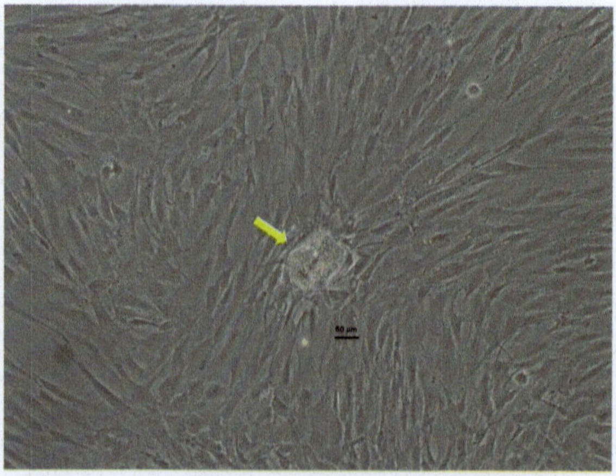

Fig. 4.3 An example of spontaneously formed specific cell cluster
Under a proper quality FBS, MSCs higher than 70% confluence spontaneously demonstrate formation of cell clusters as shown, although not at high frequency. FBS that did not generate any of the clusters is suggested to be low quality. The figure is from human BM-MSCs

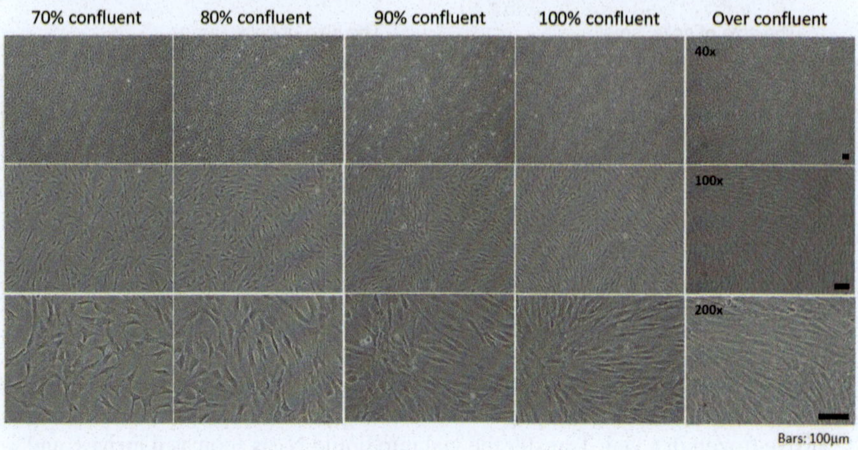

Fig. 4.4 Cell confluence in human BM-MSCs
An example of 70%, 80%, 90%, 100%, and over confluent in human BM-MSCs

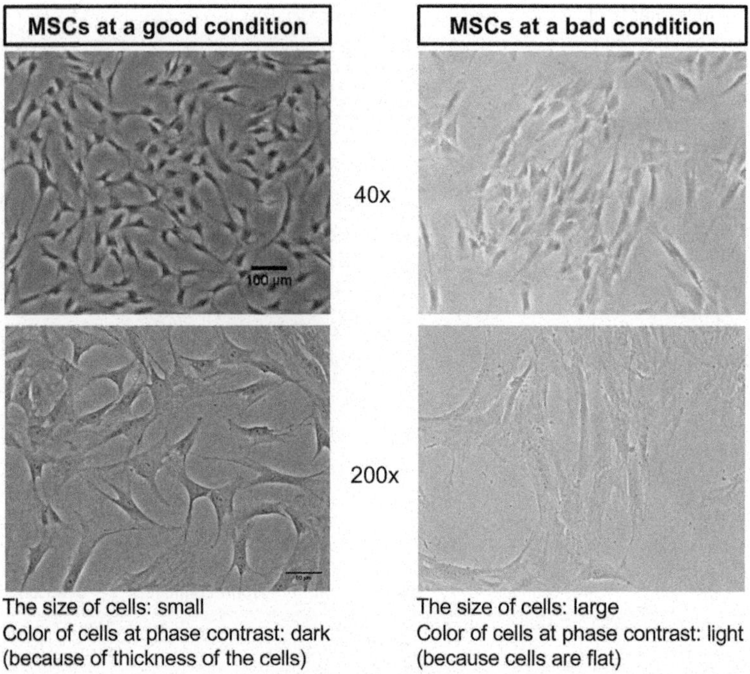

Fig. 4.5 An example of MSCs at good and bad conditions
At a good condition, the size of cells is small, and the color of cells under the phase contrast microscope is dark because of the thickness of the cells. At a bad condition, however, the size is larger and the color is lighter because cells are flat.

4.2.3 Thawing Frozen Cells

Attention! Carefully read the manufacturer's instructions prior to thawing purchased cells (BM-MSCs, NHDF, HDFa, ADSDs). Purchased cells usually arrive frozen in a vial. Transfer the vial into liquid N_2 as soon as it arrives, and keep it there until use.

- Wash and clean hands using 70% ethanol (EtOH). Use sterilized gloves if necessary.
- Set a water bath to 37 °C.
- Clean the bench by using 70% EtOH.
- Remove the frozen vial from the liquid N_2, and quickly transfer to the 37 °C water bath to thaw the cells.

- *Take care not to touch the vial cap to the water; otherwise the cells will be easily contaminated by bacteria.*

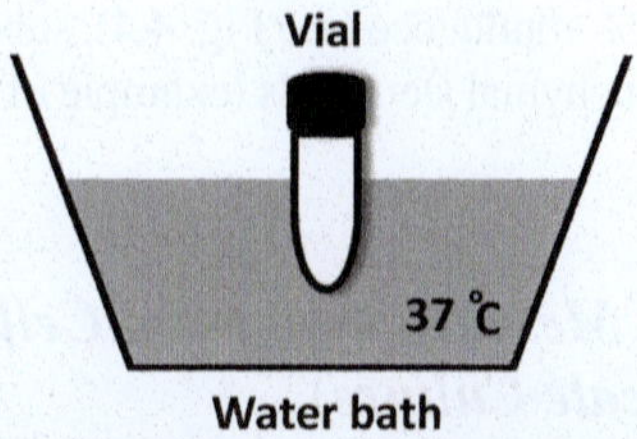

Attention! To avoid cell death, remove the vial from the water bath before it is completely thawed. The best time to remove the vial from the water bath is the point at which the solution still contains a piece of ice.

- Clean the vial by 70% EtOH, and then bring it to the clean bench.
- Carefully open the cap, and melt the piece of ice with gentle pipetting.
- Transfer cells to a 15 mL Falcon tube.
- Slowly add culture medium to the tube to make a final volume 10 mL.
- Centrifuge the 15 mL tube at 300 g for 5 min.
- Remove the supernatant, add 1 mL culture medium and loosen the cell pellet with gentle pipetting, add 9-mL culture medium, and count the total number of the cells by cell counter plate. Then plate the cells at the density indicated as follows: (1) cells just after the purchase for BM-MSCs, NHDF, HDFa, and ADSC, 1.5×10^4 cells/cm^2, and (2) cells that were once expanded after purchase and were stored in liquid N_2, 2×10^4 cells/cm^2.
- Culture the cells overnight at 37 °C, 5% CO_2.

Attention! The cells should be plated homogenously.

- At the next day of plating cells, remove the culture media, and add 10 mL fresh culture medium for medium exchange. If floating dead cells are still visible, wash a couple of times with culture media to remove all the dead cells.

Attention! In the case of cells purchased from a company, thawing the frozen cells and culturing for the first time is counted as passage 1 (P = 1) in this protocol.

Attention! If the cells were cryopreserved at P = 4 and then thawed, the cells are counted as P = 4 at this point, and passage number continues after following subculture (i.e., P = 5, P = 6, …).

- When the cells reach 90% confluence (see Fig. 4.4), subculture them as described in 2-4) Subculture of mesenchymal stem cells (example of a 10-cm dish scale culture).

4.2.4 Subculture of Mesenchymal Stem Cells (Example of a 10 cm Dish Scale Culture)

Attention! The best timing for subculture of mesenchymal stem cells (NHDFs, HDFa, BM-MSCs, and ADSCs) is at when cells reach 90% confluence. Please see how NHDF/HDFa and BM-MSCs look like at 90% confluence in Fig. 4.4.

Attention! The growth of mesenchymal stem cells is strongly suppressed by contact inhibition. Please watch the cells, and perform subculture before reaching 100% confluence.

Attention! Subculture the cells at 1:2 ratio. *Never exceed 1:3 ratio*; one ~90% confluent dish should be split into two dishes, and never into more than three dishes. If Muse cells are expanded at 1:3 or more, then the percent of Muse cells will substantially decrease.

Attention! Please use the specified culture medium recommended by the company you purchase the cells until P = 2. After P = 3, use 10% FBS in DMEM (+kanamycin) for culture.

BEST timing for subculture (*~90% confluent;* some spaces are remaining between cells)
BAD timing for subculture (*over confluent:* cells are piling up)

Protocol for Subculture

- When the mesenchymal stem cells (NHDFs, HDFa, BM-MSCs, and ADSCs) reach 90% confluence, subject the cells to subculture. Remove the culture medium, and wash the cells with 10 mL serum-free DMEM for several times.
- Add 2 ml trypsin per 10 cm dish. Rotate the dish to distribute the trypsin uniformly, and then incubate the cells at 37 °C in 5% CO_2 for 5 min.
- After 5 min incubation, check under a phase contrast microscope whether the cells have detached from the dish.

Attention! If all the cells are not detached, incubate for 5 more minutes at 37 °C in 5% CO_2, or add 1 mL trypsin, and incubate for a couple of minutes. If these treatments do not work, the trypsin itself might be deactivated. Please prepare a fresh trypsin, and redo.

- If cell detachment is confirmed, add 1 mL FBS to inactivate the trypsin reaction.
- Gently pipette the cells in the dish using a P-1000 with a blue tip for dissociation of the cells.

Attention! Do not use the P-200 yellow tip for pipetting. The opening of the yellow tip is not wide enough so that the cells will be seriously damaged by pipetting.

- Prepare a new 50 mL tube, and transfer the cells and reagents. Add 7 mL serum-free DMEM to the dish to collect the remaining cells, and transfer all the cells and reagents to the 50 mL tube.
- Centrifuge the 50 mL tube at 300 g for 5 min.
- Discard the supernatant, add 1 mL of culture medium, and loosen the cell pellet by gentle pipetting.
- Add 19 mL *culture medium* to the cells and plate into two 10 cm dishes.

Attention! Always subculture the mesenchymal stem cells *at a ratio at 1:2*. Never expand cells to 1:3 or more. Otherwise, cell growth or ratio of Muse cells will decrease. In case of a 10 cm dish, for example, expand one 10 cm dish to two 10 cm dishes.

- Incubate the cells overnight at 37 °C, 5% CO_2. Exchange the medium the next day.
- Cells are maintained by exchanging medium every 2–3 days.

Attention! Mesenchymal stem cells can be kept expanded and then used for Muse cell collection. We usually use P = 4 ~ P = 10 cultured mesenchymal stem cells for collecting Muse cells for analytical and transplantation experiments, while cells over P = 11 are not used for experiments, since the activity of Muse cells may not be fully guaranteed. We strongly recommend the readers to preserve mesenchymal stem cells in liquid N_2 if Muse cells are not in use in the near term rather than keeping subculture of the cells for a longer period.

4.2.5 *Cryopreservation of the Cells for Making Stocks*

At P = 3 or P = 4 of cultured mesenchymal stem cells (NHDFs, HDFa, BM-MSCs, and ADSCs), we usually dispense the cells and make a stock by cryopreservation.

- When mesenchymal stem cells reach 90% confluence (see Fig. 4.4), subject the cells to subculture. Remove the medium, and wash the cells several times with 10 mL serum-free DMEM.
- Add 2 ml trypsin per 10 cm dish. Rotate the dish to uniformly distribute the trypsin, and then incubate at 37 °C, 5% CO_2 for 5 min.
- After 5 min incubation, check under a phase contrast microscope to examine whether the cells have detached from the dish.
- Once cell detachment is confirmed, add 1 mL serum to inactivate the trypsin reaction.
- Gently pipette the cells and solution to dissociate cells using a P-1000 with a blue tip. *Do not use yellow tip.*
- Prepare a new 15 mL tube, and transfer all the cells and reagents. Add another 7 mL serum-free DMEM to the dish to collect the remaining cells, and transfer the cells and reagents to the 15 mL tube.
- Centrifuge the 15 mL tube at 300 g for 5 min.
- Discard the supernatant.
- Add Cellbanker 1 plus (1 mL) to the cells, and gently mix by pipetting.

Attention! The correct volume of Cellbanker 1 plus is 2~3 × 10^6 cells/mL/tube. At least 1 mL of cells + Cellbanker 1 plus should be placed into one cryotube.

- Transfer the cells in 1 mL Cellbanker 1 plus to a cryotube, and screw the lid to form a seal.
- Place the tube into the BICELL, and gradually freeze at –80 °C for 24 h.
- After freezing at –80 °C for 24 h, transfer the tube into liquid N_2, and store the cells.

Attention! The cells can be stored at −80 °C for 2~3 days without losing activity, but for a longer storage period, the cells must be stored in liquid N_2.

4.3 Procedures for Labeling Muse Cells with SSEA-3

Attention! If the purpose of Muse cell collection is to form Muse cell-derived clusters either by single-cell suspension or by methylcellulose gel, please prepare poly-HEMA-coated plates in advance of making clusters. For details, refer to the section 4.5) Poly-HEMA coating of wells and dishes. Please bear in mind that poly-HEMA coating requires overnight treatment.

4.3.1 Preparation of Secondary Antibody

- Dissolve commercially obtained secondary antibody (FITC/APC-labeled anti-rat IgM/-mouse IgG antibody) at 1.0 mg/mL in sterile water, dispense, and store at −30 °C before use.

4.3.2 Preparation of FACS Buffer

- Freshly prepare FACS buffer just before use and cool on ice constantly after preparation. Discard remainder and do not reuse.

FACS Buffer

5% BSA	5 mL
100 mM EDTA	1 mL
PBS or FluoroBrite DMEM	44 mL
Total	50 mL/50-mL tube

Attention!

5% BSA solution: dissolve BSA in PBS or FluoroBrite DMEM, sterilize with 0.22 μm filter, and store at 4 °C.

100 mM EDTA solution: dissolve in PBS or FluoroBrite DMEM, sterilize with 0.22 μm filter, and store at 4 °C.

If cell viability is decreased by using PBS, please use FluoroBrite DMEM instead.

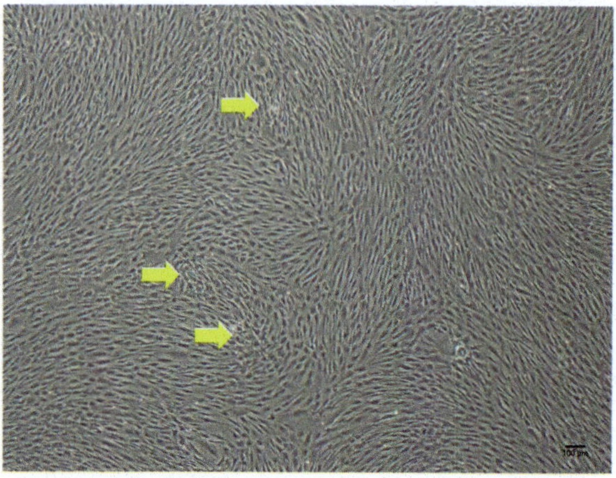

Fig. 4.6 An example of the best timing for FACS in NHDF in 100% confluent.
NHDF is at 100% confluent. Yellow arrows are loci that are about to pile up. These loci are the sign of 100% confluence

4.3.3 Preparation of Mesenchymal Stem Cells for Muse Cell Collection

Attention! Different from the subculture of mesenchymal stem cells described above, the mesenchymal stem cells *must reach 100% confluence* just before collecting Muse cells by FACS (Fig. 4.6). If mesenchymal stem cells were under or over 100% confluence, the yield of Muse cells will be substantially decreased.

BEST timing for Muse cell collection by FACS (*100% confluent: no spaces are remaining between cells)*

Attention! For FACS analysis, mesenchymal stem cells must be completely dissociated into single cells. Because mesenchymal stem cells are sticky and the cells are at 100% confluent before trypsin incubation, incubation with trypsin should be longer than the usual subculture to obtain completely dissociated cells.

Attention! When obtaining Muse cells from frozen mesenchymal stem cells, the cells must be subcultured at least once before being subjected to FACS analysis. Mesenchymal stem cells soon after thawing and plating are weak and unstable and provide a lower ratio of SSEA-3+ Muse cells. After thawing, the mesenchymal stem cells should be cultured to reach 90% confluence, subcultured at least once until they reach *100% confluence*, and then subjected to FACS analysis.

> *Attention!* When primary cultured mesenchymal stem cells are to be analyzed by FACS, the cells should be at P = 4 to P = 10. If the cells are earlier than P = 3, collected Muse cells will be weak and unstable, and if they are later than P = 10, the activity of Muse cells is not guaranteed.

- Using cells cultured in a 10 cm dish (the following procedures are for a 10 cm dish).
- Remove culture media, and wash the cells several times with PBS to completely remove serum.
- Add trypsin (0.25%)/EDTA (2 mL) and incubate cells at 37 °C in incubator for 5–10 min until cells are detached from the dish.
- Stop trypsin reaction by adding FBS (1 mL) and transfer cells and all the medium into a 15 mL tube.
- Collect remaining cells in the dish with serum-free DMEM (7 mL) and transfer them into the same 15 mL tube (total, ~10 mL).
- Centrifuge the tube with swing rotor (400 g, 5 min, room temperature).

> *Attention!* Use centrifuge with *swing rotor*.

- Remove supernatant, and resuspend cell pellet in 10 mL of PBS or FluoroBrite DMEM.
- Centrifuge the tube with swing rotor (400 g, 5 min, room temperature).
- Remove supernatant, resuspend cell pellets with FACS buffer (1 mL), and sample a part of the cells for counting.
- Add FACS buffer (9 mL) (total, 10 mL).
- Centrifuge with swing rotor (400 g, 5 min, room temperature). Count the number of cells during the above processes.
- Remove supernatant and resuspend cell pellets in FACS buffer to establish cell density of 1×10^6 cells/100 μL.

> *Attention!* Be sure that cell density does not exceed 1×10^6 cells/100 μL; otherwise cell will make aggregation. The upper limit of a cell suspension stainable in a 1.5 mL tube is 1×10^7 cells/1000 μL. If the number of cells exceeds the upper limit, split the cells into another tube.

Sample Dispensing and Examples of Cell Numbers

- Dispense cell suspension of 1×10^6 cells/100 μL into a 1.5 mL tube.

Dispensing and cell number examples

#	Sample	Dispensing volume	Number of cells
#1	Without staining	100 μL	1×10^6 cells
#2	Secondary antibody only	100 μL	1×10^6 cells
#3	Isotype control + secondary antibody	100 μL	1×10^6 cells
#4	Anti-SSEA-3 + secondary antibody	100 μL	1×10^6 cells

4.3.4 Staining with Primary Antibody

Attention! Be aware that anti-SSEA-3 antibodies released by several manufacturers do not work for labeling Muse cells. Recommended manufacturers of anti-SSEA-3 (anti-stage-specific embryonic antigen-3) is Rat IgM class clone MC-631 provided by BioLegend (Cat#330302) and Thermo Fisher Scientific (Cat#MA1-020) and Mouse IgG class clone 15B11 provided by IBL (Cat#10431). Anti-SSEA-3 antibody of clone MC-631 is also available from several other manufacturers; however, those do not work for labeling Muse cells (Fig. 4.7). Based on our preliminary data, we recommend to use rat anti-SSEA-3 antibody provided by Thermo Fisher Scientific (Cat#MA1-020) for isolation of Muse cells from umbilical cord-derived MSCs. In order to receive the latest information, please contact Prof. Dezawa (mdezawa*med.tohoku.ac.jp) because the SSEA-3-positive rate markedly depends on company and lot number. Please convert "*" into "@."

Attention! When using cells other than BM-MSCs, NHDF, HDFa, and ADSC listed above, FcR blocking is strongly recommended to be performed in the SSEA-3 antibody staining. Please incubate cells with 10% normal human serum at 4 °C for 20 min before staining with primary antibodies.

Attention! Prepare samples for isotype control. This is necessary for the accurate estimation of SSEA-3 positivity in FACS. Our laboratory use Rat IgM Isotype Control (Cat#400801, BioLegend) for BioLegend and Thermo Fisher Scientific anti-SSEA-3 antibodies and Mouse IgG2b Isotype Control (Cat#401201, BioLegend) for IBL anti-SSEA-3 antibody. Please evaluate the SSEA-3-positive cell rate based on the gating of isotype control in FACS analysis.

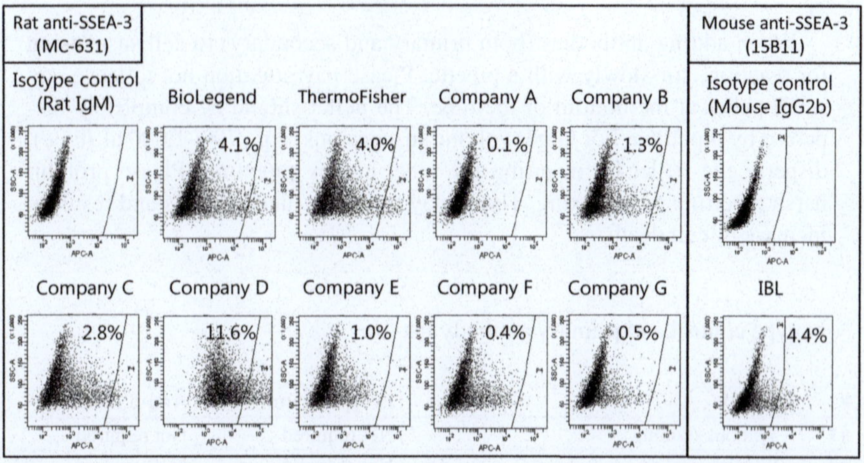

Fig. 4.7 The SSEA-3+ ratio in BM-MSCs by different companies anti-SSEA-3 antibody (as of 2018)
SSEA-3-positive ration in human BM-MSCs with isotype control, anti-SSEA-3 antibodies from BioLegend, Thermo Fisher Scientific, IBL, as well as from company A~G. While antibody from company D had a high positive rate, the negative population showed high background staining. Therefore, D is not recommended. The experimental condition is as follows:
Rat anti-SSEA-3 monoclonal antibody (MC-631) (0.5 μg/100 μL)
Mouse anti-SSEA-3 monoclonal antibody (15B11) (0.5 μg/100 μL)
Isotype control: Purified Rat IgM, κ Isotype Control (BioLegend, Cat#400801, 0.5 μg/100 μL)
Isotype control: Purified Mouse IgG2b, κ Isotype Control (BioLegend, Cat#401201, 0.5 μg/100 μL)
Human BM-MSCs (Passage 7)

- Place samples #1, 2, 3, and 4 in the table of "dispensing and cell number examples "on ice (samples should be kept on ice until analysis).
- Add isotype control at a concentration of 0.5 μg/100 μL into sample #3, stir slowly with a pipette, and incubate on ice for 1 h (stir slowly every 10 min with a pipette).
- Add anti-SSEA-3 antibody at the concentration of 0.5 μg/100 μL (for all the BioLegend, Thermo Fisher Scientific, and IBL antibodies) in sample #4, stir slowly with a pipette, and incubate on ice for 1 h (stir slowly every 10 min with a pipette).

Attention! After adding antibodies, stir slowly with a pipette to ensure a thorough reaction. A sample video is provided in the home page of Dezawa's lab for demonstrating the correct pipetting → http://www.stemcells.med.tohoku.ac.jp/protocol/movie/suspension_good.mp4

When adding antibodies (both primary and secondary) to cell suspension for reaction, stir slowly with a pipette. Please pay attention not to leave any small pellet at the bottom of the tube. The pellet should be completely dispersed by pipetting. For larger volume suspensions (e.g., 700–1,000 μL/tube), disperse the pellet by inverting the tube, not by pipetting. Please prohibit harsh pipetting or vortexing. These may seriously damage cells and result in increase of cell death.

Isotype control and primary antibody list

#	Sample	Isotype control	Anti-SSEA-3
#1	Without staining	Not required	Not required
#2	Secondary antibody only	Not required	Not required
#3	Isotype Control + secondary antibody	+ (0.5 μg/100 μL)	Not required
#4	Anti-SSEA-3 + secondary antibody	Not required	+ (0.5 μg/100 μL)

- Start washing 1 h after the incubation with antibody. Add FACS buffer to make the total volume 1 mL.
- Centrifuge (400 g, 5 min, 4 °C).

Attention! For 1.5 mL tubes, use a swing rotor for centrifugation.

- Remove supernatant, and leave ~100 μL solution in the 1.5 mL tube (FACS buffer and cell pellet at 0.1 mL total volume remains).
- After resuspending the pellet with slow stirring with a pipette, add 900 μL FACS buffer.
- Centrifuge (400 g, 5 min, 4 °C)
- Remove supernatant, and leave ~100 μL solution in the 1.5 mL tube.
- After resuspending the pellet with slow stirring with a pipette, add 900 μL FACS buffer.
- Centrifuge (400 g, 5 min, 4 °C).
- Remove supernatant, and leave ~100 μL solution in the 1.5 mL tube.
- After resuspending the pellet with slow stirring with a pipette, add 900 μL FACS buffer.
- Centrifuge (400 g, 5 min, 4 °C).
- Remove supernatant and leave ~100 μL solution in the 1.5 mL tube.
- Resuspend the pellet with slow stirring using a pipette.

4.3.5 Staining with Secondary Antibody

Attention! Non-specific staining may occur due to impurities in antibody solutions. To prevent such conditions, centrifuge secondary antibodies immediately before use (10,000 g, 3 min, 4 °C) to precipitate impurities, and use supernatants only to eliminate impurities. Store remaining secondary antibodies at 4 °C. Expiration date is within 2 weeks.

- Thaw on ice the secondary antibody that was stored at −30 °C.
- Mix the thawed secondary antibody solution with a pipette, centrifuge (10,000 g, 3 min, 4 °C), and use the supernatant for cell staining. Store the remaining secondary antibody at 4 °C.

Attention! Expiration date of secondary antibody is within 2 weeks after thawing.

- Add secondary antibody at a concentration of 1 μg/100 μL to samples #2, 3, and 4, stir slowly with a pipette, and incubate on ice for 1 h (stir slowly every 10 min with a pipette).

  Secondary antibody list

#	Sample	FITC/APC-labeled secondary antibody
#1	Without staining	Not required
#2	Secondary antibody only	+ (1 μg/100 μL)
#3	Isotype control + secondary antibody	+ (1 μg/100 μL)
#4	Anti-SSEA-3 + secondary antibody	+ (1 μg/100 μL)

- Start washing 1 h after the incubation with antibody. Add FACS buffer to make the total volume 1 mL.
- Centrifuge (400 g, 5 min, 4 °C).
- Remove supernatant, and leave ~100 μL solution in the 1.5 mL tube (FACS buffer and cell pellet at ~100 μL total volume remains).
- After resuspending the pellet with slow stirring with a pipette, add 900 μL FACS buffer.
- Centrifuge (400 g, 5 min, 4 °C).
- Remove supernatant, and leave ~100 μL solution in the 1.5 mL tube.
- After resuspending the pellet with slow stirring with a pipette, add 900 μL FACS buffer.
- Centrifuge (400 g, 5 min, 4 °C).

- Remove supernatant, and leave ~100 μL solution in the 1.5 mL tube.
- Resuspend the pellet in FACS buffer to make the cell concentration 1×10^6 cells/0.1 mL, and prepare for analyses.

Option: If the number of cells is larger than 1×10^7 cells and cell aggregates are observed in suspensions before FACS

- Resuspend the pellet in an appropriate amount of FACS buffer, and filter the cells with a cell strainer (40 μm).
- Centrifuge (400 g, 5 min, 4 °C).
- Remove supernatant and leave ~100μL solution in the 1.5mL tube.
- Resuspend the pellet in FACS buffer to make the cell concentration 1×10^6 cells/0.1 mL, and prepare for analyses.

4.4 Analysis of SSEA-3 Positivity and the Procedure for Collecting Muse Cells

Attention! A sample with a lot of dead cells is inappropriate for analysis and collection. Non-specific staining is likely to occur for dead cells, resulting in an inaccurate measurement of the Muse cell rate. Even if the sampling is forced for the further procedure, cell survival will be largely decreased. If the FSC (frontal scatting light) A vs. SSC (side scattering light) A plot demonstrated a dead cell population in FACS (Fig. 4.8), preparation of another new fresh sample is strongly recommended.

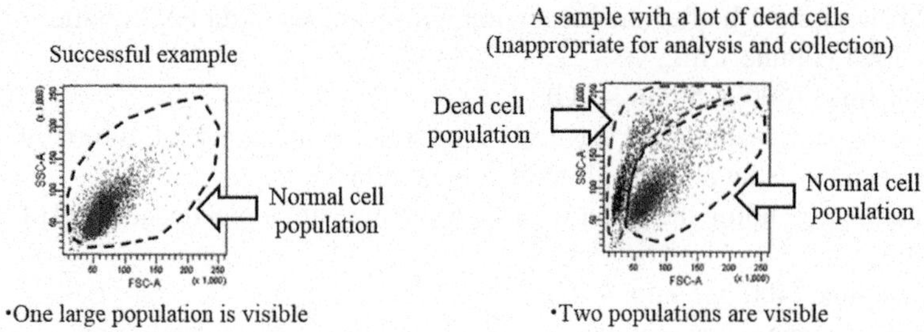

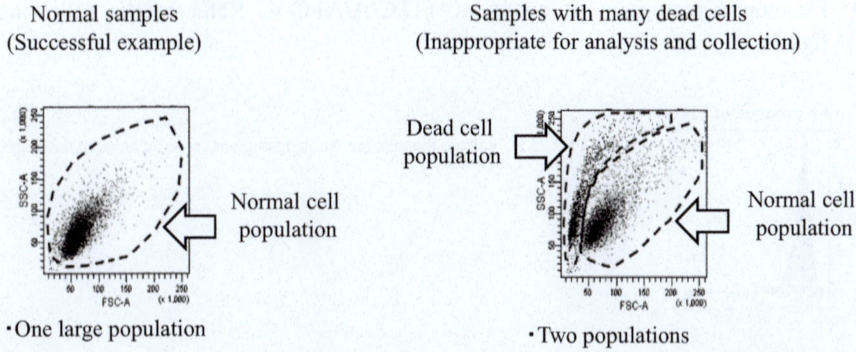

Fig. 4.8 An example of a sample with a lot of dead cells.
In a successful sample, one large population with normal cells is clearly visible. A sample with a lot of dead cells shows two cell populations, one with dead cells and another with normal cells

*4.4.1 Gate Setting and Data Acquisition (*The Following Data Was Obtained by BD FACS Aria II*)*

- Load unstained sample #1 first.
- Adjust sensitivity for SSC and FSC and perform area scaling. Refer to the following figures.

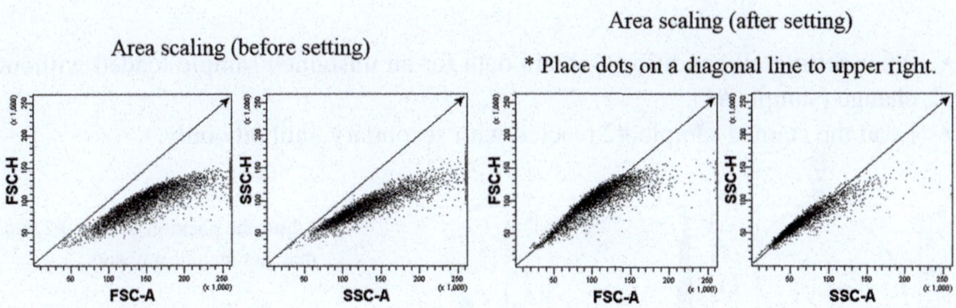

- Develop a plot of SSC-A vs. FSC-A. Refer to the following figure.

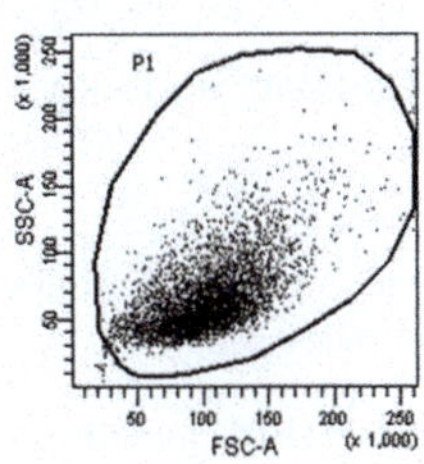

·Draw a Gate P1.

·Set SSC and FSC sensitivities to make the P1 % population more than 95%

- Develop a histogram of count vs. FITC-A/APC-A. Refer to the following figure.

An example of FITC

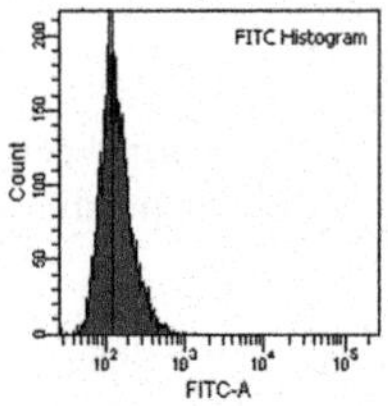

•Adjust FITC/APC voltage to make the whole histogram curve included in the graph.

- Develop a plot of SSC-A vs. FITC/APC (screen to determine SSEA-3-positive rate). Refer to the following figure.

An example of FITC

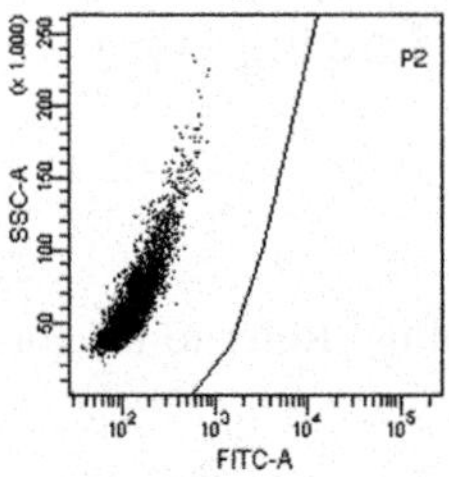

•Reflect the dots in Gate P1 to SSC-A vs. FITC/APC, and newly form Gate P2. Set Gate P2 not to include any dots.

- After setting all conditions, obtain data for an unstained sample loaded without change (sample #1).
- Load the stained sample #2 reacted with secondary antibody only.

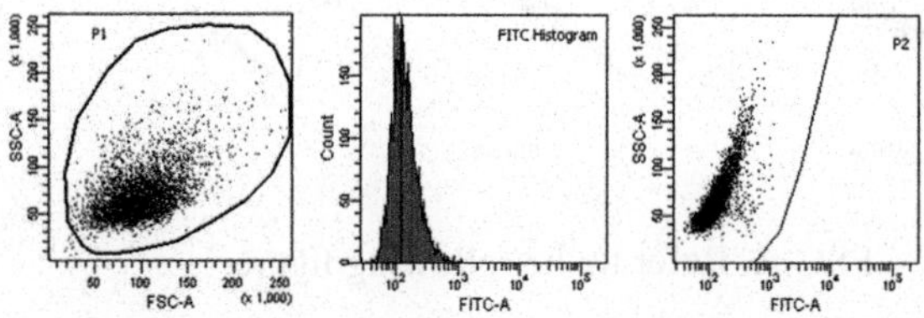

•Adjust the position of Gate P2 that may not include any dots.

•Gate P2 should be 0% for SSEA-3.

- Load the stained sample #3 reacted with isotype control + secondary antibody.

An example of FITC

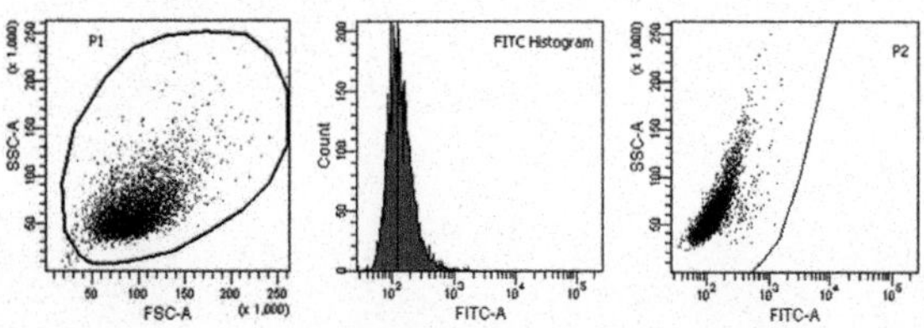

•Adjust the position of Gate P2 that may not include any dots.

•Gate P2 should be 0% for SSEA-3.

- Load the stained sample #4 stained with anti-SSEA-3 + secondary antibody.

An example of FITC

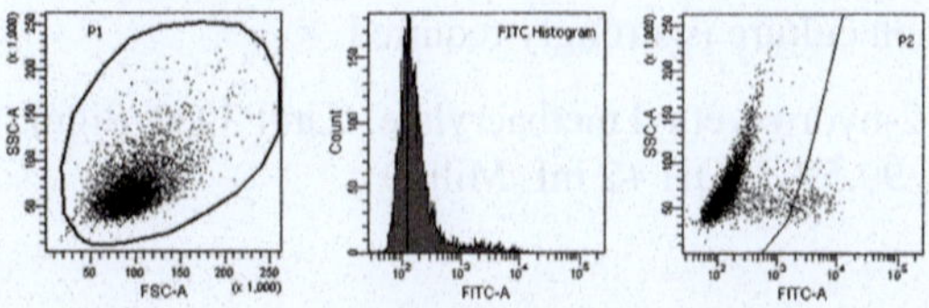

•Do not change the position of Gate P2 set at Isotype control sample #3.

•Gate P2: this case showed 3.8% of SSEA-3-positive cells in BM-MSC.

4.4.2 *Muse Cell Sorting*

- To eliminate doublets, develop the plot for FSC-W vs. FSC-H and SSC-W vs. SSC. See the figure below.

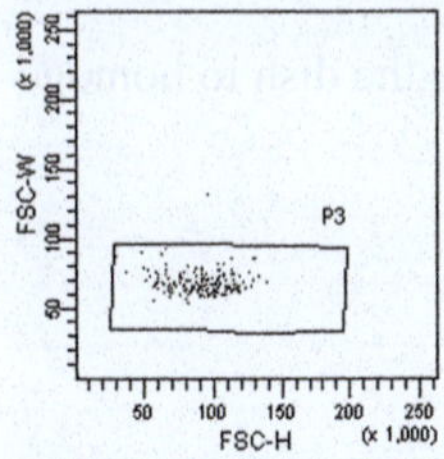

• Reflect the dots in Gate P2 to FSC-W vs FSC-H, and newly form Gate P3. This corresponds to Gate P2 of sample #4.

•Set Gate P3 not to include doublets.

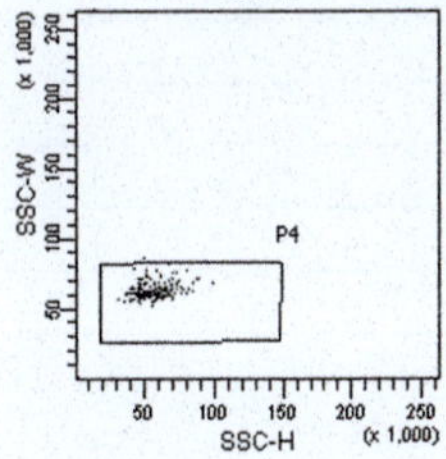

• Reflect the dots in Gate P3 to SSC-W vs SSC-H, and newly form Gate P4. This corresponds to Gate P3 of sample #4.

•Set Gate P4 not to include doublets.

- Prepare a tube containing 10% FBS/DMEM (choose low-glucose or high-glucose type according to the cell type), and set it at a sorting site for receiving sorted cells.
- Start sorting Muse cells by collecting the populations of Gate P4.

4.5 Poly-HEMA Coating of Wells and Dishes

One of the characters of Muse cells is that they form clusters similar to ES cell-derived embryoid body in suspension culture. The formation of clusters is often used for evaluation of Muse cells. There are two methods to make Muse cell-derived clusters: one is single-cell suspension culture and the other is suspension culture in methylcellulose gel. However, Muse cells easily adhere to the bottom of culture dish/well unless the dish/well is not coated with poly-HEMA, the material that completely blocks the adherence of Muse cells. Therefore, pretreatment of dish/well with poly-HEMA before suspension culture is strongly required.

5-1) Add 1.2 g poly-HEMA (poly 2-hydroxyethyl methacrylate, Cat#P3932, Sigma) into 40 mL 95% EtOH (38 mL 99.5% EtOH +2 mL MilliQ).

> *Attention!* Poly-HEMA is highly insoluble. It will not dissolve in 100% EtOH, either. Therefore, please keep in mind to fill the tube first with 95% EtOH solution, and then add the poly-HEMA.

5-2) Shake the solution for several hours at 37 °C to dissolve the poly-HEMA completely. Do not tilt the tube. The poly-HEMA at the bottom of the tube will not dissolve.

1. Add the dissolved poly-HEMA to the dish/well, and rotate the dish to homogenously cover the bottom of the dish/well.

Dish	Volume
10 cm	3.2 mL
6 cm	1.3 mL
3.5 cm	500 μL
12 well	200 μL
24 well	100 μL
48 well	70 μL
96 well	25 μL

2. Leave the cover off the dish, and allow the dish to completely dry out in a clean cabinet overnight. The door of the cabinet should be left open 10–20 cm. Do not turn on a UV light.

> *Attention!* When evaporating the poly-HEMA solution in dish/well in a clean cabinet, EtOH will be saturated in the cabinet. In this situation, dish/well will not dry completely, even when allowed to be placed in the cabinet overnight. Do not to make poly-HEMA coated dishes more than 20 dishes/plates at one time.

Attention! Dried dishes can be stored IN THE DARK, at room temperature for a couple of months.

3. Wash the poly-HEMA-coated dish at least three times by PBS before use.

4.6 Bulk Production of Muse Cell-Derived Clusters by Using Methylcellulose Gel

Attention! Preparation of *poly-HEMA coated dish/well* is required before the FACS isolation of Muse cells (see poly-HEMA coating section). It takes at least for *overnight*.

Attention! Methylcellulose (MC, MethoCult H4100, Cat#04100, STEMCELL Technologies) must be purchased before experiment.

1. Stain FACS-sorted cells with trypan blue and count the percent of live cells.
2. According to the following table, place cells + fetal bovine serum (FBS) + methylcellulose into each well. Methylcellulose has a very high viscosity. Therefore, the tip of a blue or yellow pipette tip should be cut off to facilitate sucking the methylcellulose medium.

Plate	Cell number	Cell + DMEM (μL)	FBS (μL)	2.6% MC (μL)	Total (μL)
6 well	25,000	1,700	300	1,000	3,000
12 well	10,000	705	125	420	1,250
24 well	5,000	400	70	230	700
48 well	3,000	230	40	130	400
96 well	1,000	77	13	40	130

※The final solution will contain 10% FBS and 0.9% MC in DMEM
※Only poly-HEMA coated dishes should be used

3. Slowly and gently stir the cells, FBS, and methylcellulose medium using a cell scraper. Pay attention not to scratch the poly-HEMA coating; otherwise Muse cells will stick to the dish/well. Use a phase microscope to confirm the homogeneous distribution of Muse cells.

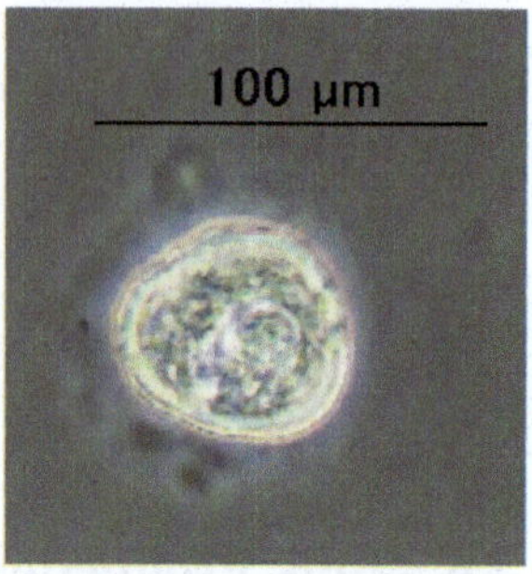

Fig. 4.9 Muse cell cluster in bulk culture
Muse cell cluster in methylcellulose culture on day 7

4. After plating, the cells should be maintained by addition of 10% FBS in low-glucose DMEM (according to the volume indicated in the table below) every 3 days. Cells should be cultured for 7–10 days, and clusters (Fig. 4.9) will be picked up for further analysis.

Plate	Volume (μL)
6 well	1,300
12 well	530
24 well	300
48 well	170
96 well	60

4.7 Generation of Muse Cell Clusters in Single-Cell Suspension Culture

Attention! 96-well *poly-HEMA coated dishes* need to be prepared before FACS isolation. It takes overnight for preparation (see Poly-HEMA coating above).

Attention! Use 10% FBS in low-glucose DMEM for culturing BM-MSCs, NHDFs and HDFs, and 15% FBS in high-glucose DMEM for ADSCs. Furthermore, be sure to add 1 ng/mL FGF-2 (bFGF) to culture BM-MSCs.

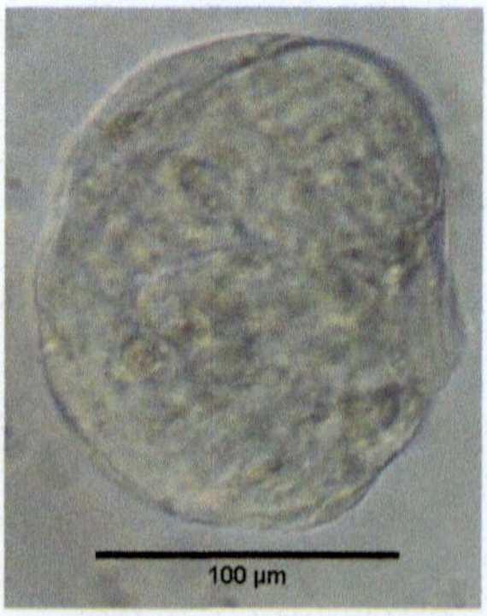

Fig. 4.10 Muse cell cluster in single-cell suspension culture Clusters formed in single-cell suspension culture at day 7

1. Stain FACS sorted cells by trypan blue and count the number of live cells.
2. For limiting dilution, calculate cell number and adjust cell solution in medium.

For example, each well needs 100 μL medium. For 96-well plate, 96 cells are suspended in 9600 μL medium. However, we routinely adjust the medium to make ~3 cells per well. Therefore, the cell solution will be ~288 cells in 9600 μL medium. The cell solution will be gently mixed and then plated 100 μL to each well. This will properly make one cell in each well after plating. Logical calculation is usually too strict to make single cell in each well.

3. Next day, observe each well under phase microscope, and check vacant well and well with multiple number of cells from counting. Those wells should be eliminated from counting the cluster formation ratio.
4. Add 30 μL of medium for each well every 3 days. Culture for 7~10 days and pick up Muse cell-derived clusters for analysis (Fig. 4.10).

4.8 Evaluation of Pluripotency of Muse Cell Clusters: Alkaline Phosphatase Reaction

Attention! Purchase Leukocyte Alkaline Phosphatase Kit (Cat#86R-1KT, Sigma).

Attention! Do not use PBS for alkaline phosphatase (ALP) reaction. PBS is used for stopping the reaction.

1. Prepare ALP solution according to the manufacture's protocol. In brief:
   Mix 10 μL of sodium nitrite solution and 10 μL of FRV-alkaline solution. Both solutions are provided in the kit. Leave for 2 min in room temperature, and then add 450 μL saline.
   ↓
   Add 10 μL naphthol AS-BI alkaline solution to the above solution (this is also provided in the kit).
2. Collect Muse cell clusters in 1.5 mL tube, and add 1 mL saline to suspend the clusters. *Do not use* PBS!
3. Centrifuge at 400 g in room temperature for 5 min. Remove the supernatant.
4. Add 1 mL saline to suspend the clusters.
5. Centrifuge at 400 g in room temperature for 5 min. Remove the supernatant.
6. Add 1 mL saline to suspend the clusters.
7. Centrifuge at 400 g in room temperature for 5 min. Remove the supernatant as much as possible.
8. Fix the clusters by 4% paraformaldehyde. → This is optional. Reaction would be stronger without fixation. We usually skip this procedure.
9. Add 200 μL ALP solution to the clusters. Incubate in 37 °C incubator for 15 min. → The manufacture's protocol instructs us to incubate in room temperature, but in our experience, 37 °C gives better reaction.
10. Add 800 μL PBS to stop the ALP reaction.
11. Centrifuge at 400 g in room temperature for 5 min. Remove the supernatant.
12. Add 1 mL PBS to suspend the clusters.
13. Centrifuge at 400 g in room temperature for 5 min. Remove the supernatant.
14. Transfer clusters to slide glass, and observe under light microscope.

4.9 Evaluation of Pluripotency of Muse Cell Clusters: Gelatin Culture for Trilineage Differentiation

Attention! Prepare gelatin-coated dish or cover slip before doing following experiment.

1. Prepare gelatin-coated dish or cover slip.

   Gelatin (Cat#G-1890, Sigma) → Stock solution is 0.1% gelatin in PBS. Sterilize by autoclaving, and use.

   For coating, load a plentiful amount of 0.1% gelatin solution in plastic wells or wells placed cover slips (we usually use 18 mm diameter round cover slip for 24 well plate) in the bottom, and incubate at 37 °C at least for 30 min.

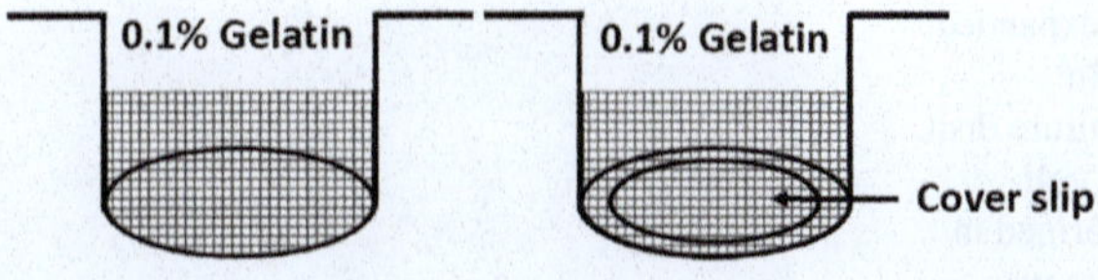

For use, aspirate gelatin solution, and directly use for experiment without washing.

After coating, pay attention *not to dry* the coated wells or cover slips.

2. Pick up Muse cell-derived clusters. Use glass capillary or micropipette (P20 scale) for picking up the clusters.

Attention! In the case of clusters formed in methylcellulose, wash clusters by low-glucose DMEM for a couple of times because methylcellulose clings to cluster disturb its adherence to dish or cover slip. In brief, supply 200 μL low-glucose DMEM into each well of 4 well plate, transfer clusters into low-glucose DMEM, and wash several times by pipetting.

3. Remove gelatin solution after incubation in 37 °C, and quickly supply 10% FBS in low-glucose DMEM into each well. Pay attention not to dry the dish or cover slip.
   Initially, the volume of medium should be a bit lesser than usual. Lesser volume makes transferred clusters easier to adhere to the bottom of dish or to the set cover slip.
   (*For example, 250 μL for 24-well scale, ~800 μL for 12 well)*
4. Transfer clusters into above well.
5. After a couple of hours, add the 10% FBS in low-glucose DMEM for volume up. For 24 well, 300~400 μL, and for 12 well, 1mL solution is preferable for the final volume.

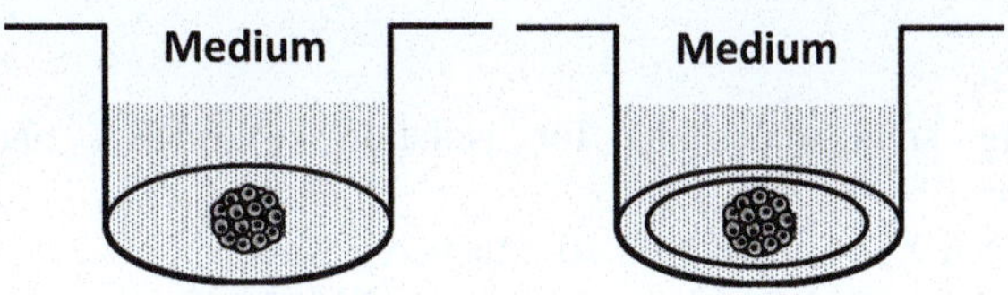

6. Clusters will adhere to the bottom of the well or cover slip by next day or latest by 3 days. The cells gradually expand out of the cluster (Fig. 4.11).
7. Culture for 1~2 weeks, and then subject the samples to RT-PCR or immunocytochemistry

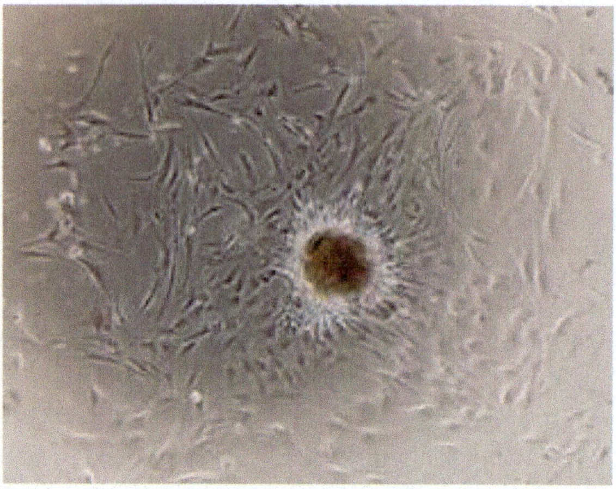

Fig. 4.11 Cells expanded from the cluster in gelatin-coated culture dish. The single Muse cell-derived cluster formed in suspension culture was transferred onto gelatin-coated culture dish. Cells expanded from the cluster at 10 days

human RT-PCR primer

β-actin	F: 5'-GGCGGACTATGACTTAGTTGCGTTACACC-3'
	R: 5'-AAGTCCTCGGCCACATTGTGAACTTTG-3'
Nkx2.5	F: 5'-GGGACTTGAATGCGGTTCAG-3'
	R: 5'-CTCCACAGTTGGGTTCATCTGTAA-3'
α-fetoprotein	F: 5'-CCACTTGTTGCCAACTCAGTGA-3'
	R: 5'-TGCAGGAGGGACATATGTTTCA-3'
MAP-2	F: 5'-ACTACCAGTTTCACACCCCCTTT-3'
	R: 5'-AAGGGTGCAGGAGACACAGATAC-3'
GATA6	F: 5'-CCTGCGGGCTCTACAGCAAGATGAAC-3'
	R: 5'-CGCCCCTGAGGCTGTAGGTTGTGTT-3'

Fig. 4.12 The list of human RT-PCR primers
Primers for human beta-actin, Nkx2.5 (mesodermal), alpha-fetoprotein (endodermal), MAP-2 (ectodermal), and GATA-6 (endodermal) are shown

<For RT-PCR>.
Use following small-scale kits for isolation of mRNA and for reverse transcription.
NucleoSpin RNA XS: Cat#740902.10, Macherey-Nagel.
SuperScript VILO cDNA Synthesis Kit: Cat#11754050, Thermo Fisher Scientific.
TaKaRa Ex Taq: Cat#RR001A, TaKaRa.
Primers: see Fig. 4.12.
<For Immunocytochemistry>
Fix the sample with 4% (vol/vol) paraformaldehyde / 0.01M PBS.

Antibodies for Use
Anti-SMA (Cat#MS-113-P0, Thermo Fisher Scientific, 1:100)
Anti-Neurofilament-M (Cat#AB1987, Merck Millipore, 1:200)
Anti-α-fetoprotein (Cat#N1501, DAKO, 1:100)
Anti-desmin (Cat#550626, BD Biosciences, 1:100)
Anti-cytokeratin 7 (Cat#MAB3226, Merck Millipore, 1:100)

Blocking solution: 20% (vol/vol) Block Ace/5% (wt/vol) (Cat#UKB40, KAC Co., Ltd.) BSA/0.3% (vol/vol) Triton X-100/0.02M D-PBS

Antibody diluent: 5% (vol/vol) Block Ace/1% (wt/vol) BSA/0.3% (vol/vol) Triton X-100/0.02M D-PBS

Chapter 5
Stem Cells and DNA Repair Capacity: Muse Stem Cells Are Among the Best Performers

Tiziana Squillaro, Nicola Alessio, Giovanni Di Bernardo, Servet Özcan, Gianfranco Peluso, and Umberto Galderisi

Abstract Stem cells persist for long periods in the body and experience many intrinsic and extrinsic stresses. For this reason, they present a powerful and effective DNA repair system in order to properly fix DNA damage and avoid the onset of a degenerative process, such as neoplastic transformation or aging. In this chapter, we compare the DNA repair ability of pluripotent stem cells (ESCs, iPSCs, and Muse cells) and other adult stem cells. We also describe personal investigations showing a robust and effective capacity of Muse cells in sensing and repairing DNA following chemical and physical stress. Muse cells can repair DNA through base and nucleotide excision repair mechanisms, BER and NER, respectively. Furthermore, they present a pronounced capacity in repairing double-strand breaks by the nonhomologous end joining (NHEJ) process. The studies addressing the role of DNA damage repair in the biology of stem cells are of paramount importance for comprehension of their functions and, also, for setting up effective and safe stem cell-based therapy.

Keywords Senescence · Apoptosis · DNA damage · DNA repair · Embryonic stem cells · Adult stem cells

---

T. Squillaro · N. Alessio · G. Di Bernardo
Department of Experimental Medicine, Campania University "Luigi Vanvitelli", Naples, Italy

S. Özcan
Genome and Stem Cell Center (GENKOK), Erciyes University, Kayseri, Turkey

G. Peluso
Institute of Agro-Environmental and Forest Biology, CNR, Naples, Italy

U. Galderisi (✉)
Genome and Stem Cell Center (GENKOK), Erciyes University, Kayseri, Turkey

Department of Experimental Medicine, Campania University "Luigi Vanvitelli", Naples, Italy

Sbarro Institute for Cancer Research and Molecular Medicine, Center for Biotechnology, Temple University, Philadelphia, PA, USA
e-mail: umberto.galderisi@unicampania.it

M. Dezawa (ed.), *Muse Cells*, Advances in Experimental Medicine and Biology 1103, https://doi.org/10.1007/978-4-431-56847-6_5

5.1 Cellular Intrinsic and Extrinsic Stress

Cells experience several types of intrinsic and extrinsic stresses for the entire duration of their life. DNA replication and cell metabolism are the principal intrinsic stressors [1]. Cells are continuously exposed to reactive oxygen species (ROS) such as hydrogen peroxide, hydroxyl-free radicals, superoxide anion radicals, and singlet oxygen. ROS derive from the imbalanced intracellular reduction of oxygen or by mitochondrial respiratory activity. Low concentrations of intracellular ROS are not deleterious and act in several signaling pathways, but a sharp increase in ROS level produces oxidative stress that damages proteins, lipids, and DNA. Oxidation of nucleotides may induce DNA mutations, and thousands of them are damaged per day per cell [2].

DNA replication is a vulnerable event in the cell's life. The double helix DNA molecule is quite stable, but during duplication, DNA unwinds and becomes single stranded. The borders between single and double strands (replication forks) present an intrinsically labile structure; moreover, these regions are histone-free. This occurrence renders replication forks open to attack by chemical or physical damaging agents that may induce DNA mutations. Telomeres are the physical ends of linear chromosomes and contain repetitive sequences. These sequences shorten after every cell replication and are another vulnerable region of DNA that may undergo mutation and alterations [3].

Hundreds of chemical and physical genotoxic agents are present in the environment and may act as extrinsic stress factors for cells. Ionizing radiations, UV rays, and high electromagnetic fields are among the most frequent and dangerous physical agents. Prooxidant molecules, factors that produce DNA deamination or alkylation, are chemical agents that are very hazardous and may induce many types of DNA damages [4].

5.1.1 How Do Cells Cope with DNA Damage?

After a stress event that induces DNA damage, cells have to eliminate and/or reduce the possibility they will undergo neoplastic transformation. A correct DNA damage sensing and repair system may recover damaged cells. Cells have different mechanisms to repair DNA (Fig. 5.1). Base pair excision repair (BER) system removes single-nucleotide mutations, such as methylation on O^8 of guanine. DNA glycosylases recognize mutated base moiety of nucleotides of the lesion and cleave the N-glycosidic bond to the damaged base, leaving an apurinic/apyrimidinic site (AP). The AP endonucleases cleave an AP site and leave a nucleotide gap that is filled by combined action of DNA polymerases and ligases. Nucleotide excision repair (NER) is the more versatile system to remove single-strand mutation. It does not recognize a mutated base; rather, it identifies distortions in the double helix DNA structure that are caused by the presence of damaged nucleotides. In this way, almost any kind of mutation can be repaired. NER may eliminate pyrimidine dimers created by UV irradiation. A protein complex runs along DNA to find distortions associated

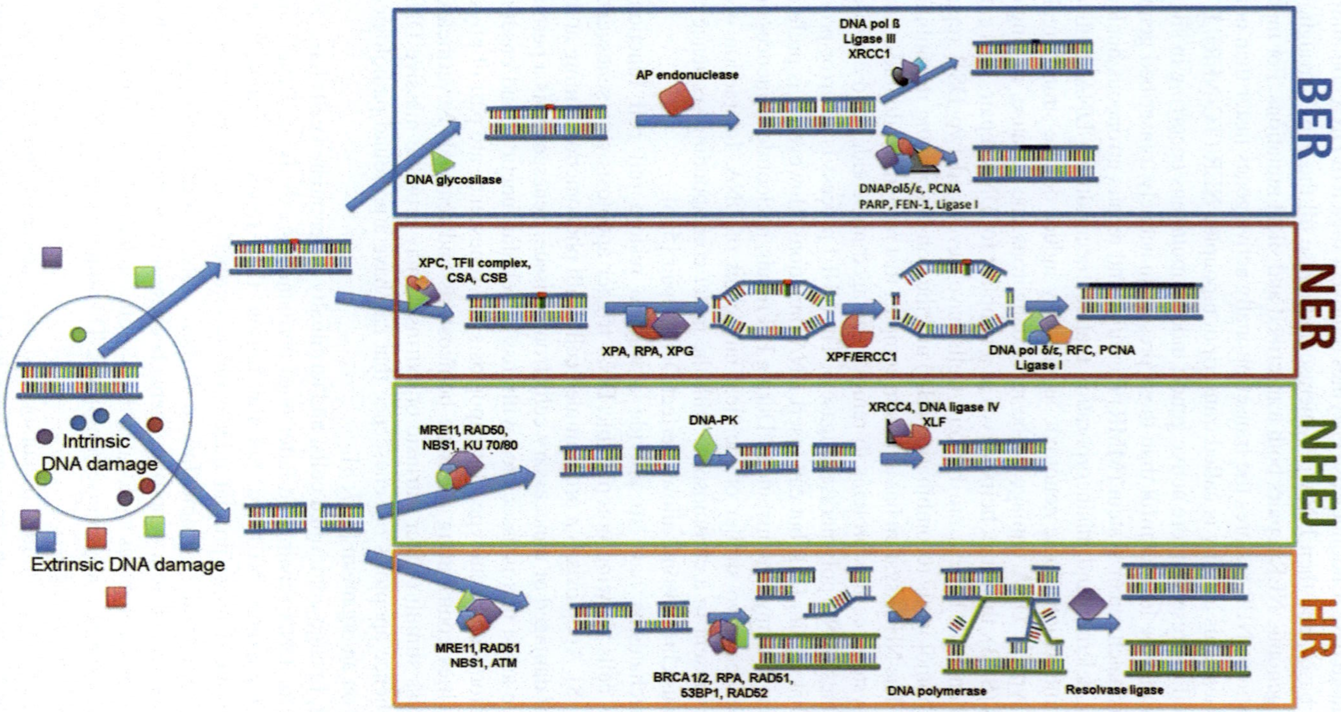

Fig. 5.1 DNA damage sensing and repair
Every day, cells receive hundreds of extrinsic and intrinsic DNA injuries that may induce single- or double-strand breaks: SSB and DSB, respectively. SSB are repaired by BER and NER, while DSB are repaired by NHEJ and HR. Figure 5.1 shows the principal factors that participate in sensing and repairing DNA. See main text for further details

with damaged nucleotides. An endonuclease activity is activated when the complex reaches a mutated base. Endonuclease enzymes remove a stretch of DNA containing the damaged bases, and then DNA polymerase and ligase fill the gap. Throughout DNA, global genomic NER repairs both transcribed and silent mutations. For many types of mutations, NER restores the transcriptionally active genes faster than non-transcribed ones. This activity is called transcription-coupled NER (TC-NER). It is a way for a cell to preserve more active genes; indeed, mutations occurring in silent DNA regions may be less harmful than those present in actively transcribed genes. The DNA mismatch repair system (MMR) identifies and repairs erroneous nucleotide insertion, deletion, and misincorporation that can occur during DNA duplication and recombination. Some genotoxic agents may induce double-strand breaks (DSB). This type of DNA damage is the most dangerous one because, in single-strand breaks (SSB), the intact helix serves as a template for DNA repair. This cannot occur when both helices are broken. Cells have developed two DSB repair systems: the homologous recombination (HR) and the non-end joining recombination (NHEJ). The NHEJ system has a protein complex that recognizes double-strand breaks; this system recruits proteins to remove damaged nucleotides and end-processing enzymes (DNA polymerases, ligases). The NHEJ process is an error-prone mechanism that repairs DSB but can insert DNA mutations. However, it is preferable to no repair in many circumstances. During HR repair, nucleotide sequences are exchanged between two similar or identical molecules of DNA. Interaction of a broken DNA present in a chromosome with the intact corresponding sequence on the homologous chromosome allows perfect DNA repair [5, 6].

Repair mechanisms have to provide a way to restore full cell functions. Alternatively, cells showing unrepairable DNA trigger apoptosis or a senescence process [1, 7, 8]. Apoptosis is a programmed cell death phenomenon aiming at the elimination of damaged or unnecessary cells from a tissue. Senescence is a permanent cell cycle arrest that is associated with loss of cellular functions and onset of pro-inflammatory cell phenotype. Both apoptosis and senescence are part of physiological activities contributing to tissue homeostasis and renewal. Nevertheless, persistent stress stimuli (either intrinsic or extrinsic) may promote massive DNA damage and hence apoptosis or senescence that can have a profound pathological consequence on an organism [1, 7, 8].

Stem cells, lineage-committed cells, and terminally differentiated cells have different needs for DNA repair. The latter ones permanently exit the cell cycle and replicate their genomes no more. For this reason, their repair mechanisms are focused on fixing damage present mainly on transcribed genes rather than constantly scanning the entire genome. Indeed, in these cells, a considerable part of the DNA is well protected from genotoxic agents because it is in compact heterochromatin. Only transcribed genes are located in euchromatin that is less densely packed to allow access to RNA polymerase complexes. This status, anyway, renders DNA more vulnerable to damages.

On the other hand, committed precursors and stem cells are cycling cells that can experience DNA replication errors and have less heterochromatin than differentiated cells. In addition, given their long life, stem cells may suffer from several rounds of intrinsic and extrinsic stresses.

5.2 DNA Repair in Embryonic Stem Cells and Induced Pluripotent Stem Cells

Embryonic stem cells (ESCs) originate from the inner cell mass of mammalian embryo blastocyst. ESCs are pluripotent stem cells and can differentiate into all cell types of the three primary germ layers: ectoderm, endoderm, and mesoderm [9]. The stability of the ESC genome is highly desirable because a mutation can have profound consequences on the organism's development. Indeed, the mutation rate of ESCs is far lower than in differentiated cells. As an example, the mutation frequency for the hypoxanthine-guanine phosphoribosyltransferase gene (*Hprt*) is lower than 10^{-8} per base pair per year, while in mouse embryonic fibroblasts (MEFs), it is around 10^{-5}. The frequency for the adenine phosphoribosyltransferase gene (*Aprt*) is 10^{-6} in ESCs and 10^{-4} in MEFs [10–12].

The low frequency of mutation is related to an efficient DNA repair system. Following ionizing irradiation, peroxide hydrogen, or psoralen treatments, ESCs are more proficient in repairing DNA by BER and NER than somatic cells. This is related to an increased expression of genes involved in DNA repair [13]. ESCs can also cope with DSB. It is reasonable to hypothesize that ESCs prefer to repair DSB by the high-fidelity HR rather than by the error-prone NHEJ, given the importance of a stable genome for these cells. In agreement with this hypothesis, scientists proved that ESCs spend roughly 70/75% of their life in S-phase when the HR is active [14]. Others showed that the resolution of RAD51 foci, involved in the HR pathway, is highly active in differentiated cells (astrocytes) and not in stem cells (neural progenitor cells and ESCs) [15]. In a model of induced DSB by the conditional expression of an endonuclease, Francis and Richardson showed that somatic cells repair DNA either by HR or NHEJ, while in ESCs the activity of NHEJ is negligible [16].

Nevertheless, there are many studies reporting that ESCs express proteins belonging to the NHEJ pathway. Adam and coauthors evidenced that ATM-independent, high-fidelity NHEJ predominates in human ESCs [17]. This system seems to be DNA-PKs independent, while the classic NHEJ is not. The factors and features of NHEJ repair change through differentiation: it becomes more error prone as differentiation proceeds. The existence of a high-fidelity NHEJ may explain why ESCs could use it safely. Others, however, challenged the idea that ESCs may use such a system. Bogomazova and coauthors showed evidence that in the G_2 phase, the human ESCs use the error-prone NHEJ [18]. These authors treated ESCs with X-rays and found several repairs that occurred through chromatid exchanges. These derived from NHEJ that misrejoined DNA breaks. ESCs with misrepaired DNA were prone to apoptosis, and authors hypothesize that ESCs use NHEJ repair and consequent apoptosis as a strategy to preserve genome stability. Indeed, cells that cannot properly repair their DNA by HR are quickly eliminated through programmed cell death.

Adult somatic cells can be induced to reprogram their biological features and become induced pluripotent stem cells (iPSCs). These cells show high similarity with ESCs since they can produce mesodermal, endodermal, and ectodermal derivatives. The iPSCs, like ESCs, are mitotically active and can self-renew. Currently, iPSCs are used in clinical trials for cell therapy or are used as in vitro models for the studies of several diseases [19, 20].

An in-depth analysis of Fan and collaborators suggested that ESCs and iPSCs use overlapping strategies to cope with DNA damage. Both upregulate the BER and NER following oxidative stress and activate the HR and NHEJ pathways after heavy genotoxic injuries. These authors also showed that ESCs ensure genomic stability by having a low apoptotic threshold in response to DNA damage. The iPSCs evidenced a partial apoptotic response [21].

5.2.1 DNA Repair in Adult Somatic Stem Cells

Among the different types of adult stem cells, the DDR has been deeply investigated only in hematopoietic stem cells (HSCs) and neural stem cells (NSCs). In general, stem cells were more proficient in repairing DNA with respect to mature cells. Nevertheless, in some types of damage or specific conditions, scientists did not find significant differences between stem cells and their progeny.

Bracker and collaborators studied the DNA repair capacity of HSCs in comparison with that of committed and mature cells belonging to the lymphohematopoietic system. They noted a decline in repair capacity during commitment and differentiation. The elimination of DNA adducts, the resealing of repair gaps, and the resistance to DNA-reactive drugs were higher in HSCs compared to progenitors and differentiated cells. This was in agreement with increased expression of DDR genes in HSCs [22].

Other researchers evaluated NER in several human acute myeloid leukemia cell lines, before and after differentiation into macrophage-like cells. They found that repair of cisplatin crosslinks in mature cells was robustly reduced compared to progenitors. While some UV-induced damages were repaired with the same efficiency [23], HSCs and their differentiated derivatives showed no difference in the capacity to repair methylation DNA damage since treatment of cells with methylating agent N-methyl-N-nitrosourea (MNU) induced similar levels of apoptosis in stem cells and their progeny. This is in spite of a general trend toward the high expression of several DNA repair genes in HSCs compared to their differentiated counterparts. The overexpressed genes belonged to the NER, BER, MMR, and DSB repair pathways [24].

Of interest, several studies evidenced that the expression of genes involved in DNA repair declines with age. This phenomenon is accompanied by a progressive accumulation of DNA damage and leads to adult stem cell exhaustion. This is supposed to be one of the principal aging mechanisms. Studies of Nijnik and collaborators strengthen this hypothesis. Mice hypomorphic ligase IV mutation showed diminished double-strand break repair by NHEJ. This mutation induces progressive exhaustion of HSCs and bone marrow cellularity during aging and negatively affects stem cell function in tissue culture and transplantation [25].

A study carried out on NSCs and their differentiated derivatives showed that these stem cells were highly proficient in removing 8-oxoguanine from DNA by means of BER. This damage is one the most common DNA injuries induced by an oxidative attack and accumulates at high levels in the genome. The high capacity of NSCs in repairing this damage compared to differentiated cells suggested that NSCs are highly susceptible to oxidative stress [26].

NER activity in repairing UV-induced DNA lesions was analyzed in mature human neurons and their precursor NT2 cells. Removal of UV damage was lower in neurons than in the NT2 cells [27].

A finding of Nowak and collaborators evidenced that following irradiation of a developing mouse brain with a 2Gy dose of gamma rays, neural progenitors went into apoptosis, while neurons did not and survived. Analysis of gamma histone H2AX, a marker of the DNA damage repair process, indicated that the level of DNA damage was equivalent in the two different cell types. Kinetics of gamma H2AX staining evidenced that the DDR system repaired DNA damage in neural progenitor's DNA more slowly than in neurons. This was in agreement with high radiosensitivity of progenitor cells. Also, in this case, the suicide strategy of neural precursors appears to be a strategy to preserve genome stability [28].

5.3 Repair Capacity of Muse Cells

The mesenchymal stromal cells (MSCs) are found in the stroma of almost every organ. Their presence in bone marrow, umbilical cords, and adipose tissues is higher than in other organs. MSCs are heterogeneous because they are composed of distinct cell populations: stem cells, committed progenitors, and mature cells. Many investigators reported that stem cells of MSCs can give origin to progeny having a mesodermal phenotype such as osteocytes, chondrocytes, adipocytes, and muscle cells [29]. In recent years, Dezawa and collaborators identified a pluripotent stem cell population within MSCs that they named multilineage-differentiating stress enduring (Muse) cells, given their stress tolerance. These cells can differentiate in endodermal, ectodermal, and mesodermal derivatives. They express genes of pluripotency such as OCT3/4, SOX2, and NANOG. Muse cells present on their surface the pluripotency antigen SSEA3 that is used to isolate them from MSCs [30, 31]. Recently, Muse cells are considered endogenous reparative stem cells that contribute to tissue repair at serious damage as well as to daily minute repair for maintenance of tissue homeostasis. Since damaged tissue is hostile microenvironment for cells and the main function of Muse cells is to repair tissues, their capacities for high DNA repair and resistance to senescence/apoptosis seem reasonable and rational [32].

Several studies proved that the therapeutic potential of Muse cells is comparable to that of ESCs without the ethical issues that these last cells pose to public opinion. Furthermore, the use of Muse cells may challenge the current applications of iPSCs because Muse cells are naturally found in mammalian organs, while iPSCs are artificially created. In this scenario, it is of paramount importance to evaluate if Muse

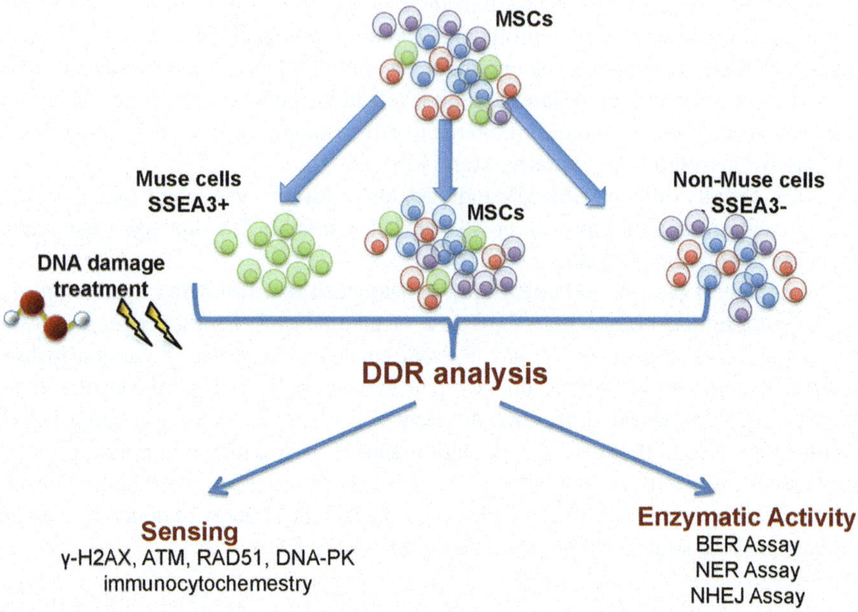

Fig. 5.2 Experimental workflow
We performed research on bone marrow MSCs and their SSEA3 positive and negative fractions, Muse, and non-Muse cells, respectively. Cells were treated with H_2O_2 and UV rays to induce DNA injuries. 1, 6, and 48 h following stress treatments, cells were collected for DDR analysis. We evaluated DNA damage sensing by immunochemistry detection of ATM, RAD51, DNA-PK, and γ-H2AX. We also analyzed the enzymatic activity of DNA repair enzymes by BER, NER, and NHEJ biological assays

cells have a high ability to cope with DNA damage as several reports have demonstrated for ESCs and partially for iPSCs.

Our research group evaluated DNA repair capacity of Muse cells in comparison with MSCs [33]. We analyzed the activation of the DNA damage checkpoint and repair system following the induction of chemical (H_2O_2) and physical (UV rays) stresses (Fig. 5.2).

We studied three different cell populations: naïve MSCs and their SSEA3 positive (Muse cells) and negative (non-Muse cells) subpopulations.

Muse cells showed better protection from chemical and physical damages than non-Muse cells and MSCs. This resulted in resistance to senescence and apoptosis, which are phenomena that cells trigger when damaged DNA is unrepairable.

DNA damage triggers cell cycle arrest and the induction of DNA repair machinery. The ataxia-telangiectasia mutated kinase (ATM) coordinates DNA repair activities by recruiting to damaged foci several DNA repair enzymes. ATM is one of the first proteins activated following genomic injuries, and if DDR works properly, ATM activation is switched off some hours later [34]. In Muse cells, we observed an increase in ATM activation soon after H_2O_2 treatment or UV irradiation and a decline to basal level by 48 h following stresses. The ATM activation of MSCs occurred at a level higher than in Muse cells and did not decline to basal values even 2 days after induction of DNA damages. In non-Muse cells, the activation of ATM did not occur properly.

Our investigation further continued by studying the expression of RAD51 and DNA-PK, which are two fundamental downstream effectors of ATM. RAD51 is part of HR, while DNA-PK is a component of NHEJ [35].

The treatment of Muse cells with H_2O_2 or UV rays produced an increase of RAD-51(+) and DNA-PK(+) quickly after injuries and then a return to basal levels. In MSCs, the activation of RAD51 and DNA-PK took place accurately, but again, no decline to basal level was detected. Of interest, in non-Muse cells, this activation did not occur. These results suggest that only Muse cells repaired DNA damages correctly. The analysis γ-H2AX staining validated this hypothesis.

Following the activation of ATM, the histone protein H2AX is activated by phosphorylation (γ-H2AX). The phosphorylated H2AX contributes to recruitment of DNA repair factors. Soon after DNA injury, the presence of γ-H2AX foci indicated regions with damaged DNA that is undergoing repair. The permanence of active γ-H2AX foci evidences that DNA has been unrepaired or misrepaired [36].

Muse cells presented several H2AX foci soon after treatments with H_2O_2 or UV rays. Two days post-stresses, we still observed H2AX foci in cells that were in the G_1 phase, but this percentage was far lower than those detected in non-Muse cells and in MSCs.

The DDR system relies on sensing factors that identify damaged regions of DNA and then recruit enzymes involved in repair of different types of injuries. We evaluated the enzymatic activities involved in repairing SSB by BER and NER and DSB by NHEJ. The proficiency of Muse cells in correcting SSB by NER and BER was equivalent to those observed in MSCs and non-Muse cells, while NHEJ was significantly higher in Muse cells.

The high activity of NHEJ in Muse cells could be related to the fact that this is the only mechanism that repairs DSB in every cell cycle phase. Further investigations should demonstrate whether, in Muse cells, NHEJ is error prone or a high-fidelity system.

After DNA damage cells activate a quick response by triggering the repair proteins already present in the nucleus. Then, cell may switch on a late response to cope future harmful stimuli. This can be obtained inducing changes in mRNA expression of genes implicated in DNA repair.

After DNA injury, Muse cells changed the mRNA expression of several genes involved in NER, BER, mismatch repair (MMR), NHEJ, and homologous recombination (HR) to cope with DNA repair. These changes occurred soon after damage and persisted 48 h later. Non-Muse cells showed minimal modifications of gene expression only soon after DNA damage (1 h) [33]. Changes in the mRNA expression of genes belonging to DNA damage and repair pathways suggest that in Muse cells is active an efficient adaptive response mechanism. This phenomenon implies that a minimal priming dose of a DNA damaging agent can protect cells against a larger second dose given several hours or days later by the activation of a repair process.

5.4 Conclusions

Mechanisms that properly cope with environmental stress are a typical feature of stem cells. Muse cells present a quick and effective DDR. The rapidity in detecting damaged foci and fixing damages may explain the effectiveness of this process. The study of DDR is of paramount importance for comprehension of stem cell functions and for setting up effective and safe stem cell-based therapy. In summary, Muse cells appear to have high DDR capacity as already evidenced for ESCs and iPSCs, and this is important for therapeutic applications. However, ESCs and iPSCs are not endogenous, and they don't reside in normal living body. They are consistently exogenous cells. Muse cells are naturally existing endogenous stem cells, normally distributing in the connective tissue of every organ, peripheral blood, and the bone marrow. They may represent a better alternative to ESCs and iPSCs in cell therapy.

References

1. Tower J (2012) Stress and stem cells. Wiley Interdiscip Rev Dev Biol 1(6):789–802
2. Richardson C, Yan S, Vestal CG (2015) Oxidative stress, bone marrow failure, and genome instability in hematopoietic stem cells. Int J Mol Sci 16(2):2366–2385
3. Ruzankina Y, Asare A, Brown EJ (2008) Replicative stress, stem cells and aging. Mech Ageing Dev 129(7–8):460–466
4. Chakarov S et al (2014) DNA damage and mutation. Types of DNA damage. Biodiscovery 11:1–10
5. Branzei D, Foiani M (2008) Regulation of DNA repair throughout the cell cycle. Nat Rev Mol Cell Biol 9(4):297–308
6. Dexheimer TS (2013) DNA repair pathways and mechanisms. In: Mathews LA, Cabarcas SM, Hurt E (eds) DNA repair of cancer stem cells. Springer, New York, pp 19–32
7. Alessio N et al (2016) Mesenchymal stromal cells having inactivated RB1 survive following low irradiation and accumulate damaged DNA: Hints for side effects following radiotherapy. Cell Cycle 16(3):251–258
8. Zanichelli F et al (2012) Dose-dependent effects of R-sulforaphane isothiocyanate on the biology of human mesenchymal stem cells, at dietary amounts, it promotes cell proliferation and reduces senescence and apoptosis, while at anti-cancer drug doses, it has a cytotoxic effect. Age (Dordr) 34(2):281–293
9. Ilic D, Ogilvie C (2017) Concise review: human embryonic stem cells-what have we done? What are we doing? Where are we going? Stem Cells 35(1):17–25
10. Hong Y et al (2007) Protecting genomic integrity in somatic cells and embryonic stem cells. Mutat Res 614(1–2):48–55
11. Lin Q, Donahue SL, Ruley HE (2006) Genome maintenance and mutagenesis in embryonic stem cells. Cell Cycle 5(23):2710–2714
12. Tichy ED, Stambrook PJ (2008) DNA repair in murine embryonic stem cells and differentiated cells. Exp Cell Res 314(9):1929–1936
13. Maynard S et al (2008) Human embryonic stem cells have enhanced repair of multiple forms of DNA damage. Stem Cells 26(9):2266–2274
14. Savatier P et al (2002) Analysis of the cell cycle in mouse embryonic stem cells. Methods Mol Biol 185:27–33
15. Adams BR et al (2010) Dynamic dependence on ATR and ATM for double-strand break repair in human embryonic stem cells and neural descendants. PLoS One 5(4):e10001

16. Francis R, Richardson C (2007) Multipotent hematopoietic cells susceptible to alternative double-strand break repair pathways that promote genome rearrangements. Genes Dev 21(9):1064–1074
17. Adams BR et al (2010) ATM-independent, high-fidelity nonhomologous end joining predominates in human embryonic stem cells. Aging (Albany NY) 2(9):582–596
18. Bogomazova AN et al (2011) Error-prone nonhomologous end joining repair operates in human pluripotent stem cells during late G2. Aging (Albany NY) 3(6):584–596
19. Lengner CJ (2010) iPS cell technology in regenerative medicine. Ann N Y Acad Sci 1192:38–44
20. Yamanaka S, Blau HM (2010) Nuclear reprogramming to a pluripotent state by three approaches. Nature 465(7299):704–712
21. Fan J et al (2011) Human induced pluripotent cells resemble embryonic stem cells demonstrating enhanced levels of DNA repair and efficacy of nonhomologous end-joining. Mutat Res 713(1–2):8–17
22. Bracker TU et al (2006) Stringent regulation of DNA repair during human hematopoietic differentiation: a gene expression and functional analysis. Stem Cells 24(3):722–730
23. Hsu PH, Hanawalt PC, Nouspikel T (2007) Nucleotide excision repair phenotype of human acute myeloid leukemia cell lines at various stages of differentiation. Mutat Res 614(1–2):3–15
24. Casorelli I et al (2007) Methylation damage response in hematopoietic progenitor cells. DNA Repair (Amst) 6(8):1170–1178
25. Nijnik A et al (2007) DNA repair is limiting for haematopoietic stem cells during ageing. Nature 447(7145):686–690
26. Hildrestrand GA et al (2007) The capacity to remove 8-oxoG is enhanced in newborn neural stem/progenitor cells and decreases in juvenile mice and upon cell differentiation. DNA Repair (Amst) 6(6):723–732
27. Nouspikel T, Hanawalt PC (2000) Terminally differentiated human neurons repair transcribed genes but display attenuated global DNA repair and modulation of repair gene expression. Mol Cell Biol 20(5):1562–1570
28. Nowak E et al (2006) Radiation-induced H2AX phosphorylation and neural precursor apoptosis in the developing brain of mice. Radiat Res 165(2):155–164
29. Galderisi U, Giordano A (2014) The gap between the physiological and therapeutic roles of mesenchymal stem cells. Med Res Rev 34(5):1100–1126
30. Wakao S et al (2011) Multilineage-differentiating stress-enduring (muse) cells are a primary source of induced pluripotent stem cells in human fibroblasts. Proc Natl Acad Sci U S A 108(24):9875–9880
31. Wakao S et al (2012) Regenerative effects of mesenchymal stem cells: contribution of muse cells, a novel pluripotent stem cell type that resides in mesenchymal cells. Cells 1(4):1045–1060
32. Dezawa M (2016) Muse cells provide the pluripotency of mesenchymal stem cells: direct contribution of muse cells to tissue regeneration. Cell Transplant 25:849–861
33. Alessio N et al (2018) Stress and stem cells: adult muse cells tolerate extensive genotoxic stimuli better than mesenchymal stromal cells. Oncotarget 9(27):19328–19341
34. Freeman AK, Monteiro AN (2010) Phosphatases in the cellular response to DNA damage. Cell Commun Signal 8:27
35. Shibata A et al (2011) Factors determining DNA double-strand break repair pathway choice in G2 phase. EMBO J 30(6):1079–1092
36. Fu S et al (2012) gamma-H2AX kinetics as a novel approach to high content screening for small molecule radiosensitizers. PLoS One **7**(6):e38465

Chapter 6
Immunomodulatory Properties and Potential Therapeutic Benefits of Muse Cells Administration in Diabetes

Marcelo Javier Perone, María Laura Gimeno, and Florencia Fuertes

Abstract It is well established the link between inflammation and the development of insulin resistance and pathogenesis of type 2 diabetes. Type 1 diabetes is an autoimmune disease characterized by the destruction of insulin-producing pancreatic β cells mediated by autoreactive T lymphocytes and pro-inflammatory agents. Therefore, developing new strategies to efficiently control dysregulated inflammation could have substantial benefits in the treatment of diabetes. Recently, a novel population of non-tumorigenic pluripotent stem cells, named multilineage-differentiating stress-enduring (Muse) cells, was discovered. Muse cells secrete significant amounts of TGF-β1, a key cytokine governing down-modulation of T lymphocytes and macrophages. In this chapter, we discuss the immunomodulatory properties of Muse cells as well as the molecular mechanism of TGF-β1 as mediator of Muse cell action. We also describe the role of certain cytokines/growth factors highly expressed in Muse cells as potential mediators of their effects. Finally, we provide evidence of the beneficial effects of adipose tissue-derived Muse cells in an experimental mice model of type 1 diabetes.

Keywords Stem cells · Inflammation · Tissue regeneration · TGF-β1 · Interleukins · T lymphocytes · Macrophages · Adipose-derived stem cells

6.1 Introduction

Mesenchymal stromal cells (MSCs) are a group of multipotent stem cells from adult tissues with capacity to regulate the immune system, participating in both the innate and adaptive responses [1–7]. These cells have been described as capable of inhibiting both T cell proliferation and cytokine secretion [1, 3, 6]. They are also able to drive T helper balance towards a regulatory phenotype, laying the ground for

M. J. Perone (✉) · M. L. Gimeno · F. Fuertes
Instituto de Investigación en Biomedicina de Buenos Aires (IBioBA) – CONICET – Partner Institute of the Max Planck Society, Buenos Aires, Argentina

M. Dezawa (ed.), *Muse Cells*, Advances in Experimental Medicine and Biology 1103, https://doi.org/10.1007/978-4-431-56847-6_6

tolerogenic environment. Other immunomodulatory abilities already attributed to MSCs are inhibition of activation and proliferation of dendritic cells [4], as well as blocking the proliferation of natural killer (NK) cells, being both key actors of the innate immune response [3, 4, 6]. These effects are thought to be a consequence of soluble factors released by MSCs; however, the precise mechanisms of stimulation, production of active agents and their secretion by MSCs are still not fully understood. MSCs release soluble factors with major impact on the regulation of the immune system including hepatocyte growth factor (HGF), transforming growth factor-β1 (TGF-β1), prostaglandin E2 (PGE2) and interleukin-10 (IL-10) among others [5, 7]. The non-classic MHC I molecule, HLA-G, expressed in the surface of MSCs, has also been described as a mediator of their immunosuppressive effects [3, 6, 8]. Based on their immunomodulatory properties, the use of MSCs has been thought to be an excellent therapeutic alternative for immune-mediated diseases.

In contrast to induced pluripotent stem cells (iPSCs), MSCs became good candidates for cell regeneration therapy both due to their easy accessibility and their lack of tumorigenic activity, making them a good therapeutic alternative for immune-mediated diseases. However, one of the main reasons why MSCs regenerative therapies have not been successful so far is because of the low rate of cell implantation [9, 10]. That is why, in the past few years, researchers have tried to encounter other candidates for cell therapy.

In 2010, a subset of pluripotent stem cells was described and named multilineage-differentiating stress-enduring (Muse) cells, isolated from adult tissues [11]. Currently, Muse cells can be isolated from mesenchymal tissues mainly skin, bone marrow and adipose tissue [11, 12]. These conspicuous stem cells have been characterized for their SSEA-3 expression (a well-known surface marker for human embryonic stem cells). They also express pluripotency markers, such as octamer-binding transcription factor 3/4 (Oct3/4), sex-determining region Y-box 2 (Sox-2) and Nanog [11–13]. Furthermore, Muse cells have high capacity of migrating to injured areas, regenerating damage tissues and fostering function recovery.

Remarkably, Muse cells express low levels of HLA-DR, a major histocompatibility class (MHC) II molecule, suggesting the potential use of Muse cells for allogeneic transplantation. Muse cells telomerase activity is very low, and it has been proven their incapacity of undergoing teratogenesis in vivo [11, 13–16]. Based on all these properties, Muse cells are ideal candidates for tissue regeneration and cell therapy.

6.2 Diabetes

During the last two decades, several attempts have been made to employ the administration of MSCs for the treatment of diabetes. Diabetes is a highly prevalent endocrine-metabolic disease with a constant growing rate, affecting nearly half a billion people worldwide. It is characterized by hyperglycaemia as a result of whether a lack of insulin production in the pancreas (type 1 diabetes) or peripheral insulin resistance and insufficient insulin production (type 2 diabetes), in addition to genetic and environmental components. Type 1 and type 2 diabetes are distinguished by their etiopathogenesis, where the former is characterized by an autoimmune-mediated destruction of β cells in the pancreatic islets, while type 2 diabetes comes out as a consequence of insulin resistance, which leads to β cells inability to cope with high blood glucose and, eventually, their death by apoptosis [17].

In type 1 diabetic patients, autoantigens produced by damaged β cells (regardless of the primary stimulus) are recognized by antigen-presenting cells (APCs) and activate pro-inflammatory responses. In this context, MSCs and Muse cells may either modulate the action of the uncontrolled immune system or protect pancreatic β cells from death.

Dendritic cells and macrophages belong to the first line of immune cells infiltrating pancreatic islets in type 1 diabetes, followed by B and T lymphocytes. At early disease stages, benign T helper type 2 (Th2) cells can be found within islets, followed by a wave of Th1 lymphocytes generating a pro-inflammatory environment [18]. Later on, cytotoxic T lymphocytes (CD8+) contribute to the pathogenesis establishing direct contact and damaging insulin-producing cells through the release of perforin and granzyme molecules, among other agents [17, 19]. Also, β cell damage is promoted by pro-inflammatory cytokines released by macrophages and activated T lymphocytes, such as IL-1β, TNF-α and IFN-γ [20, 21], and it is further accelerated by superoxide production [17]. These pro-inflammatory stimuli, together with radical oxygen and nitrogen species generated by the intracellular oxidative stress of β cells, will lead to their demise and inability of the remaining β cells to cope appropriately with blood glucose levels.

The pathophysiological mechanisms of type 2 diabetes differ from type 1 diabetes, although recent evidence demonstrated the participation of selected components of the immune system as well [19]. Autoreactive CD4+ T cells have been found in the blood of type 2 diabetic patients, and their cytokine production (particularly IFN-γ and IL-10) is similar to that observed in type 1 diabetic subjects. Insulin resistance in type 2 diabetes determines higher blood glucose levels which, combined with reduced β cell function, are translated into poor insulin output and subsequent dysregulated glucose homeostasis [22].

The nonobese diabetic (NOD) mouse is frequently employed as a spontaneous model of type 1 diabetes, in which the immune system plays an essential role in the pathogenesis of the disease. This mouse strain is characterized by multiple defects in regulatory pathways of the immune system [23]. Infiltration of leukocytes in the pancreatic islets – insulitis – can be observed as early as 4 weeks of age in females, and incidence of diabetes commonly occurs in 60–80% of 30-week-old female NOD mice. In this chapter, we describe the effects of adipose tissue-derived Muse cells modulating diabetogenic T lymphocytes and their beneficial administration in diabetic NOD mice.

6.3 Immunomodulatory Action of Muse-AT Cells

Muse cells are generally obtained by cell sorting based on pluripotent surface marker SSEA-3 expression [11]. However, we obtained Muse cell-enriched fraction from human lipoaspirates (Muse-AT cells) by simple, fast and affordable procedure, based on the original method reported by Heneidi et al. [12]. The lipoaspirate was treated with severe cellular stress conditions combining long-term collagenase incubation, lack of nutrients, low temperature and hypoxia and resulted in the enrichment stage-specific embryonic antigen-3 (SSEA-3)-positive cells for up to ~60% of [13].

Recent reports have described the immunoregulatory capacity of Muse-AT cells. The fact that Muse-AT cells isolated from adipose tissue secrete high levels of TGF-β1 is a major milestone supporting the immunomodulatory properties of these conspicuous stem cells [13]. TGF-β1 is a cytokine which plays a critical role in inflammation and other biological processes, depending on the microenvironment surrounding damaged tissues. We demonstrated that the expression level of TGF-β1 secreted by Muse-AT cells in culture increases spontaneously over time, reaching the highest levels between 5 and 10 days, probably as a response to the stringent stress culture conditions triggered by clustering formation in suspension culture.

The immunomodulatory capacity of Muse-AT cells is mainly mediated by secreted TGF-β1 [13]. T cells and macrophages secrete pro-inflammatory cytokines upon stimulation in vitro. Pro-inflammatory IFN-γ and TNF-α secreted by T cells and macrophages, respectively, boost the immune response. To investigate whether Muse-AT cells might influence the response of macrophages, we first employed the mouse macrophage cell line RAW 264.7. LPS activates and dramatically induces the secretion of TNF-α by RAW264.7 cells [24]. Muse-AT cells co-cultured with RAW 264.7 macrophages significantly reduced TNF-α secretion upon LPS stimulation. In addition, using a transwell co-culture system to find out whether soluble factors were the mediators of the observed inhibition, we discovered a significant decrease of TNF-α production by RAW264.7 cells upon LPS stimulation. These results indicate that soluble mediators secreted by Muse-AT cells are, in part, responsible for their immunomodulatory effects. Similarly, Muse-AT cell conditioned-media significantly reduced TNF-α secretion by LPS-stimulated RAW264.7 cells. We could also confirm the presence of TGF-β1 in Muse-AT cell conditioned-media as the one responsible for the observed effects on RAW 264.7 cells. Co-culture experiments were performed in the presence of SB-431542 (SB), a small molecule inhibitor that blocks TGF-β1 action. Under these circumstances, the inhibitory effect of Muse-AT cell conditioned-media was reverted, indicating that TGF-β1 plays a critical role as a mediator of Muse-AT cell immunomodulation. The same results were obtained when a more reliable source of macrophages was used. Indeed, Muse-AT cell conditioned-media reduced TNF-α secretion by LPS-stimulated freshly isolated murine macrophages, and the incorporation of SB into the media restored TNF-α secretion.

We then explored the ability of Muse-AT cells on T-lymphocyte responses. For this purpose, an in vitro antigen-specific T-lymphocyte assay was employed. NOD BDC2.5 CD4+ T cells specifically recognize a chromogranin-A peptide in the context of major histocompatibility class II complex (H-2 g7) of the NOD mice. BDC2.5 CD4+ T lymphocytes mainly differentiate into a Th1 phenotype secreting high amounts of pro-inflammatory cytokines, e.g. IFN-γ, upon antigen stimulation and have been extensively characterized regarding pathogenicity in autoimmune diabetes [25]. Thus, BDC2.5 CD4+ T cells responded with a high secretion of INF-γ upon antigen challenge [26, 27]. Muse-AT cell conditioned-media diminished IFN-γ secretion by antigen-stimulated BDC2.5 CD4+ T lymphocytes in a dose-dependent manner. Interestingly, conditioned-media from Muse-AT cells generated more efficient results in blocking INF-γ secretion when compared to conditioned-media obtained from culturing MSCs isolated from adipose tissue. When comparing TGF-β1 expression from cells isolated from a single individual, Muse-AT cells express higher levels than adipose MSCs, suggesting this cytokine may be responsible for the observed immune downregulating effect.

To further confirm whether the immune-regulatory potency exerted by Muse-AT cells on T cells might depend on their TGF-β1 expression levels, we blocked its signalling pathway using SB. When SB was applied to T cell cultures in the presence of Muse-AT cell conditioned-media, the secretion of IFN-γ by antigen-specific stimulated BDC2.5 CD4+ T lymphocytes was restored. Similar results were observed when a neutralizing anti-TGF-β1 antibody was added. These results confirmed the role of TGF-β1 as a key mediator of Muse-AT immunomodulatory activities in macrophages and T lymphocytes. In addition, we discovered upregulation of the classic anti-inflammatory IL-10 in the presence of Muse-AT cell conditioned-media in BDC2.5 CD4+ T-lymphocyte cultures [13].

6.4 Mechanistic Insights into Muse-AT Cells Activity

Based on all the results described above, we analysed the TGF-β1 signalling pathway of Muse-AT cells. Analysing several preparations of human Muse-AT cells, we have found that these cells express TGF-βII receptor, the cognate receptor for TGF-β1, on their surface membrane (Gimeno et al. unpublished), suggesting that this cytokine may promote autocrine/paracrine effects on the cells. What could be the role of TGF-β1 signalling in Muse-AT cells? This pleiotropic cytokine has been proposed as a stemness regulator of haematopoietic stem cells (HSCs); interestingly, these cells require stimulation with activin-like kinase 5 (TGF-β type I receptor) ligands to maintain a quiescent state within the bone marrow niche. In accordance to this, we observed that increments of TGF-β1 levels in Muse-AT cell clusters also matched with very low proliferation cell activity, while very low TGF-β1 levels were observed in highly proliferative ASCs. Thus, our observations led us to speculate that TGF-β1 autocrine/paracrine loop might help maintain Muse-AT cells at low proliferative rate [11]. Freshly isolated Muse-AT cells express

TGF-βII-receptor, suggesting that they might show an early response to TGF-β1 ligands (see Fig. 6.1). Regarding this, Ylöstalo et al. proposed that aggregation of MSCs into spheroids induces a type of cellular stress that results in intracellular signalling leading to upregulation of anti-inflammatory effectors [28]. In vivo, activation might occur by the action of soluble factors released from injured tissues [29]. Accordingly, we found that Muse-AT cells spontaneously expressed high levels of TGF-β1 when cultured in non-adherent conditions (e.g. after 5–10 days). Cell aggregation might trigger this expression. Furthermore, when cells are transplanted and reach damaged tissues, it can be expected to find low levels of TGF-β1 in the microenvironment, which in turn will allow cell differentiation and tissue repair.

TGF-β binds to its specific receptor initiating intracellular signalling cascades and activating several phosphorylated mediators. Among them, SMAD2 has been described as an important mediator of the TGF-β1 anti-inflammatory pathway [30–

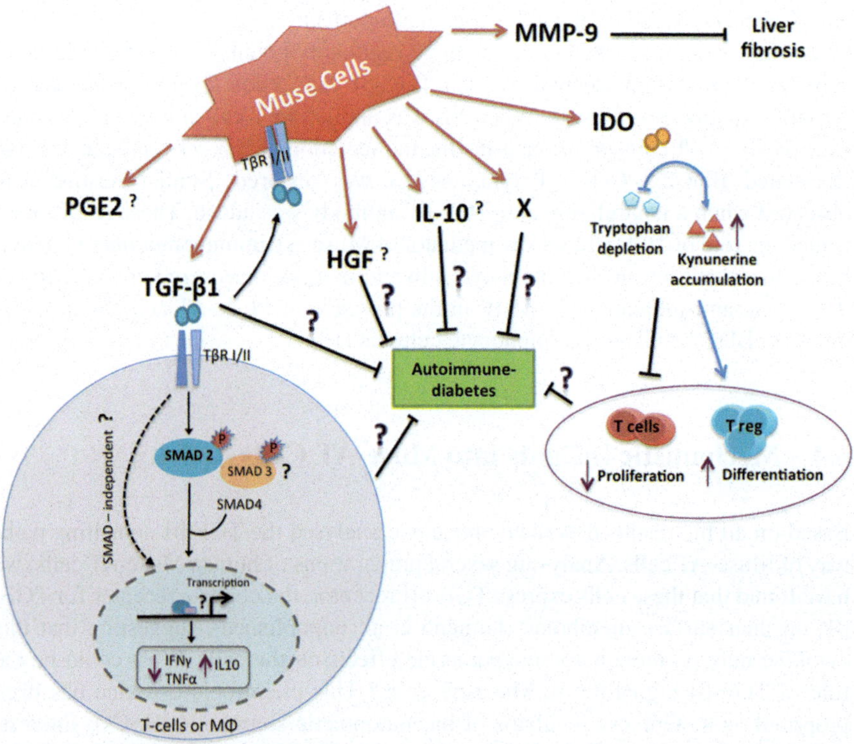

Fig. 6.1 Schematic diagram illustrating the agents secreted by Muse-AT cells, their impact on immune cells response and possible therapeutic effects. Secreted-TGF-β1 by Muse-AT cells binds to its cognate receptor (TβR I/II) on T lymphocytes and macrophages. A putative intracellular TGF-β1 signalling pathway involves the expression of phosphorylated-SMAD proteins regulating the secretion of indicated cytokines. Also, putative action of biologics in indicated diseases is represented with "?". X represents unknown factors. *T reg* T regulatory lymphocytes, *MΦ* macrophages

32]. We analysed phosphorylated SMAD2 (pSMAD2) in T lymphocytes treated with Muse-AT cells conditioned-media. For this purpose, we employed BDC2.5 CD4+ T lymphocytes, as described [25, 27]. When BDC2.5 CD4+ T lymphocytes were cultured in Muse-AT cell conditioned-media, pSMAD2 levels increased in an antigen-specific dependent manner evaluated by Western blot [13] (see Fig. 6.1).

Upon antigen stimulation, naïve CD4+ T lymphocytes have the ability to differentiate into several Th effector subsets. NOD BDC2.5 CD4+ T cells have a marked Th1 bias [25]. Surprisingly, the presence of Muse-AT cells conditioned-media in antigen-stimulated BDC2.5 CD4+ T cells augmented the secretion and expression of IL-10 and its mRNA, respectively. Therefore, Muse-AT cells change the balance between the hallmark Th1 cytokine (IFN-γ) and regulatory IL-10 in T lymphocytes.

To further evaluate the intracellular mechanisms that govern T cells regulation by TGF-β1-mediated Muse-AT cells conditioned-media, we evaluated the master transcription factor TBX21 (T-box transcription factor, also known as T-bet) which governs Th1 differentiation and IFN-γ expression [33]. Our Western blot analysis indicated that Muse-AT cell conditioned-media reduced the expression of T-bet by antigen-specific stimulated BDC2.5 CD4+ T lymphocytes, suggesting that low levels of IFN-γ expression might be due, in part, to reduced synthesis of T-bet.

Are Muse-AT cells able to modify the expression of any other master transcription factor involved in T-lymphocyte differentiation? Although the levels of GATA-3 mRNA, a master transcription factor for Th2 differentiation, were not influenced by Muse-AT cell conditioned-media in BDC2.5 CD4+ T lymphocytes, we should not rule out the possibility that different diabetogenic clones may respond differentially to the conditioned-media. On the other hand, our preliminary data suggest the existence of other biologics expressed by Muse-AT cells which might influence immune responses, probably through direct action on T lymphocytes and macrophages. Therefore, a comprehensive knowledge of the microenvironment generated by Muse-AT cells is further needed.

TGF-β1 is also capable of maintaining the undifferentiated state of MSCs [34]. The Smad3 signalling pathway, mediated by the nuclear translocation of β-catenin, is required for the proliferation of MSCs. Complementary to this, it has been observed that osteoblast maturation is induced when endogenous TGF-β1 is inhibited [35]. Moreover, differentiation towards an osteogenic phenotype was inhibited in the presence of TGF-β1 in the medium, according to published results.

Muse cells seem to have a dual therapeutic effect: apart from their immunomodulatory activity, they can regenerate damaged tissue and restore its function, for example, inhibiting fibrosis. This is in accordance with the ability of Muse cells to produce matrix metalloprotease 9 (MMP-9), which has been proven to have an anti-inflammatory role [36]. High levels of expression of MMP-9 by Muse cells could explain their capacity to degrade and remove extracellular molecules, promoting fibrolysis or suppressing fibrosis and contributing to healing of the damaged tissue [37]. Once Muse cells reach the damaged tissue, they start differentiating into new cells for tissue repair. We can speculate that, during this process, a decrease in TGF-β1 production by Muse-AT cells would facilitate synthesis of metalloproteinases, therefore promoting fibrolysis. Furthermore, TGF-β1 also induces the expression of TIMP, a metalloproteinase inhibitor [38]; it would be interesting to analyse transcription levels of this protein in differentiated cells derived from Muse-AT cells and to compare TGF-β1 and MMP-9/TIMP levels. Crosstalk between TGF-β1 and metalloproteinases could explain the mechanism through which Muse-AT cells control fibrosis. It has also been described that TGF-β1 induces differentiation into myofibroblasts, therefore contributing to fibrosis [31]. It also induces the expression of fibronectin, a characteristic protein of the extracellular matrix (ECM) and mediator of cell proliferation [39, 40], preventing the degradation of ECM by regulating transcriptional levels of metalloproteinases and its inhibitor [38]. Taking all this into consideration, Muse-AT cells would act as promoters of fibrosis due to their expression of TGF-β1. However, treatment with Muse cells tested in murine models of liver fibrosis and nephropathy proved to be effective and showed no development of fibrosis [37, 41].

The complete molecular mechanism and cytokine interplay are not completely understood. If TGF-β1 secretion is decreased after differentiation of Muse cells, other biologics may be involved in its maintenance of the stem state. For example, inhibition of PGE2 synthesis in human umbilical cord mesenchymal stem cells (hUC-MSCs) co-cultured with human peripheral blood monocytes (PBMCs) resulted in the loss of stem cells properties, evaluated as proliferation of hPBMCs and secretion of IFN-γ [7]. Interestingly, PGE2 levels were increased when stimulated hPBMCs and hUC-MSCs were incubated together, indicating that a pro-inflammatory stimulus, possibly the secretion of IFN-γ and IL-1β, was needed to induce secretion of the PGE2 by MSCs. When NK cells were co-cultured with MSCs, a synergistic effect between indoleamine 2,3-dioxygenase (IDO) and PGE2 was observed to inhibit the cytotoxic phenotype of these immune cells [5]. Remarkably, as in the case of hPBMCs and hUC-MSCs, PGE2 secretion by MSCs was augmented when cultured together with NK cells but not when cultured alone.

IDO is a critical enzyme that catalyses the first and rate limiting step of tryptophan catabolism along the kynurenine pathway, which is involved in immune tolerance, preventing autoimmunity or immunopathology that could result from uncontrolled and overreacting immune responses [42]. When IDO catalyses tryptophan, this essential amino acid is locally depleted, while its catabolites accumulate, including kynurenine and its derivatives. As a result, tryptophan shortage inhibits T-lymphocyte division [43], and accumulation of tryptophan catabolites induces T cell apoptosis and differentiation into regulatory T cells [44–46]. Muse cells were also reported to express IDO in the similar level to MSCs, indicating that the pro-inflammatory environment observed in autoimmune diseases may trigger IDO expression in Muse cells and contribute to the abrogation of reactive immune cells. This could be another mechanistic process explaining the immunosuppressive phenotype of Muse cells [41].

Not only soluble factors are involved in modulating the immune system. As it has been demonstrated [47], bone marrow stromal cells (BMSCs) suppress T-lymphocyte proliferation, and this inhibition is less effective when cells are cultured together but not allowed to have cell-to-cell contact (transwell system). Interaction between

MSCs and leukocytes may require the expression of several leukocyte chemokines by the former, such as CXCL9, CXCL10 and CXCL11 [48]. In mice, VCAM-1 and ICAM-1 are some of the surface molecules expressed by MSCs, also participating in inhibition of T-lymphocyte proliferation.

Another mechanism that might contribute to the regeneration of the tissue is the ability of Muse cells to migrate to damaged tissue with high efficiency. This property enhances their potential use as regenerative and anti-inflammatory agents, since they act locally in the affected area. The CXCR4-SDF-1 axis is one of the pathways involved in homing of stem cells into the damaged tissue [49, 50], and it seems to be one that drives Muse cell migration. This axis was abrogated when cells were incubated with a CXCR4 antagonist, proving its participation in Muse cells migration in a model of liver fibrosis [37]. In addition to CXCR4-SDF-1 axis, recent study suggested that sphingosine-1-phosphate (S1P) is more specific migratory factor for Muse cells in acute myocardial infarction patients [51]. The next question that arouse from these observations was related to the fate of Muse cells once they have arrived in the damaged tissues. It has been proven that, when administered intravenously, MSCs can transmigrate through the endothelial barrier and reach the lesion site [52, 53]. This migration may be generated by a chemokine gradient released from the injured tissue.

In a mouse model of focal segmental glomerulosclerosis damaged kidney cells, Muse cells showed high efficacy of homing, tissue regeneration and function recovery. Interestingly, the renal tissue of the Muse cell-treated group showed significantly high levels of expression of IGF-1, VEGF-A, VEGF-B, VEGF-C, VEGF-D and EGF known factors that not only protect kidney cells from damage but also participate in the generation of new blood vessels, a desirable characteristic which may lengthen tissue survival [41]. Previous reports indicate that TGF-β1 is a recognized regulator of angiogenesis, both during developmental stages and pathophysiological conditions [54, 55]. Binding of TGF-β1 to its type II receptor (TβRII) recruits type I receptors termed activin receptor-like kinase (ALKs). Pro-angiogenic responses in endothelial cells are elicited by downstream phosphorylation of Smad 1/5/8. Exploring this signalling pathway promoted by Muse cells could uncover new properties related to neovascularization and open new horizons for cell therapy.

6.5 Muse-AT Cells as Drivers of Diabetes Improvement

Strategies aim to preserve β cell function in type 1 diabetes, improve glycaemic control and decrease microvascular complications and hypoglycaemic events [56]. Despite intensive efforts, several clinical attempts have failed to preserve residual β cell mass for long periods of time, and progressive loss of β cells is ineluctable in most cases of type 1 diabetes. MSCs contribute to tissue repair and have demonstrated potential capacity against autoimmunity development [57]. For example, MSCs are capable of (i) suppressing antigen- or alloantigen-challenged T cell proliferation and (ii) upregulating the activity and number of T reg cells [58, 59]. This immunomodulatory activity is commonly associated to a decrease in inflammatory cytokine production [60]. Several agents produced by MSCs have been described as mediators of immunomodulation including heme oxygenase-1 (HO-1), IDO, MMPs and nitric oxide (NO). In this regard, infusion of bone marrow MSCs improves β cell function in diabetic NOD mice and even in patients [61].

A step forward in the characterization of Muse-AT cells would be to determine whether the immunomodulatory properties observed in vitro might translate into therapeutic benefits, reducing the incidence and/or ameliorating diabetes in mice. As stated earlier in this chapter, type 1 diabetes is characterized by a dysregulated immune system contributing to disease progression. Therefore, our findings regarding the ability of Muse-AT cells to down-regulate diabetogenic BDC2.5 Th1 T lymphocytes encourage efforts to prove whether these cells have potential for diabetes treatment. Thus, if Muse-AT cells can abrogate the pro-inflammatory surge, a more suitable environment would be generated, and survival of insulin-producing cells within islets would be extended.

We tested this hypothesis employing overt diabetic NOD mice, as a model of spontaneous autoimmune diabetes. NOD mice were considered diabetic when spontaneously reached glycaemia levels above 250 mg/dl during two consecutive days. Diabetic animals were divided into two groups: (1) one of them received intraperitoneally a single injection of 1 × 106 Muse-AT cells, and (2) the control group was injected with the same volume of sterile PBS (100 μl). Our preliminary results indicated that glucose levels from the control group dramatically increased, reaching glycaemia >500 mg/dl after a week, while the Muse-AT cell group showed oscillating blood glucose levels between 202 and 500 mg/dl during the 7-week period of observation (Gimeno et al., unpublished). While body weight diminished abruptly (between 2 and 5 g) within the 2 weeks of follow-up in the control group, showing clear signs of cachexia, all mice belonging to the Muse-AT cell group maintained a stably body weight during 7-week post-treatment. These results suggest that Muse-AT cells may control the autoimmune process after onset of spontaneous diabetes. Current ongoing experiments in our laboratory aim to improve these observations by several means, including increasing the total number of injected Muse-AT cells, repeating the administration once a week during several weeks and/or changing the route of administration, e.g. i.v. Also, therapy targeting major players involved in the immune-mediated destruction of β cells, such as the use of sub-therapeutic doses of CD3 antibodies, in combination with Muse-AT cells would be an interesting approach to tackle autoimmune diabetes.

Once the in vivo experiment is optimized, it would be interesting to elucidate the mechanism by which Muse-AT cells improve glycaemic control in immune-mediated diabetes. In this regard, it would be possible that Muse cells can differentiate into insulin-producing cells in vivo. It has already been established that Muse cells are able to migrate to damaged liver and differentiate into hepatocytes in a model of murine liver fibrosis [37]. Therefore, migration of Muse cells damage islets might be possible to explore. In this scenario, the anti-inflammatory environment promoted by Muse-AT cell population could also protect β cells from immune-driven demise and, even more, nourish their functionality.

The prevalence of chronic wounds related to diabetes is a major focus of diabetes care and hospitalization, as many as one in four patients with diabetes develops diabetic foot ulcer [62]. Kinoshita et al. employed Muse cells in an experimental mice model of type 1 diabetes skin ulcers [63]. Skin ulcers generated in SCID mice suffering type 1 diabetes showed delayed wound healing compared with nondiabetic SCID mice. Interestingly, injection of Muse cells around the wound significantly accelerated wound healing. The mechanism by which Muse cells promotes wound healing remains to be elucidated. A possible explanation may be that the high amount of growth factors secreted by Muse cells, particularly under hypoxic conditions, might be responsible for their accelerated skin repair process.

6.6 Muse Cells: Source of Factors Involved in Cell Survival and Immunoregulation

Whether Muse-AT cells secrete soluble factors able to directly promote β cell proliferation and/or survival is a matter of interest. However, there is a lack of information so far. A microarray analysis study revealed the expression of growth factors such as PDGF-A, EGF and SDF-1 under hypoxic conditions; however, this study was limited to only one sample [63]. Researchers found secreted growth factors, including PDGF-BB, TGF-β, bFGF and TNF-α in large quantities, particularly under hypoxic conditions. EGF- and PDGF-R signalling contributes to β cell mass expansion during high-fat mass diet and pregnancy in mice [64] and controls age-dependent proliferation of β cells in mice and humans [65]. Growth factors secreted could explain, at least in part, the repair effect of Muse-AT cells regarding β cell damage during diabetes.

Recently, a comprehensive secretome study of Muse cells was reported. In brief, Muse cells secretome was enriched in biologics that might have potential in conferring stemness preservation, cell survival under stress conditions and immunomodulatory properties [66]. It is worth highlighting the expression by Muse cells of several 14-3-3 isoforms that have the capacity to inactivate pro-apoptotic BAD, proteins belonging to protein kinase A pathways with putative roles in autocrine/paracrine signalling and proteins involved in cellular proliferation and lipid metabolism, such as liver X receptor (LXR) and farnesoid X receptor (FXR) with multiple metabolic roles. Lastly, some components of the complement system are expressed by Muse cells, among other proteins with immune-associated activities, such as alpha-2 macroglobulin and pregnancy zone protein, together with the expression of large number of interleukins and factors involved in the regulation of extracellular matrix remodelling with recognized immune functions [66]. All these biologics are worth investigating to further characterize Muse and Muse-AT cells and their therapeutic potential.

References

1. Yañez R, Lamana ML, García-Castro J, Colmenero I, Ramírez M, Bueren JA (2006) Adipose tissue-derived mesenchymal stem cells have in vivo immunosuppressive properties applicable for the control of the graft-versus-host disease. Stem Cells 24:2582–2591
2. Le Blanc K, Rasmusson I, Sundberg B, Götherström C, Hassan M, Uzunel M, Ringdén O (2004) Treatment of severe acute graft-versus-host disease with third party haploidentical mesenchymal stem cells. Lancet 363:1439–1441
3. Selmani Z, Naji A, Zidi I, Favier B, Gaiffe E, Obert L, Borg C, Saas P, Tiberghien P, Rouas-Freiss N et al (2008) Human leukocyte antigen-G5 secretion by human mesenchymal stem cells is required to suppress T lymphocyte and natural killer function and to induce CD4+CD25highFOXP3+ regulatory T cells. Stem Cells 26:212–222
4. Ramasamy R, Fazekasova H, Lam EW-F, Soeiro I, Lombardi G, Dazzi F (2007) Mesenchymal stem cells inhibit dendritic cell differentiation and function by preventing entry into the cell cycle. Transplantation 83:71–76
5. Spaggiari GM, Capobianco A, Abdelrazik H, Becchetti F, Mingari MC, Moretta L (2008) Mesenchymal stem cells inhibit natural killer-cell proliferation, cytotoxicity, and cytokine production: role of indoleamine 2,3-dioxygenase and prostaglandin E2. Blood 111(3):1327–1333
6. Yen BL, Chang CJ, Liu K-J, Chen YC, Hu H-I, Bai C-H, Yen M-L (2009) Brief report--human embryonic stem cell-derived mesenchymal progenitors possess strong immunosuppressive effects toward natural killer cells as well as T lymphocytes. Stem Cells 27:451–456
7. Chen K, Wang D, Du WT, Han Z-B, Ren H, Chi Y, Yang SG, Zhu D, Bayard F, Han ZC (2010) Human umbilical cord mesenchymal stem cells hUC-MSCs exert immunosuppressive activities through a PGE2-dependent mechanism. Clin Immunol 135:448–458
8. Spaggiari GM, Capobianco A, Becchetti S, Mingari MC, Moretta L (2006) Mesenchymal stem cell-natural killer cell interactions: evidence that activated NK cells are capable of killing MSCs, whereas MSCs can inhibit IL-2-induced NK-cell proliferation. Blood 107:1484–1490
9. Hodgetts SI, Beilharz MW, Scalzo AA, Grounds MD (2000) Why do cultured transplanted myoblasts die in vivo? DNA quantification shows enhanced survival of donor male myoblasts in host mice depleted of CD4+ and CD8+ cells or Nk1.1+ cells. Cell Transplant 9:489–502
10. Mingliang R, Bo Z, Zhengguo W (2011) Stem cells for cardiac repair: status, mechanisms, and new strategies. Stem Cells Int 2011:1–8. https://doi.org/10.4061/2011/310928
11. Kuroda Y, Kitada M, Wakao S, Nishikawa K, Tanimura Y, Makinoshima H, Goda M, Akashi H, Inutsuka A, Niwa A et al (2010) Unique multipotent cells in adult human mesenchymal cell populations. Proc Natl Acad Sci 107:8639–8643
12. Heneidi S, Simerman AA, Keller E, Singh P, Li X, Dumesic DA, Chazenbalk G (2013) Awakened by cellular stress: isolation and characterization of a novel population of pluripotent stem cells derived from Hu-man adipose tissue. PLoS One 8:e64752
13. Gimeno ML, Fuertes F, Barcala Tabarrozzi AE, Attorressi AI, Cucchiani R, Corrales L, Oliveira TC, Sogayar MC, Labriola L, Dewey RA, Perone MJ (2017) Pluripotent nontumorigenic adipose tissue-derived Muse-AT cells have immunomodulatory capacity mediated by transforming growth factor-β1. Stem Cells Transl Med 6:161–173
14. Wakao S, Kitada M, Kuroda Y, Shigemoto T, Matsuse D, Akashi H, Tanimura Y, Tsuchiyama K, Kikuchi T, Goda M et al (2011) Multiline-age-differentiating stress-enduring (Muse-AT) cells are a primary source of induced pluripotent stem cells in human fibroblasts. Proc Natl Acad Sci 108:9875–9880
15. Simerman AA, Dumesic DA, Chazenbalk GD (2014) Pluripotent Muse-AT cells derived from human adipose tissue: a new perspective on regenerative medicine and cell therapy. Clin Transl Med 3:12
16. Ogura F, Wakao S, Kuroda Y, Tsuchiyama K, Bagheri M, Heneidi S, Chazenbalk G, Aiba S, Dezawa M (2014) Human adipose tissue possesses a unique population of pluripotent stem cells with nontumorigenic and low telomerase activities: potential implications in regenerative medicine. Stem Cells Dev 23:717–728

17. Burrack AL, Martino T, Fife BT (2017) T cell-mediated Beta cell destruction: autoimmunity and alloimmunity in the context of type 1 diabetes. Front Endocrinol 8:343. https://doi.org/10.3389/fendo.2017.00343
18. Rothe H, Hausmann A, Casteels K, Okamura H, Kurimoto M, Burkart V, Mathieu C, Kolb H (1999) IL-18 inhibits diabetes development in nonobese diabetic mice by counterregulation of Th1-dependent destructive insulitis. J Immunol 163:1230–1236
19. Sarikonda G, Pettus J, Phatak S, Sachithanantham S, Miller JF, Wesley JD, Cadag E, Chae J, Ganesan L, Mallios R, Edelman S, Peters B, von Herrath M (2013) CD8 T-cell reactivity to islet antigens is unique to type 1 while CD4 T-cell reactivity exists in both type 1 and type 2 diabetes. J Autoimmun 50:77–82. https://doi.org/10.1016/j.jaut.2013.12.003
20. Suarez-Pinzon WL, Rabinovitch A (2001) Approaches to type 1 diabetes prevention by intervention in cytokine immunoregulatory circuits. Int J Exp Diabetes Res 2(1):3–17
21. Lundh M, Bugliani M, Dahlby T, Chou DH, Wagner B, Ghiasi SM, De Tata V, Chen Z, Lund MN, Davies MJ, Marchetti P, Mandrup-Poulsen T (2017) The immunoproteasome is induced by cytokines and regulates apoptosis in human islets. J Endocrinol 233(3):369–379
22. Kahn SE, Cooper ME, Del Prato S (2014) Pathophysiology and treatment of type 2 diabetes: perspectives on the past, present, and future. Lancet 383(9922):1068–1083
23. Leiter EH, Schile A (2013) Genetic and pharmacologic models for type 1 diabetes. Curr Protoc Mouse Biol 3(1):9–19
24. Lichtman SN, Wang J, Lemasters JJ (1998) LPS receptor CD14 participates in release of TNF alpha in RAW 264.7 and peritoneal cells but not in Kupffer cells. Am J Phys 275(1 Pt 1):G39–G46
25. Stadinski BD, Delong T, Reisdorph N, Reisdorph R, Powell RL, Armstrong M, Piganelli JD, Barbour G, Bradley B, Crawford F, Marrack P, Mahata SK, Kappler JW, Haskins K (2010) Chromogranin a is an auto-antigen in type 1 diabetes. Nat Immunol 11:225–231
26. Judkowski V, Pinilla C, Schroder K, Tucker L, Sarvetnick N, Wilson DB (2001) Identification of MHC class II-restricted peptide ligands, including a glutamic acid decarboxylase 65 sequence, that stimulate diabetogenic T cells from transgenic BDC2.5 nonobese diabetic mice. J Immunol 166:908–917
27. Castro CN, Barcala Tabarrozzi AE, Winnewisser J, Gimeno ML, Antunica Noguerol M, Liberman AC, Paz DA, Dewey RA, Perone MJ (2014) Curcumin ameliorates autoimmune diabetes. Evidence in accelerated murine models of type 1 diabetes. Clin Exp Immunol 177:149–160
28. Ylöstalo JH, Bartosh TJ, Coble K, Prockop DJ (2012) Human mesenchymal stem/stromal cells cultured as spheroids are self-activated to produce prostaglandin E2 that directs stimulated macrophages into an anti-inflammatory phenotype. Stem Cells 30(10):2283–2296
29. Nemeth C, Nunnally M, Bitan Y, Nunnally S, Cook RI (2009) Between choice and chance: the role of human factors in acute care equipment decisions. J Patient Saf 5(2):114–121
30. Weiss A, Attisano L (2013) The TGF beta superfamily signalling pathway. Wiley Interdiscip Rev Dev Biol 2(1):47–63
31. Morikawa M, Derynck R, Miyazono K (2016) TGF-β and the TGF-β family: context-dependent roles in cell and tissue physiology. Cold Spring Harb Perspect Biol 8(5):a021873
32. Ayyaz A, Attisano L, Wrana JL (2017) Recent advances in understanding contextual TGFβ signalling. F1000Research 6:749
33. Szabo SJ, Kim ST, Costa GL, Zhang X, Fathman CG, Glimcher LH (2000) A novel transcription factor, T-bet, directs Th1 lineage commitment. Cell 100:655–669
34. Jian H, Shen X, Liu I, Semenov M, He X, Wang XF (2006) Smad3-dependent nuclear translocation of beta-catenin is required for TGF-beta1-induced proliferation of bone marrow-derived adult human mesenchymal stem cells. Genes Dev 20(6):666–674
35. Maeda S, Hayashi M, Komiya S, Imamura T, Miyazono K (2004) Endogenous TGF-beta signalling suppresses maturation of osteoblastic mesenchymal cells. EMBO J 23(3):552–563
36. Fingleton B (2017) Matrix metalloproteinases as regulators of inflammatory processes. Biochim Biophys Acta 1864(11 Pt A):2036–2042

37. Iseki M, Kushida Y, Wakao S, Akimoto T, Mizuma M, Motoi F, Asada R, Shimizu S, Unno M, Chazenbalk G, Dezawa M (2017) Human Muse-AT cells, nontumorigenic pluripotent-like stem cells, have liver regeneration capacity through specific homing and cell replacement in a mouse model of liver fibrosis. Cell Transplant 26:821–840
38. Edwards DR, Murphy G, Reynolds JJ, Whitham SE, Docherty AJ, Angel P, Heath JK (1987) Transforming growth factor beta modulates the expression of collagenase and metalloproteinase inhibitor. EMBO J 6(7):1899–1904
39. Ignotz RA, Massagué J (1986) Transforming growth factor-beta stimulates the expression of fibronectin and collagen and their incorporation into the extracellular matrix. J Biol Chem 261(9):4337–4345
40. Roberts AB, Sporn MB, Assoian RK, Smith JM, Roche NS, Wakefield LM, Heine UI, Liotta LA, Falanga V, Kehrl JH (1986) Transforming growth factor type beta: rapid induction of fibrosis and angiogenesis in vivo and stimulation of collagen formation in vitro. Proc Natl Acad Sci U S A 83(12):4167–4171
41. Uchida N, Kushida Y, Kitada M, Wakao S, Kumagai N, Kuroda Y, Kondo Y, Hirohara Y, Kure S, Chazenbalk G, Dezawa M (2017) Beneficial effects of systemically administered human Muse-AT cells in adriamycin nephropathy. J Am Soc Nephrol 28(10):2946–2960
42. van Baren N, Van den Eynde BJ (2015) Tryptophan-degrading enzymes in tumoral immune resistance. Front Immunol 6:34
43. Munn DH, Shafizadeh E, Attwood JT, Bondarey I, Pashine A, Mellor AL (1999) Inhibition of T cell proliferation by macrophage tryptophan catabolism. J Exp Med 189(9):1363–1372
44. Fallarino F, Grohmann U, Vacca C, Bianchi R, Orabona C, Spreca A, Fioretti MC, Puccetti P (2002) T cell apoptosis by tryptophan catabolism. Cell Death Differ 9(10):1069–1077
45. Terness P, Bauer TM, Röse L, Dufter C, Watzlik A, Simon H, Opelz G (2002) Inhibition of allogeneic T cell proliferation by indoleamine 2,3-dioxygenase-expressing dendritic cells: mediation of suppression by tryptophan metabolites. J Exp Med 196(4):447–457
46. Fallarino F, Grohmann U, You S, McGrath BC, Cavener DR, Vacca C, Orabona C, Bianchi R, Belladonna ML, Volpi C, Santamaria P, Fioretti MC, Puccetti P (2006) The combined effects of tryptophan starvation and tryptophan catabolites down-regulate T cell receptor zeta-chain and induce a regulatory phenotype in naive T cells. J Immunol 176(11):6752–6761
47. Di Nicola M, Carlo-Stella C, Magni M, Milanesi M, Longoni PD, Matteucci P, Grisanti S, Gianni AM (2002) Human bone marrow stromal cells suppress T-lymphocyte proliferation induced by cellular or nonspecific mitogenic stimuli. Blood 99(10):3838–3843
48. Haddad R, Saldanha-Araujo F (2014) Mechanisms of T-cell immuno-suppression by mesenchymal stromal cells: what do we know so far? Biomed Res Int 2014:1–14. https://doi.org/10.1155/2014/216806
49. Kucia M, Jankowski K, Reca R, Wysoczynski M, Bandura L, Allendorf DJ, Zhang J, Ratajczak J, Ratajczak MZ (2004) CXCR4-SDF-1 signalling, locomotion, chemotaxis and adhesion. J Mol Histol 35(3):233–245
50. Hirbe AC, Morgan EA, Weilbaecher KN (2010) The CXCR4/SDF-1 chemokine axis: a potential therapeutic target for bone metastases? Curr Pharm Des 16(11):1284–1290
51. Tanaka T, Nishigaki K, Minatoguchi S, Nawa T, Yamada Y, Kanamori H, Mikami A, Ushikoshi H, Kawasaki M, Dezawa M, Minatoguchi S (2017) Mobilized Muse-AT cells after acute myocardial infarction predict cardiac function and remodelling in the chronic phase. Circ J 82(2):561–571
52. Nitzsche F, Müller C, Lukomska B, Jolkkonen J, Deten A, Boltze J (2017) Concise review: MSC adhesion cascade-insights into homing and transendothelial migration. Stem Cells 35(6):1446–1460
53. Bayo J, Marrodán M, Aquino JB, Silva M, García MG, Mazzolini G (2014) The therapeutic potential of bone marrow-derived mesenchymal stromal cells on hepatocellular carcinoma. Liver Int 34(3):330–342
54. Zhao M, Hu Y, Jin J, Yu Y, Zhang S, Cao J, Zhai Y, Wei R, Shou J, Cai W, Liu S, Yang X, Xu GT, Yang J, Corry DB, Su SB, Liu X, Yang T (2017) Interleukin 37 promotes angiogenesis through TGF-β signalling. Sci Rep 7(1):6113

55. Pardali E, Goumans MJ, ten Dijke P (2010) Signalling by members of the TGF-beta family in vascular morphogenesis and disease. Trends Cell Biol 20:556–567
56. Bonfanti R, Bazzigaluppi E, Calori G, Riva MC, Viscardi M, Bognetti E, Meschi F, Bosi E, Chiumello G, Bonifacio E (1998) Parameters associated with residual insulin secretion during the first year of disease in children and adolescents with type 1 diabetes mellitus. Diabet Med 15(10):844–850
57. Abdi R, Fiorina P, Adra CN, Atkinson M, Sayegh MH (2008) Immunomodulation by mesenchymal stem cells: a potential therapeutic strategy for type 1 diabetes. Diabetes 57(7):1759–1767
58. Di Ianni M, Del Papa B, De Ioanni M, Moretti L, Bonifacio E, Cecchini D, Sportoletti P, Falzetti F, Tabilio A (2008) Mesenchymal cells recruit and regulate T regulatory cells. Exp Hematol 36(3):309–318
59. Ghannam S, Bouffi C, Djouad F, Jorgensen C, Noël D (2010) Immuno-suppression by mesenchymal stem cells: mechanisms and clinical applications. Stem Cell Res Ther 1(1):2
60. Oh W, Kim DS, Yang YS, Lee JK (2008) Immunological properties of umbilical cord blood-derived mesenchymal stromal cells. Cell Immunol 251(2):116–123
61. Li L, Hui H, Jia X, Zhang J, Liu Y, Xu Q, Zhu D (2016) Infusion with human bone marrow-derived mesenchymal stem cells improves β-cell function in patients and nonobese mice with severe diabetes. Sci Rep 6:37894
62. International Diabetes Federation (2017) International diabetes federation atlas, 7th edn. Brussels, Belgium
63. Kinoshita K, Kuno S, Ishimine H, Aoi N, Mineda K, Kato H, Doi K, Kanayama K, Feng J, Mashiko T, Kurisaki A, Yoshimura K (2015) Therapeutic potential of adipose-derived SSEA-3-positive Muse-AT cells for treating diabetic skin ulcers. Stem Cells Transl Med 4(2):146–155
64. Hakonen E, Ustinov J, Mathijs I, Palgi J, Bouwens L, Miettinen PJ, Otonkoski T (2011) Epidermal growth factor (EGF)-receptor signalling is needed for murine beta cell mass expansion in response to high-fat diet and pregnancy but not after pancreatic duct ligation. Diabetologia 54(7):1735–1743
65. Chen H, Gu X, Liu Y, Wang J, Wirt SE, Bottino R, Schorle H, Sage J, Kim SK (2011) PDGF signalling controls age-dependent proliferation in pancreatic β-cells. Nature 478(7369):349–355
66. Alessio N, Özcan S, Tatsumi K, Murat A, Peluso G, Dezawa M, Galderisi U (2017) The secretome of MUSE-AT cells contains factors that may play a role in regulation of stemness, apoptosis and immunomodulation. Cell Cycle 16(1):33–44

Chapter 7
The Role of the Mitochondria in the Evolution of Stem Cells, Including MUSE Stem Cells and Their Biology

James E. Trosko

Abstract From the transition of single-cell organisms to multicellularity of metazoans, evolutionary pressures selected new genes and phenotypes to cope with the oxygenation of the Earth's environment, especially via the symbiotic acquisition of the mitochondrial organelle. There were many new genes and phenotypes that appeared, namely, stem cells, low-oxygen-micro-environments to house these genes ("niches"), new epigenetic mechanisms to regulate , selectively, the gene repertoire to control proliferation, differentiation, apoptosis, senescence and DNA protection mechanisms, including antioxidant genes and DNA repair. This transition required a critical regulation of the metabolism of glucose to produce energy for both the stem cell quiescent state and the energy-requiring differentiated state. While the totipotent-, embryonic-, pluripotent-, and a few adult organ-specific stem cells were recognized, only relatively recently, because of the isolation of somatic cell nuclear transfer (SCNT) stem cells and "induced pluripotent stem" cells, challenges to the origin of these "iPS" cells have been made. The isolation and characterization of human MUSE stem cells and more adult organ-specific adult stem cells have indicated that these MUSE cells have many shared characteristics of the "iPS" cells, yet they do not form teratomas but can give rise to the trigeminal cell layers. While the MUSE cells are a subset of human fibroblastic cells, they have not been characterized, yet, for the mitochondrial metabolic genes, either in the stem cell state or during their differentiation processes. A description of other human adult stem cells will be made to set future studies of how the MUSE stem cells compare to all other stem cells.

Keywords WARBURG hypothesis · Human adult stem cells · Gap junctional intercellular communication · Oxidative stress

---

J. E. Trosko (✉)
The Department of Pediatrics/Human Development, Center For Integrative Toxicology, College of Human Medicine, Michigan State University, East Lansing, MI, USA
e-mail: james.trosko@hc.msu.edu

M. Dezawa (ed.), *Muse Cells*, Advances in Experimental Medicine and Biology 1103, https://doi.org/10.1007/978-4-431-56847-6_7

> Many of the events involved in the emergence of complex life have been driven by the need of individual cells to rid themselves of toxic excesses of oxygen and calcium. This involved the concurrent shedding of other elements that happened to be abundant where and when oxygen or calcium reached toxic concentrations, iron, silicon, phosphorus, and carbon, in particular. This shedding into extracellular space of oxygen-expensive molecules of the collagen family induced 3-D tissue like structures to come into being, necessitated cell-to-cell cooperation, and led to the emergence of the metazoan. J. M. Saul [1].

> Unicellular organisms are "immortalized" by reproducing through binary fission, whereas eukaryotes evolved sexual reproduction as a means of communicating their genetic information from one generation to the next. John Torday [2].

7.1 Introduction: The Interplay of Physical, Chemical, and Biological Evolution – From Single Cells to Multicellular Metazoans, to Cancer, and to the Aging Process

It was Theodosius Dobzhansky who stated: "Nothing in biology makes sense except in the light of evolution" [3]. In view of the amazing advances in all branches of science, due to the power of sophisticated technologies, insightful concepts/paradigms, and amazing experimental results, we are left with many unanswered questions, ranging from the evolution of living organisms and the origin of diseases in human beings. Because, modern scientific advances require detailed reductionistic training in sub-branches of each discipline, the detailed advances in these sub-branches are not often seen to be linked to either neighboring sub-disciplines, let alone the "overall big picture" of all life's processes.

Although this analysis of stem cells should be on multilineage-differentiating stress-enduring (MUSE) cells, because of the lack of mitochondrial studies to date, many characteristics of all other known types of stem cells will be made on some other human organ-specific adult stem cells. This analysis should provide a guideline or framework for future analyses on these MUSE cells. Therefore, the goal will be made to see what is known about other human adult stem cells that pertains to the comparisons between the embryonic stem cells (ES), induced pluripotent stem cells ("iPS"), and multilineage-differentiating stress-enduring (MUSE) cells, as they might give insight to the origin, biology, functions, and relationship to health and stem cell-related diseases, as well as to their potential for stem cell therapy.

From the field of developmental biology and embryology, current characterizations of various types of embryonic, "induced pluripotent stem cells" ("iPS") [4], somatic cell nuclear transfer (SCNT) cells [5], "multilineage-differentiating stress-enduring" (MUSE) cells [6], and organ-specific adult stem cells, as well as in situ identification of adult stem cells in various tissues, have given new insights to their potential roles in the carcinogenic process.

From the time of origin of the planet Earth, with its inhospitable physical conditions (high temperatures, high radiation levels, toxic chemicals in the atmospheric/

aqueous environments, lack of needed nutrients, etc.), life, as we know it today, could not exist. When several of these physical characteristics of the earth changed, simple one-celled organisms appeared in an anoxic atmosphere. The life-affirming "blueprint" of nucleic acids, which, by its ability to code for biochemical reactions and cellular structures, allowed for the preservation or perpetuation of those particular species. Important enzymes, such as for DNA replication and DNA repair, as well as antioxidant production, detoxification of free radicals and sugar metabolism for maintenance of life and reproduction, were coded in these DNA molecules.

Clearly, in view of Dobzhansky's quote, if these important genetically coded enzymes, selected to protect "error-free" replication of DNA or "error-free" repair of damaged DNA, were, themselves, able to prevent any mutations during replication of DNA or during repair of DNA damage, then that individual organism and its offspring would never have been able to survive. That is because the inevitable changes in the environment would have made its genome incapable of adapting to the new environment. On the other hand, if these genes that coded enzymes formed too many mutations during normal DNA replication or the repair of DNA damage, then the organism and its offspring would have produced too many life-jeopardizing mutations for either the individual or species to survive the changing environment. Since life has evolved from the single-cell organism in an anoxic environment to the normoxic multicellular organism, such as *Homo-sapiens*, it should be obvious that a balance of mutation production of both somatic and germ line cells has occurred, as well as a new means, i.e., epigenetic mechanisms, to regulate gene expression. That means, within a population of a given species, there are individuals that carry either or both somatic and germ line mutations, which would allow them to survive changes in the environment for which most members of that species could not survive.

The relative short range of extremes in temperature, specific ratios of ambient gases, the change in seasons, and diurnal cycles of light selected specific biological evolutionary mechanisms led to the generation of energy from available nutrients, for life to begin. While the detailed changes in paleo-geochemistry of the Earth are not known in great detail, there was a moment in its history when the first single-cell organisms, together with nutrients, including a food-generating energy sources, glucose, had the metabolizing capabilities to metabolize glucose to ATP via glycolysis and the release of the energy for life's maintenance and cell replication. Here is a key phenotypic observation for the ultimate evolution of the metazoan to come. These single-cell organisms replicated by the process of "symmetrical" or binary cell division. That is to say, the two daughters, formed by the cell division of its mother, ostensibly, inherited the identical genetic codes and phenotypes of their mother. It might be carried to an extreme characterization to state that the original mother was kept in an "immortalized" state by this process, in that the daughters did not "mortalized" or terminally end its ability to replicate. In effect, this transition from an anoxic environment to a normoxic one provided the evolutionary force to selectively create the "stem cell" and its unique microenvironment in a multicellular organism, the low-oxygen "niche," to protect this new cell type.

This biological evolutionary change that occurred forced a major change in the physicochemical environments that, in turn, forced a major evolutionary change, leading to the formation of multicellular organisms. When,through a sequence of mutational changes, single-cell phytoplanktons could extract energy from the sun via photosynthesis to produce sugars and release, as a by-product, oxygen. Ironically, this oxygenated world became toxic to anaerobic single-cell organisms. In order to survive, these organisms had to seek anaerobic microenvironments to survive. Again, while the details of evolutionary events that had to occur are not known, a new microorganism, the mitochondrion, appeared that had the ability to metabolism sugar via oxidative phosphorylation and to produce ATP much more efficiently than via glycolysis [7]. To put this approach into context, single -cell organisms survived, albeit the low efficiency of ATP production, but because of the low risk of reactive oxygen species damage to their DNA. Because the historic details are still not delineated, an evolutionary event occurred by which a symbiotic union occurred, in which a fusion between the mitochondrion organism and an unknown eukaryotic cell allowed the formation of a multicellular organism. In effect, the mitochondria symbiotic fusion with this early single-cell organism leads to the selection of a new type of cell, the "stem cell." This new cell type led to the creation of "multicellularity." Unlike the single-cell organism of the anoxic environmental state, which could only divide symmetrically to produce two daughters that inherited their mother's ability to adapt to changing environments via mutagenesis, this "stem cell" acquired the ability to divide either by symmetrical cell division, similar to the single-cell organism or also by asymmetrical cell division to form one daughter similar to its mother, but another cell that could differentiate to a phenotype that acquired new means to adapt via epigenetic mechanisms rather than by mutation [8].

7.2 The Transition of a Single-Cell Organism to the First Multicellular Metazoan Was Associated with the Formation of Asymmetric Cell Division, Appearance of Gap Junctional Intercellular Communication, Differentiation of Cells, Germinal and Somatic Stem Cells, Chronic Diseases, and Aging

The first multicellular organism in the oxygenated environment formed an adherent social collection of mitochondrial-containing cells, now had a means, i.e., oxidative phosphorylation, to generate more energy from sugars. Included in this new oxygenated environment was the acquisition of genes that could biosynthesize the molecules of the cholesterol and collagen family, which provided the individual cells with a membrane quite unlike that of prokaryotes. This new type of cell membrane allowed for coupled interactions between the external and internal environments of a cell, as well as between the individual cells of an adherent colony. Cholesterol in the membrane allowed it to be more flexible, allowing increased gas exchange due to the thinning of the cell membrane (respiration), facilitating both endocytosis and exocytosis (metabolism) and enhanced directed cell movement [9]). The collagen family provided not only the "glue" to hold individual cells together as a coexisting, co-interacting tissue, organ, and organisms but allowed specific collagen-like molecules to provide unique signals to the cells anchored to these molecules. The future challenge in the origin and evolutionary function of the MUSE cells is to find the earliest appearance of these cells in some evolutionary appearance of a primitive multicellular organism.

With the negative side effects of oxidative phosphorylation, namely, the generation of a number of reactive oxygen species (ROS)/reactive nitrogen species (RNS) [10], the coevolution of genes and a new cellular strategy were required to cope with them. Oxidative stress was the inevitable consequence of this new metabolic process to survive in an aerobic environment. In addition, to antioxidant defense mechanisms, ROS signaling, rather than ROS cellular damage, evolved as a means to regulate gene expression (epigenetically) [11]. Moreover, new genes were acquired to utilize them to act as adaptive signal transducer and gene regulators to induce or repress whole batteries of genes to respond to maintaining a balanced redox state. In other words, oxidative stress, by controlling specific gene expression, controls new cellular behaviors, besides just cell proliferation. This provided new cellular phenomena to occur, namely, cellular differentiation, apoptosis, senescence, as well as selective adaptive gene expression [12, 13].

Biological evolution of multicellularity of both the individual and the species depended on maintaining the adaptive integrity of genomic and mitochondrial DNA. Therefore, genes and a cellular strategy (stem cells; tissue-amplifying cells and terminally differentiated cells) were selected that protected these DNA's from the highly reactive ROS's [14]. This series of genes had to be coevolved (a) to protect the genomic DNA from ROS-induced macromolecular damage (endogenous antioxidants). The probable reason for the selection of the nuclear membrane was (a) to sequester the genomic DNA from the ROS's generated by the mitochondrial metabolism and (b) to repair the inevitable damage that might occur (DNA repair mechanisms). In particular, extracellular matrices and the "niche" in this simple multicellular organism were needed to sequester the unique specialized cells, the germinal and somatic stem cells [15–18].

These genes to protect the germ and somatic stem cells were selected in order to provide a microenvironment to help maintain the conditions of stemness. Basically, one could speculate that the genes, which coded for the niches to keep the stem cells in an anoxic microenvironment, helped to maintain the metabolic state of a single-cell organism. Specifically, by a low-oxygenated and low-nutrient microenvironment in the niche, the metabolism of these cells would prevent DNA-damaging metabolites from inducing mutations and would keep these stem cells from constant proliferation. By keeping these stem cells in this quiescent state, the genome could be kept in a state to protect the species germinal and somatic stem cells from risks to germ line and somatic stem cell-related diseases. This included the stem cells' ability to divide in a symmetrical or binary manner. However, these germinal and somatic stem cells could escape this "immortalized" state by factors that breached these niches to allow these stem cells to be exposed to hyperoxic states. Under these hyperoxic conditions, new genes that could afford a new means of cell division in the stem cells, namely, asymmetrical cell division, would respond to external stimuli, to allow one daughter to maintain the capacity of stemness and the other daughter could differentiate. It is here that mitochondriogenesis occurred and cell differentiation occurred.

To give witness to this, the introduction of "mortality" (or terminal differentiation) during the transition for an anaerobic to an aerobic state, several interesting observations have been made on single-cell organisms, which, when exposed to an oxygenated environment, caused the bacterium to enter into a "primitive differentiated" phenotype. These bacteria replicated its DNA but could not septate (Fig. 7.1, Panel A [19]). When the single anaerobic ciliated cells were exposed to oxygenated medium, they clumped together as if they might protect some interior cells from the

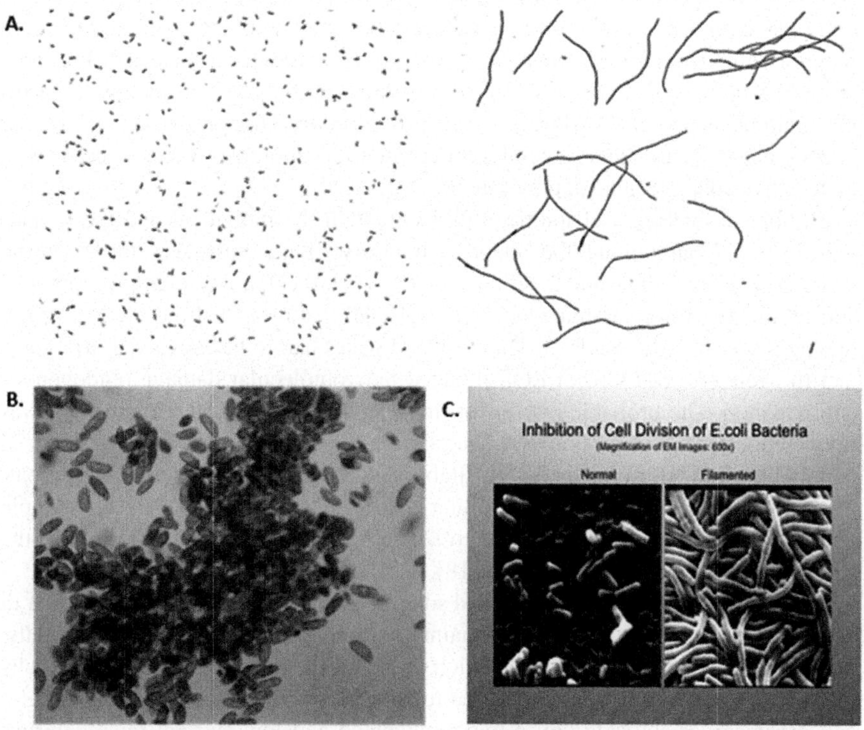

Fig. 7.1 *Panel A*. filamentation of aerobically grown Hpx-mutants of *E. coli* cells. Cells were grown in Luria broth, anaerobically (**a**) or aerobically (**b**). Magnification: ×400. [19] Permission granted By Proc Natl. Acad. Sci., (PNAS). *Panel B*. J.M. Saul, Lethaia, 2008: Clumping of anaerobic ciliates in oxygenated water [15]. Permission granted. *Panel C. E.coli*, grown in traditional growth medium, showing normal morphology. However, when *E.coli* were grown in the same medium, but with a submerged platinum electrode, the *E.coli* had their DNA replicated, but they did not septate. This observation led to Dr. Barnett Rosenberg's discovery of the anticancer drug, cisplatin. Permission grant by Paul Rosenberg of the Board of Barros Foundation

toxic oxygen (Fig. 7.1, Panel B [15]). The dramatic consequence of bacteria exposed to the leached cisplatin from electrodes in their culture medium induced oxidative stress, which, also, led the bacteria to replicate their DNA but preventive septation of the bacteria (Fig. 7.1, Panel C; [20]).

7.3 A Reverse Faustian Bargain: Anaerobic Bacteria Do Not Get Cancer or Age

While the appearance of the mitochondria in a eukaryote cell allowed for the emergence of a multicellular metazoan, which, now, had acquired new adaptive genotypes/phenotypes to survive this oxygenated environment with the appearance of new cells (nerve, muscle, liver, pancreatic, retinal, etc.) within this society of cells, it did come with a price, because not only did the individual metazoan eventually age and die but also the individual acquired various acute and chronic diseases, such as birth defects and cancer. Many classic observations have been made to suggest the link between "cancer and aging" [21] but also between the delicate interactive relationship between the development of the embryogenesis, correct gene expression and normal development, and all kinds of developmental problems when those communication processes go wrong.

The late Van R. Potter, a cancer biochemist, stated:

> …, cancer is a problem in regulatory dysfunction, which in this case, results in a failure to **orchestrate** the available repertory of gene capabilities in a manner appropriate to the whole organism at any given time [22].

He, also, noted that:

> The cancer problem is not merely a cell problem; it is a problem of cell interaction, not only within tissues, but also with distal cells in other tissues. But in stressing the whole organism, we must also remember that the integration of normal cells with the welfare of the whole organism is brought about by molecular messages acting on molecular receptors [23].

On the other hand, Leonard Hayflick, an expert on cellular aging, noted:

> …since senescence does occur in most living organisms, it is supposed that the genetic program which **orchestrates** the development of an individual is incapable of maintaining it indefinitely [24].

What is striking in both of their definitions of cancer and aging is the central role of homeostasis, or the roles that various forms of cellular communication play in both phenomena of the metazoan. As indicated previously, single-cell organisms could communicate via extracellular secreted factors that triggered intracellular signaling within the organism to regulate individual cellular behavior with a community of organisms. However, with the appearance of the collagen and cholesterol family of molecules, membrane and extracellular matrices afforded new modes of homeostatic regulation of battery of genes within different cell types in this three-dimensional multicellular organism. The new sources of communication sources included the introduction of the 20 + connexin genes [25] that helped contiguous cells to communicate ions and various small regulatory molecules to help synchronize metabolic and electrotonic functions between segregated cells within the metazoan [26]. At the same time, extracellular secreted molecules, themselves, provided not only "anchorage" of cell types via cell adhesion but also provided unique intracellular signaling to dictate specific gene responses [27]. First, this was articulated, beautifully, by Clem Markert [28]:

> Cells interact and communicate during embryonic development and through inductive stimuli mutually direct the divergent courses of their differentiation. Very little cell differentiation is truly autonomous in vertebrate organisms. The myriad cell phenotypes present in mammals, for example, must reflect a corresponding complexity in the timing, nature, and amount of inductive interactions. Whatever the nature of inductive stimuli may be, they emerge as a consequence of specific sequential interactions of cells during embryonic development.
>
> The first embryonic cells, blastomeres, of mice and other mammals are all totipotent. During cleavage and early morphogenesis these cells come to occupy different positions in the three-dimensional embryo. Some cells are on the outside, some inside. The different environments of these cells cause the cells to express different patterns of metabolism in accordance with their own developing programs of gene function. These patterns of metabolism create new chemical environments for nearby cells and these changed environments induce yet new programs of gene function in responding cells. Thus a progressive series of reciprocal interactions is established between the cellular environment and the genome of each cell. These interactions drive the cell along a specific path of differentiation until a stable equilibrium is reached in the adult. Thereafter little change occurs in the specialized cells and they become remarkably refractory to changes in the environment. They seem stably locked into the terminal patterns of gene function characteristic of adult cells. The genome seems no longer responsible to the signals that were effective earlier in development.
>
> Of course, changes can occur in adult cells that lead to renewed cell proliferation and altered differentiation as seen in neoplasms, both benign and malignant, but such changes are very rare indeed when one considers the number of cells potentially available for neoplastic transformation. Possibly, mutations in regulatory DNA of dividing adult cells can occasionally lead to new and highly effective programs gene function that we recognize as neoplastic or malignant. However, most genetic changes in adult cells can probably lead to cell death since random changes in patterns of gene activity are not likely to be beneficial.

What needs to be done with the future characterization of MUSE cells is to characterize the role of the connexin genes and functional gap junctional intercellular communication during their differentiation into the trigeminal layers and the derivative differentiated daughter cells. This evolutionary advance can be viewed from a philosophical perspective, in that single-cell organisms that survive via glycolysis in the anaerobic environment by symmetrical cell division can be considered "immortal" and does not "die" or get cancer or "age." However, in acquiring the ability to metabolize via oxidative phosphorylation in an aerobic environment, the metazoan cell, in essence, acquired the risk of a "reverse Faustian Bargain." Consequently, it is implied, that by gaining or by maintaining "immortality" or losing "mortality" of a stem cell in an adult metazoan, development might be disrupted, leading to death or teratogenic consequences. The question to be answered is: "What happens if it is the MUSE cells that are affected?" In the myth of the Faustian Bargain, the individual trades his soul for immortality. In that case of the individual metazoan germinal and somatic stem cells, a potential risk was made in its having kept its ability for symmetrical cell division for "immortality" (for the species and for future growth and repair of somatic tissue), but with the risk of acquiring its inability to divide asymmetrically in order to terminally differentiate. The price the individual paid for immortality of stem cells for the survival of the species was the inevitable risk of birth defects, cancer, and other diseases of aging and of the aging process itself for the individual [29]. However, what might be the consequences for MUSE cells, if blocked from differentiation, during development or post birth?

7.4 Role of Stem Cells or Differentiated Cells in Carcinogenesis and Aging

…cancer has countless secondary causes but….only one prime cause. Summarized in a few words, the prime cause of cancer is the replacement of the respiration of oxygen in normal body cells by a fermentation of sugar [1].

Ultimately, to get to the understanding of how the transition of energy acquisition for life and perpetuation of the species, an interesting observation was made by Warburg [30], in that he observed that, while normal cells metabolize via mitochondrial-dependent oxidative phosphorylation, tumor cells seemed to revert back to glycolysis during the carcinogenic process. However, at the time of his observations, the mechanisms of carcinogenesis were not known nor were the concepts of stem cells or, for that matter, "cancer stem cells" known. It is the basic assumption of this analysis of mitochondria and stem cells is that, only by an examination of what we believe we know about these concepts today, can we put the Warburg observations into an evolutionary perspective. Moreover, while some answers can be obtained from other human stem cells, we need to await analysis of the Warburg analyses in MUSE cells.

From the field of developmental biology and embryology, current characterizations of various types of embryonic [31, 32], "induced pluripotent stem cells" ("iPS") [4], somatic cell nuclear transfer (SCNT) cells [5], MUSE cells [6], organ-specific adult stem cells [33], as well as in situ identification of adult stem cells in various tissues, have given new insights to their potential roles in the carcinogenic process. In the case of MUSE cells, it is established that these MUSE cells do not form teratomas when injected back into adult animals, unlike the ES and "iPS" cells. The MUSE cells behave as do other organ-specific adult stem cells. What needs to be shown, in future studies, is whether MUSE cells can be transformed, neoplastically, as are organ-specific adult stem cells.

While the complex process of carcinogenesis is far from complete understanding, there seems to be a convergence of opinions that are generally, if not universally, shared. First, with the exception of teratoma formation [34], carcinogenesis seems to be explained by a "multistage," "multi-mechanism," and "initiation/promotion progression" process [35, 36]. Second, there seems to be general agreement that all cancers originate from a single cell or the monoclonal origin of cancer [37, 38]. Third, by the time a tumor has achieved an ability to invade tissues and to metastasize to distal sites, while the individual cells within the tumor exhibit multiple different genotypes and phenotypes, they did all derive from that single founding cell. With today's use of various sophisticated molecular/cell techniques and markers, the concept of "cancer stem cells" (those cancer cells within a tumor, that can sustain the tumor growth) has emerged [39, 40].

Several hypotheses, concerning the origin of cancer, involved, on one extreme, the fact that organ-specific stem cells are the origin of the first cells that, when exposed to an agent that can damage DNA, has a finite probability to form an irreversible change via an error in DNA repair or by an "error of DNA replication" to form a mutation in a finite number of genes that blocks its ability to divide asymmetrically ("initiation" step) (Fig. 7.2).

Probably the clearest example of this happening is with ultraviolet light and the human example is that of the skin cancer prone genetic syndrome of xeroderma pigmentosum [41]. In addition, an irreversible mutation could also be formed by an error of replication, in that every time a stem cell is stimulated to proliferate (i.e., the *promotion* process), there is always a finite chance of a mutation occurring. However, if that initiated adult organ-specific stem cell is prevented from proliferating either by secreted negative growth factors or by contact inhibition [42], then that premalignant initiated stem cell will never have a chance to acquire addition genetic/phenotypic or "epigenetic" alterations to acquire all the "hallmarks" of cancer [43, 44].

The opposite hypothesis exists that differentiated somatic cells can "dedifferentiated" or be "reprogrammed" to start the carcinogenic process [45]. One of the sources that preceded the recent demonstration and interpretation is that one can "reprogram"differentiated somatic cells to become embryonic-like or "induced pluripotent stem" cells ("ips") [4, 46]. The major observation that set the prevailing paradigm that cancer starts when a normal mortal cell is "immortalized" by some event. This idea comes from many sources, such as the ability of cells from tumors to grow indefinitely when placed in in vitro (i.e., HeLa cells), whereas normal fibroblasts and epithelial cells, in vitro, have a finite life span [24]. In addition, with human normal fibroblasts, attempts to "immortalized" these cells have basically been unsuccessful [47–49]. When Land et al. [50] genetically engineered normal fibroblasts with the myc oncogene, they then could neoplastically transform these

Fig. 7.2 (continued) adult stem cells to partially differentiate into cancer nonstem cells. This, together with either addition mutations or stable epigenetic changes, might allow a given initiated adult stem cell to have autonomous, invasive properties of a malignant cell. From: Trosko, J.E., and Tai, M.H., "adult stem cell theory of the multistage, multi-mechanism theory of carcinogenesis: Role of inflammation on the promotion of initiated cells." In: Infections and Inflammation: Impacts on Oncogenesis, T. Dittmar, K.S. Zaenker, and A. Schmidt, eds., S. Karger AG, Publisher, Contributions to Microbiology, Vol. 13 Infection and Inflammation: Impacts on Oncogenesis]. Permission granted from Karger AG

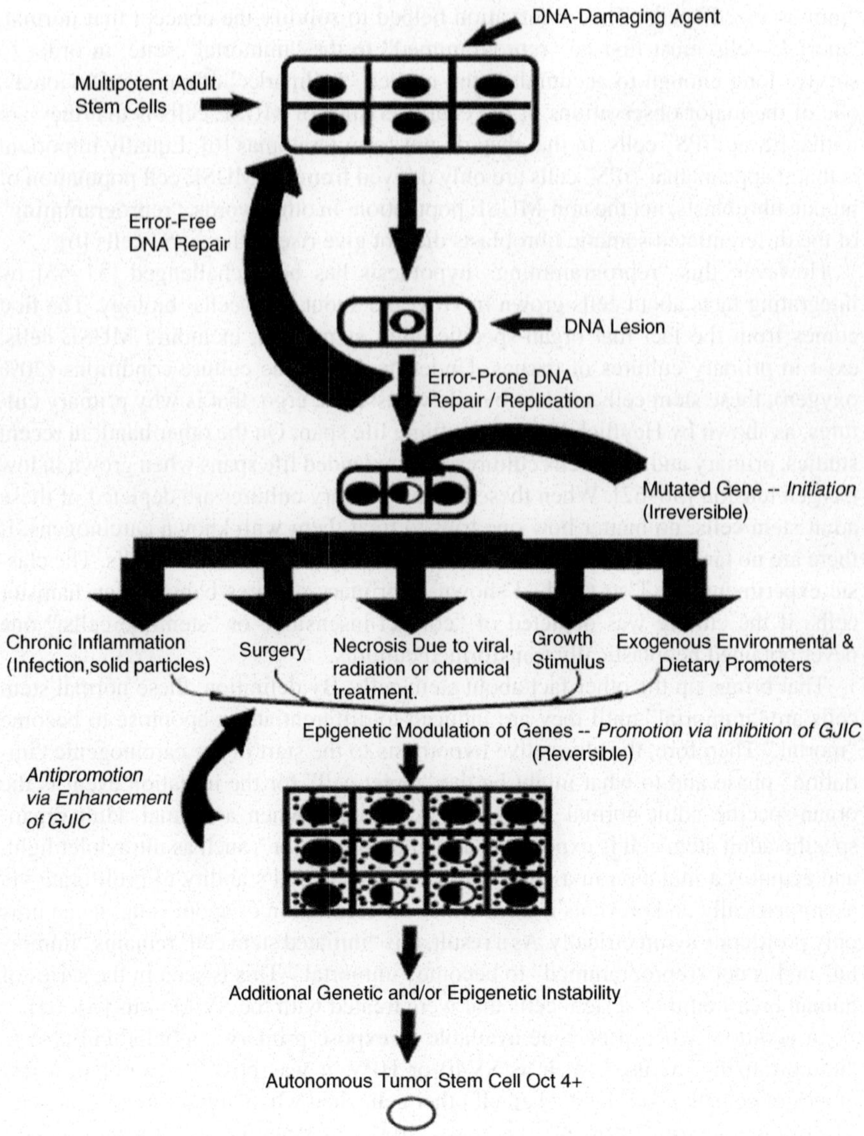

Fig. 7.2 A diagrammatic heuristic scheme to depict the postulated mechanisms of the initiation and promotion phase of carcinogenesis. DNA lesions, induced by physical mutagens or by errors in DNA replication, are substrates in adult stem cells (Oct-4+) that can be fixed if they are not removed in an error-free manner prior to DNA replication. Promotion includes conditions such as chronic inflammation induced by infectious agents and solid particles, surgery or wounding, necrotic cell death, normal growth stimuli caused by growth factors and hormones, and exogenous epigenetic natural and synthetic epigenetic molecules, in which a pluripotent, but surviving, initiated adult stem cell (Oct-4+) can escape the nonproliferative state. The buildup of initiated cells allows them to "resist" the antimitotic influence of neighboring non-initiated cells. In addition, the changing microenvironment within the growing benign tumor will cause some of the initiated

"immortalized" cells. This observation helped to solidify the concept that normal, "mortal" cells must first be "reprogrammed" to the "immortal" state, in order to survive long enough to accumulate the needed "hallmarks" of cancer. Obviously, one of the major observations of the characteristics of MUSE cells is that they are unlike ES or "iPS" cells, in that they do not form teratomas [6]. Equally important is that it appears that "iPS" cells are only derived from the MUSE cell population of human fibroblasts, not the non-MUSE population. In other words, "reprogramming" of the differentiated somatic fibroblasts did not give rise to the "iPS" cells [6].

However, this "reprogramming" hypothesis has been challenged [51–55] by integrating facts about cells grown in vitro and about stem cells' biology. The first comes from the fact that organ-specific adult stem cells, including MUSE cells, exist in primary cultures of tissues. Under normal tissue culture conditions (20% oxygen), these stem cells are lost by a few passages, ergo, that is why primary cultures, as shown by Hayflick [24]) had a finite life span. On the other hand, in recent studies, primary and stem cells cultures have extended life spans when grown in low oxygen tension [56–62]. When these human primary cultures are depleted of these adult stem cells, no matter how one tries to treat them with known carcinogens, if there are no target stem cells, one cannot obtain any "immortalized" cells. The classic experiments by T'so 63] had shown in primary cultures baby Syrian hamster cells, if the culture was depleted of "contact-insensitive or "stemlike cells," one never obtained neoplastically transformed cultures.

That brings up the other fact about stem cells. By definition, these normal stem cells are "immortal" until they are induced to differentiate or apoptose to become "mortal." Therefore, the alternative hypothesis to the start of the carcinogenic "initiation" phase and to what might be the "target cell" for the initiation event is the organ-specific adult normal "immortal" stem cell. When a normal adult organ-specific adult stem cell is exposed to an "initiating factor", such as ultraviolet light, and acquires a mutation in a gene that regulates the cell's ability to proliferate via asymmetrically and prevents it from either differentiation or apoptosing, it can now only proliferate symmetrically. As a result, this "initiated stem cell' remains "immortal" and is not "reprogrammed" to become "immortal." This is seen in the series of human breast epithelial stem cells that were treated with the SV40 virus [64, 65].

In addition, when it became available to expose primary epithelial cultures to "immortalizing" viruses, such as SV40 or HPV, it was possible to obtain a few clones of cells that survived when all other cells died when going through "crises." The original interpretation was that these viruses "reprogrammed" a few of the differentiated epithelial cells. That is how they acquired the term "immortalizing" viruses. However, in those primary epithelial cultures, a few adult organ-specific adult stem cells existed. While these viruses infected all the cells, only did the viral genes (Large T, E6, E7) rendered the p53 and RB proteins in the few stem cells inactive [66]. This prevented the stem cells from differentiating or by dying from apoptosis. The number of surviving "immortalized" clones was small (approximating a

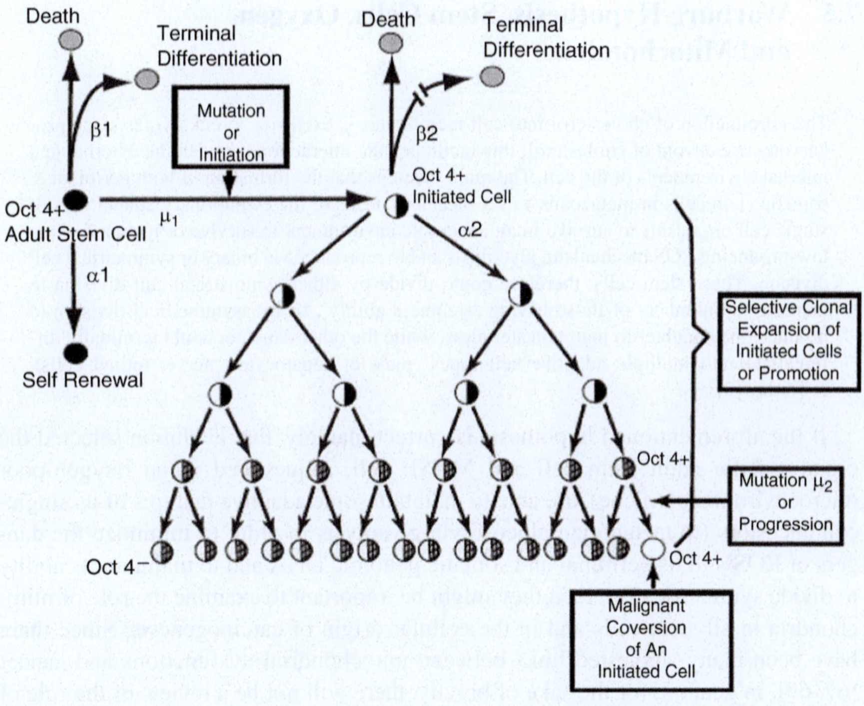

Fig. 7.3 In this diagram, a normal adult stem cell is shown dividing asymmetrically to form one daughter that is committed to ultimately terminally differentiate. The other daughter is designated to be identical to its mother adult stem cell (Oct-4+). If that adult stem cell is exposed to some condition that prevents asymmetrical cell division, but does not suppress the Oct-4 expression, it is operationally an initiated cell. That is, if mitotically stimulated to divide, it divides symmetrically to form two initiated, nonterminally differentiated cells. Initiation is, then, defined as the process that prevents an "immortal" normal adult stem cell to terminally differentiate or become "mortal." These adult-initiated stem cells are still Oct-4-positive or benign cancer stem cells. As these initiated Oct-4+ cells are stimulated to proliferate and resist apoptosis, the growing benign tumor microenvironment changes, some of these initiated Oct-4 + cells can partially differentiate into "cancer nonstem cells" [Oct-4 negative]. Eventually, additional stable mutational or epigenetic events occur, providing the benign Oct-4+ cancer stem cells to become invasive, metastatic "cancer stem cells" [55]. Permission granted by Nova Science

low mutation frequency; a few "reprogrammed" differentiated cells or the few adult stem cells in the original primary culture). If this interpretation is correct, the term, "immortalizing" viruses is incorrect and very misleading. They should be termed "mortalizing inhibitory viruses." Another major implication for understanding the role certain viruses play in human carcinogenesis is that after exposure to these viruses, any adult organ-specific stem cells that might have been infected (breast, liver, cervix, brain, etc.) has been kept from "mortalizing" and, therefore, can live long enough to acquire other changes to become a cancer. (Fig. 7.3).

7.5 Warburg Hypothesis, Stem Cells, Oxygen, and Mitochondria

The introduction of cholesterol into cell membranes is exclusive to eukaryotes since prokaryotes are devoid of cholesterol; this facilitates the interaction between the external and internal environments of the cell. The main theme is that the formation of both germinal & somatic stem cells in metazoans was a means to preserve the evolutionary adaptations of single cell organisms to survive in an anaerobic environment to survive or metabolize via low-producing ROS mechanism, glycolysis and to reproduce via binary or symmetrical cell division. These stem cells, therefore could divide by either symmetrical cell division to increase the numbers of those having 'stemness ability', or by asymmetrical division to produce one daughter to maintain stemness, while the other daughter could terminally differentiate into multiple adaptive cell types (muscle; hepatocytes, nerve, retinal cells). J. Torday [2]

If the aforementioned hypothesis is correct, namely, that evolution selected the organ-specific adult stem cell and MUSE cell, sequestered in an oxygen-poor microenvironment (niche), are able to maintain some adaptive features of its single-cell ancestors (to metabolize glucose via glycolysis in order to minimize the dangers of ROS's to its germinal and somatic genomic DNA and to maintain its ability to divide symmetrically), then they might be important to examine the role of mitochondria in all stem cells and in the cellular origin of carcinogenesis. Since there have been many suggested links between mitochondrial dysfunctions and cancer [67–69], in general, for the sake of brevity, there will not be a review of the role of mitochondrial mutations in cancer as an "initiating" event. This is based on the assumption that any DNA damaging event that might cause a mutation to initiate an adult-specific stem cell, it would happen in the genomic DNA and not in the few mitochondria or the mitochondrial DNA found in these stem cells. The DNA damage caused by mitochondrial metabolism would take place in the progenitor and terminally differentiated cells, where these cells have many mitochondria compared to the few in the stem cells. In addition, in those progenitors and differentiated cells with many mitochondria, the mutations in the individual mitochondrion are random, and each mitochondrion would have different mutations. On the other hand, since new experimental results on the biology of stem cells, the stem cell hypothesis of cancer and the resurrected concept of the "cancer stem cells" do seem to have credible support, it might be relevant to examine how normal and cancer stem cells metabolize glucose in the concept of the Warburg hypothesis. Equally relevant is the role of mitochondria, the role of the Warburg metabolism in MUSE cells.

If during the transition of an anaerobic to an aerobic environment conditions were created for the selection of multicellularity by the symbiotic relationship of the mitochondrion and a primitive pre-eukaryotic cell [70, 71] to produce more ATP for additional adaptive phenotypes of growth control, cell differentiation, and the formation of germinal and somatic stem cells, then unique features of stem cell biology, as opposed to their differentiated offspring, might provide insights to the role of stem cells and cancer. This should include the possibility during the asymmetric cell division of stem cells that there is an asymmetric apportioning of mitochondria of "older" versus newly synthesized mitochondria [72]. Moreover, given the fact that MUSE cells exist in the population of adult tissues and that they can be induced to differentiate into the trigeminal layers without forming teratomas, the Warburg metabolic studies and mitochondriogenesis must be in future studies of MUSE cell biology.

One of the early characteristics of normal and cancerous tissues was shown to be how the cancer tissues seemed to acquire or to revert back to glycolysis during the carcinogenic process. Most descriptions of this Warburg effect use these terms as though "normal "tissues metabolize glucose via oxidative phosphorylation because as multicellular tissues, the somatic, differentiated cells contained mitochondria. However, at the time of these early metabolic studies, and even today, there was no discussion of the fact that even "normal" tissues contained a mixture of a few adult stem cells, many progenitor or tissue-amplifying cells and terminally differentiated cells. Those metabolic studies were done on mixtures. In cancer tissues, we know, today, that there is also a complex mixture of cells…a few normal stromal cells, invasive cells, "cancer stem cells," and partially differentiated "cancer stem cells." To our knowledge, no study to date has examined the Warburg metabolic activities of these individual cell types within a tumor. Also, while the stem cell versus the "dedifferentiation" or "reprogramming" hypotheses have not been universally decided, the evidence concerning the adult-organ-specific stem cells as the origin of carcinogenesis has been presented for blood cancers, breast cancer, intestinal cancers, etc. [65, 73–76]. Therefore, it might serve as a way to test the stem versus dedifferentiated hypotheses of cancer to examine the role of mitochondria and glucose metabolism in normal organ-specific adult stem cells.

To put this approach into context, while single-cell organisms survived, albeit the low efficiency of ATP production, was the low risk of reactive oxygen species damage to their DNA. With the emergence of mitochondria, increased ATP production needed for multicellularity came with the increased risk for ROS-induced macromolecular damage as the by-product of oxidative phosphorylation in the differentiated somatic cells of the multicell organisms. The selection of sequestered unique germ and somatic stem cells that existed in low-oxygen microenvironments might have favored that these stem cells would not need many mitochondria to survive. In addition, while many assume that stem cells replicated more rapidly that their progenitor cells, it seems that under normal conditions, these stem cells are rather quiescent.

By understanding the fundamental biochemical processes up front, glucose can be oxidized and converted to pyruvate by either glycolysis, followed by fermentation, to become lactate or by complete oxidation with mitochondrial respiration [77]. Since it was shown that normal embryonic stem cells have significantly fewer mitochondria than their differentiated offspring [78–80], one might predict that stem cells would metabolize sugar via glycolysis. In addition, normal mesenchymal stem cells metabolized glucose by anaerobic glycolysis than by oxidative phosphorylation [81]. Furthermore, in a glucose-restricted model, increased proliferation ability, increased antioxidant defense ability, and increased aerobic metabolism were shown in mesenchymal stem cells [82]. In limited life span progenitor cells, which metabolize glucose via the tricarboxylic acid cycle and oxidative phosphorylation, their mitochondria acquire many mutations in the mitochondrial genome. However, in terms of cancer, since these cells are not the targets for the initiation step of carcinogenesis, these cells will not be "reprogrammed" to become immortal, and they will either senesce or die.

Several interesting experiments were done to examine both the number and quality of the mitochondrial genome in embryonic, "induced pluripotent stem" ("iPS") cells, and their differentiated daughters. These experiments were prompted by the question: "What happens to the mitochondria during their 'reprogramming' of the differentiated somatic cells in the 'induced pluripotent stem cells'?" Several studies to analyze this question used embryonic stem, the isolated "iPS" cells, and the differentiated cells, from which the "iPS" cells were derived [79–83]. The aims were to examine the quantity and quality of the mitochondria in the embryonic stem cells, the "iPS" cells, and the original fibroblast population from which the "iPS" were derived. The results demonstrated that the quality and quality of the embryonic and "iPS" cells' mitochondria were almost identical, but that they were very different from the original differentiated somatic fibroblasts. Given the demonstration that "iPS" cells were isolated from a small subset of human fibroblast that were the MUSE cells [6], future studies on mitochondrial metabolism in these MUSE cells ,prior to and subsequent to their differentiation into the trigerminal layer cells and their generation of "iPS" cells, are underway.

In addition, there were fewer mitochondria in both the embryonic and "iPS" cells. Clearly, this suggested that both the stem cells metabolized via aerobic glycolysis, but unlike the differentiated somatic fibroblasts, which had many mitochondria, these cells metabolized sugar via oxidative phosphorylation. Equally important to note is that the embryonic and "iPS" stem cells had few mitochondrial mutations compared to the differentiated somatic cells which acquired many mitochondrial mutations.

The interpretation of these studies is critically important. It related to the question: "are 'iPS' cells really 'reprogrammed' or 'dedifferentiated' somatic differentiated fibroblast cells" or "are they simply selected adult stem cells found in the population of primary fibroblasts?" If "iPS" cells are "reprogrammed" at the genomic DNA level by epigenetic mechanisms, and they lost mitochondria numbers during the "reprogramming" process, it could be interpreted as meaning that the "reprogramming" at the genomic level caused the alteration of numbers of mitochondria. It also would require that only the nonmutated mitochondria survived this "reprogramming" process. To accept this interpretation, one must deal with the fact that "reprogramming" by epigenetic mechanisms is possible for genomic DNA and one cannot "reprogram" mitochondrial mutations by epigenetic means. To end up with unmutated mitochondria in the "iPS" cells, back mutagenesis would have had to occur in those few mitochondria. However, that would seem to be highly unlikely for those few mitochondria to have "back-mutated" the many different kinds of mutations in the mitochondria of the somatic fibroblasts.

Alterative interpretation to the facts of these experiments is that, if the "iPS" cells were derived from the few rare adult stem cells in the primary cultures, the amount of ROS's would be minimal from anaerobic glycolysis or from oxidative phosphorylation from the few mitochondria. Given these two interpretations, "iPS" were more likely to have been derived from the few adult stem cells in the original primary population. Evidence has indicated that these "iPS" cells contain epigenetic fingerprints of the original tissue from which they were derived [84, 85] In brief, one might be able to "reprogram," epigenetically, genomic DNA, but one cannot "reprogram," epigenetically, mutated mitochondrial DNA. This was also noted by Wako et al. [6] and Trosko [86].

After the claim of cloning of Dolly and the isolation of embryonic stem cells, multiple techniques to isolate "stemlike" cells by very different approaches (SCNT, "iPS," organ-specific adult stem cells, MUSE cells, etc.) have been developed. In order to generated additional experimental data to test the "the "reprogramming" versus selection of pre-existing adult stem cells, the recently discovered and characterized MUSE cells [6] clearly demonstrated that the "iPS" cells were derived from the MUSE subpopulation in the primary human fibroblast cultures but not from the non-MUSE population in these primary cultures.

7.6 Disruption of Quorum Sensing During the Evolution of Multicellularity Between Stem Cells, Cell-Cell Communication, and Symmetrical and Asymmetrical Cell Division and During Early Development and Diseases Later in Life

During the evolution of multicellularity, the emergence of totipotent, germinal, embryonic, pluripotent, MUSE, and organ-specific adult stem cells, many new genes, new cellular functions, and phenotypes had to appear in order that a complex integration of intra- and extracellular communication mechanisms had to appear. The niche was needed to provide a low-oxygen microenvironment for the stem cells to provide a low risk for both mitochondrial and genomic DNA damage and mutations. After exposure to normoxic conditions to trigger stem cell asymmetric cell division and for the differentiation of various specialized cells, mitochondriogenesis was needed to produce increased amounts of ATP and energy. New biochemical mechanisms regulate specific limited patterns of genes from the total genome for specific cell differentiation, namely, those intracellular signaling pathways that must do two important functions, namely, (a) epigenetic regulation of specific gene regulation and (b) modulation of cell-cell communication by either secreted molecules or by gap junctions [42]. This integration of the three basic forms of cell communication in metazoans provides the means to regulate homeostatic control of cell proliferation, differentiation, apoptosis, senescence, and adaptive responses of differentiated cells in this tightly adhering society of cells.

Quorum sensing was the primitive adaptive means by which single-cell organisms alerted others in their non-adhering population via secreted molecules [87]. A breakdown in this quorum sensing in single-cell organisms could actually, under certain conditions, lead to toxin or drug resistance [88]. With the evolution of quorum sensing mechanisms in metazoans, which included secreted molecules, such as growth factors, hormones, and cytokines, that acted on specific receptors to trigger specific intracellular signaling pathways to up- or downregulate gap junctional intercellular communication [89], a means to control normal homeostatic regulation of cell function would lead to normal embryonic, fetal, neonatal, adolescent, and maturation development. Any acute disruption of this quorum sensing during development could lead to death or birth defects or chronic disruption could lead to many diseases later in life [90]. Specific inhibition of this integration of extra-, intra- and intercellular communication of metazoan quorum sensing probably explains drug resistance or cancer stem cells' resistance to chemotherapy [90].

7.7 Conclusion

The basic assumption of this study has been that the evolutionary emergence of mitochondria enabled the acquisition of genes and multicellularity to create a low-oxygen niche to house a new type of cell, the germinal and various somatic stem cells, which had the ability to divide both symmetrically and asymmetrically to form various differentiated cells. Based on previous observations that the number of mitochondria that exists in embryonic stem and a few adult organ-specific stem cell is comparatively low compared to differentiated cells, since no systematic analyses have been on all stem cell types, it will be important to analyze the adult MUSE stem cells. Included in future studies on MUSE cells will be the roles of glycolysis and oxidative phosphorylation in the MUSE cells compared to ES and "iPS" cells, prior to and after differentiation of these cells. In addition, while MUSE cells have been shown not to form teratomas, it will be important to rule in or out their ability to form adult cancer types.

References

1. Saul JM (2010) A geologist speculates on gemstones, origin of gas and oil, moonlike impact scars on the earth, the emergence of animals and cancer. Les 3 Colonnes, Paris
2. Torday J (2014) Evolutionary biology redux. Perspect Biol Med 56:455–484
3. Dobzhansky T (1975) Nothing in biology makes sense except in the light of evolution. Am Biol Teach 35:125–129
4. Takahashi K, Yamanaka S (2006) Induction of pluripotent stem cells from mouse embryonic and adult fibroblast cultures by defined factors. Cell 126:663–676
5. Cibelli JB (2014) Human somatic cell nuclear transfer is alive and well. Cell Stem Cell 14(6):699–701
6. Wakao S, Kitada M, Kuroda Y et al (2011) Multilineage-differentiating stress-enduring (Muse) cells are a primary source of induced pluripotent stem cells in human fibroblasts. Proc Natl Acad Sci USA 108(24):9875–9880. https://doi.org/10.1073/pnas

7. Trosko JE, Kang KS (2012) Evolution of energy metabolism, stem cells and cancer stem cells: how the Warburg and Barker hypotheses might be linked. Int J Stem Cells 5:39–56
8. Trosko JE (2007) Gap junction intercellular communication as a 'Biological Rosetta Stone' in understanding, in a systems manner, stem cell behavior, mechanisms of epigenetic toxicology, chemoprevention and chemotherapy. J Membr Biol 218:93–100
9. Torday J, Rehan VK (2012) Evolutionary biology, cell-cell communication and complex disease. Wiley-Blackwell, Hoboken
10. Perva S, Taneja R, Ghaffan S (2009) Oxidative stress regulation of stem and progenitor cells. Antioxid Redox Signal 11:2777–2789
11. Upham BL, Trosko JE (2009) Carcinogenic tumor promotion, induced oxidative stress signaling, modulated gap junction function and altered gene expression. Antioxid Redox Signal 11:297–308
12. Brigelius-Flohé R, Flohé L (2011) Basic principles and emerging concepts in the redox control of transcription factors. Antioxid Redox Signal 15:2335–2381
13. Rigoulet M, Yoboue ED, Devin A (2011) Mitochondrial ROS generation and its regulation: mechanisms involved in H(2)O(2) signaling. Antioxid Redox Signal 14(3):459–468. https://doi.org/10.1089/ars.2010.3363
14. Dang CV (2012) Links between metabolism and cancer. Genes Dev 26:9. https://doi.org/10.1101/gad.189365.112
15. Saul JM (2008) Did detoxification processes cause complex life to emerge? Lethaia 42:179–184
16. Nursall JR (1959) Oxygen as prerequisite to the origin of metazoan. Nature 183:1170–1172
17. Ozbek S, Balayubramanian PG, Chiquet-Ehrismann R, Tucker RP et al (2010) The evolution of excellular matrix. Mol Biol Cell 21:4300–4305
18. Mohyeldin A, Garzon-Muvdi T, Quinones-Hinojosa A (2010) Oxygen in stem cell biology: a critical component of stem cell niche. Cell Stem Cell 7:150–161
19. Park S, You X, Imlay JA (2005) Substantial DNA damage from submicromolar intracellular hydrogen peroxide detected in Hpx- mutants of Escherichia coli. Proc Natl Acad Sci USA 102:9317–9322
20. Rosenberg B, VanCamp L, Trosko JE et al (1969) Platinum compounds: a new class of potent antitumour agents. Nature 222:385–386
21. Finkel T, Manuel Serrano M, Blasco MA (2007) The common biology of cancer and ageing. Nature 448:767–774. https://doi.org/10.1038/nature05985
22. Potter VR (1978) Phenotypic diversity in experimental hepatomas: concept of partially blocked ontogeny. Br J Cancer 38:1–23
23. Potter VR (1973) Biochemistry of cancer. In: Holland J, Frei E (eds) Cancer medicine. Lea and Febiger Publishers, Philadelphia, pp 178–192
24. Hayflick L (1965) The limited in vitro lifespan of human diploid cell strains. Exp Cell Res 37:614–636
25. Cruciani V, Mikalsen SO (2005) The connexin gene family in mammals. Biol Chem 386:325–332
26. Evans WH, Martin PEM (2002) Gap junctions: structure and function. Mol Membr Biol 19:121–136
27. Horiguchi K, Kouki T, Fujiwara K et al (2011) The extracellular matrix component laminin promotes gap junction formation in the rat anterior pituitary gland. J Endocrinol 208:225–232
28. Markert CL (1984) Genetic control of cell interactions in chimeras. Dev Genet 4:267–279
29. Carruba G, Trosko JE The long evolutionary journey of cancer from ancestor to modern humans. In press
30. Warburg OH (1956) On the origin of cancer cells. Science 123:309–314
31. Shamblott MJ et al (1998) Derivation of pluripotent stem cells from cultured human premortal germ cell. Proc National Acad Sci USA 95(23):13726–13731
32. Thomson JA et al (1998) Embryonic stem cell lines derived from human blastocyst. Science 282(5391):1145–1147

33. Trosko JE, Chang CC, Wilson MR et al (2000) Gap junction and the regulation of cellular functions of stem cells during development and differentiation. Methods 20:245–264
34. Minsk B, Illensee K (1975) Normal genetically mosaic mice produced from malignant teratocarcinoma cells. Proc Natl Acad Sci USA 72:3585–3589
35. Weinstein IB, Gattoni CS, Kirschmeier P et al (1984) Multistage carcinogenesis involves multiple genes and multiple mechanisms. J Cell Physiol Suppl 3:127–137
36. Pitot H, Dragon YP (1991) Facts and theories concerning the mechanisms of carcinogenesis. FASEB J 5:2280–2286
37. Nowell PC (1976) The clonal evolution of tumor cell populations. Science 194:23–28
38. Fialkow PJ (1976) Clonal origin of human tumors. Annu Rev Med 30(1976):135–176
39. Atena M, Reza AM, Meehran G (2014) A review on the biology of cancer stem cells. Stem Cell Discov 4:83–89
40. Reya T, Morrison SJ, Clarke MF et al (2001) Stem cells, cancer, and cancer stem cells. Nature 414:105–111
41. Cleaver JE (1978) Xeroderma pigmentosum: genetic and environmental influences in skin carcinogenesis. Int J Dermatol 17:435–444
42. Trosko JE (2003) The role of stem cells and gap junctional intercellular communication in carcinogenesis. J Biochem Mol Biol 36:43–48
43. Hanahan D, Weinberg RA (2000) The hallmarks of cancer. Cell 100:57–70
44. Hanahan D, Weinberg RA (2011) Hallmarks of cancer: the next generation. Cell 144:646–674
45. Sell S (1993) Cellular origin of cancer: differentiation or stem cell maturation arrest? Environ Health Perspect 101:15–26
46. Yamanaka S (2012) Induced pluripotent stem cells: past, present, and future. Cell Stem Cell 10(6):678–684
47. DiPaolo JA (1983) Relative difficulties in transforming human and animal cells in vitro. J Natl Cancer Inst 70:3–8.84
48. Kakunaga T (1978) Neoplastic transformation of human diploid fibroblast cells by chemical carcinogens. Proc Natl Acad Sci USA 75:1334–1338
49. Rhim JS (1993) Neoplastic transformation of human cells in vitro. Crit Rev Oncog 4:312–335
50. Land H, Parada LF, Weinberg RA (1983) Tumorigenic conversion of primary embryo fibroblasts requires at least two cooperating oncogenes. Nature 304:596–602
51. Trosko JE (2008) Human adult stem cells as the target cells for the initiation of carcinogenesis and for the generation of "cancer stem cells". Int J Stem Cells 1:8–26
52. Trosko JE (2008) Commentary: re-programming or selecting adult stem cells. Stem Cell Rev 4:81–88
53. Trosko JE (2009) Cancer stem cells and cancer non-stem cells: from adult stem cells or from re-programming of differentiated somatic cells. Vet Pathol 46:176–193
54. Trosko JE (2006) From adult stem cells to cancer stem cells: Oct-4 gene, cell-cell communication, and hormones during tumor promotion. Ann NY Acad Sci 1089:36–58
55. Trosko JE (2014) Induction of iPS cells and of cancer stem cells: the stem cell or reprogramming hypothesis of cancer? Anat Rec (Hoboken) 297:161–173
56. Estrada JC, Albo C, Benguría A et al (2012) Culture of human mesenchymal stem cells at low oxygen tension improves growth and genetic stability by activating glycolysis. Cell Death Differ 19(5):743–755. https://doi.org/10.1038/cdd.2011.172
57. Parrinello S, Samper E, Krtolica A et al (2003) Oxygen sensitivity severely limits the replicative lifespan of murine fibroblasts. Nat Cell Biol 5:741–747
58. Chen Q, Fischer A, Reagan JD et al (1995) Oxidative DNA damage and senescence of human diploid fibroblast cells. Proc Natl Acad Sci USA 92:4337–4341
59. Busuttil RA, Rubio M, Dolle ME et al (2003) Oxygen accelerates the accumulation of mutations during the senescence and immortalization of murine cells in culture. Aging Cell 2:287–294

60. Fehrer C, Brunauer R, Laschober G et al (2007) Reduced oxygen tension attenuates differentiation capacity of human mesenchymal stem cells and prolongs their lifespan. Aging Cell 6:745–757
61. Li TS, Marban E (2010) Physiological levels of reactive oxygen species are required to maintain genomic stability in stem cells. Stem Cells 28:1178–1185
62. Forsyth NR, Musio A, Vezzoni P et al (2006) Physiologic oxygen enhances human embryonic stem cell clonal recovery and reduces chromosomal abnormalities. Cloning Stem Cells 8:16–23
63. Nakano S, Ueo H, Bruce SA et al (1985) A contact-insensitive subpopulation in Syrian hamster cell cultures with a greater susceptibility to chemically induced neoplastic transformation. Proc Natl Acad Sci USA 82:5005–5009
64. Kao CY, Nomata K, Oakley CS et al (1995) Two types of normal human breast epithelial cells derived from reduction mammoplasty: phenotypic characterization and response to SV40 transfection. Carcinogenesis 16:531–538
65. Tai M-H, Chang CC, Kiupel M et al (2005) Oct-4 expression in adult stem cells: Evidence in support of the stem cell theory of carcinogenesis. Carcinogenesis 26:495–502
66. Ahuja D, Sa'enz-Robles MT, Pipas JM (2005) SV40 large T antigen targets multiple cellular pathways to elicit cellular transformation. Oncogene 24(52):7729–7745
67. Wallace DC (2012) Mitochondria and cancer. Nat Rev Cancer 12(10):685–698. https://doi.org/10.1038/nrc3365
68. Zong W-X, Rabinowitz JD, White E (2016) Mitochondria and cancer. Mol Cell 61:667–676
69. Ralph SJ, Rodriguez-Enriquez S, Neuzil J et al (2010) The causes of cancer revisited: "mitochondrial malignancy" and ROS-induced oncogenic transformation – why mitochondria are targets for cancer therapy. Mol Asp Med 31(2):145–170
70. Woolfson A (2015) An improbable journey. Nature 520:6176
71. Spang A, Saw JH, Jorgensen SL et al (2015) Complex archaea that bridge the gap between prokaryotes and eukaryotes. Nature 521:173–179
72. Suomalainen A (2015) Asymmetric rejuvenation. Nature 521:296–298
73. Cozzio A, Passegue E, Ayton PM (2003) Similar MLL associated leukemias arising from self-renewing stem cells and short-lived myeloid progenitors. Genes Dev 17:3029–3035
74. Barker N, Ridgway RA, van Es JH et al (2009) Crypt stem cells as the cells-of-origin of intestinal cancer. Nature 457:608–611. https://doi.org/10.1038/nature076
75. Till JE (1982) Stem cells in differentiation and neoplasia. J Cell Physiol 1:3–11
76. Greaves MF (1986) Differentiation-linked leukemogenesis in lymphocytes. Science 234:697–704
77. Gatenby RA, Gillies RJ (2004) Why do cancers have high aerobic glycolysis? Nat Rev Cancer 4(11):891–899
78. Armstrong L, Tilgner K, Saretzki G et al (2010) Human induced pluripotent stem cell lines show stress defense mechanisms and mitochondrial regulation similar to those of human embryonic stem cells. Stem Cells 28:661–673
79. Prigione A, Fauler B, Lurz R et al (2010) The senescence-related mitochondrial/oxidative stress pathway is repressed in human induced pluripotent stem cells. Stem Cells 28:721–733
80. Nesti C, Pasquali L, Vaglini F et al (2007) The role of mitochondria in stem cell biology. Biosci Rep 27:154–171
81. Chen CT, Shih YR, Kuo TK et al (2008) Coordinated changes of mitochondrial biogenesis and antioxidant enzymes during osteogenic differentiation of human mesenchymal stem cells. Stem Cells 26:960–968
82. Lo T, Ho JH, Yang MH et al (2011) Glucose reduction prevents replicative senescence and increases mitochondrial respiration in human mesenchymal stem cells. Cell Transplant 20:813–825
83. Armstrong L, Tilgner K, Saretzki G et al (2010) Human induced pluripotent stem cell cell lines show stress defense mechanisms and mitochondrial regulation similar to those of human embryonic stem cells. Stem Cells 28:661–673

84. Kim K, Doi A, Wen B et al (2010) Epigenetic memory in induced pluripotent stem cells. Nature 467:285–290
85. Hochedlinger K (2010) Cell type of origin influences the molecular and functional properties of mouse induced pluripotent stem cells. Nat Biotechnol 28:848–855
86. Trosko JE (2018) Modulation of cell-cell communication and epigenetic mechanisms as a shared cellular mechanism in diverse childhood brain diseases, such as cancer and autism. EC Neurol 10(3):134–156
87. Miller M, Bassler BL (2001) Quorum sensing in bacteria. Annu Rev Microbiol 55:165–199
88. Lambert G, Estevez-Salmeron L, Oh S et al (2011) An analogy between the evolution of drug resistance in bacterial communities and malignant tissues. Nat Rev Cancer 11:375–382
89. Trosko JE (2016) Evolution of microbial quorum sensing to human global quorum sensing: an insight to how gap junctional intercellular communication might be linked to global metabolic disease crisis. Biology 5(2):E29
90. Trosko JE (2017) Commentary: how evolution of quorum sensing must fit into the understanding of the origin, prevention and treatment of cancer. J Cancer Treat Diagn 2(1):26–30

Chapter 8
Acute Myocardial Infarction, Cardioprotection, and Muse Cells

Shinya Minatoguchi, Atsushi Mikami, Toshiki Tanaka, Shingo Minatoguchi, and Yoshihisa Yamada

Abstract Acute myocardial infarction (AMI) is a common cause of morbidity and mortality worldwide. Severe MI leads to heart failure due to a marked loss of functional cardiomyocytes. First-line treatment for AMI is to reperfuse the occluded coronary artery by PCI as soon as possible. Besides PCI, there are several therapies to reduce the infarct size and improve the cardiac function and remodeling. These are drug therapies such as pharmacological pre- and postconditioning, cytokine therapies, and stem cell therapies. None of these therapies have been clinically developed as a standard treatment for AMI. Among many cell sources for stem cell therapies, the Muse cell is an endogenous non-tumorigenic pluripotent stem cell, which is able to differentiate into cells of all three germ layers from a single cell, suggesting that the Muse cell is a potential cell source for regenerative medicine. Endogenous Muse cell dynamics in the acute phase plays an important role in the prognosis of AMI patients; AMI patients with a higher number of Muse cells in the peripheral blood in the acute phase show more favorable improvement of the cardiac function and remodeling in the chronic phase, suggesting their innate reparative function for the heart. Intravenously administered exogenous Muse cells engrafted preferentially and efficiently to infarct border areas via the S1P-S1PR2 axis and differentiated spontaneously into working cardiomyocytes and vessels, showed paracrine effects, markedly reduced the myocardial infarct size, and delivered long-lasting improvement of the cardiac function and remodeling for 6 months. These findings suggest that Muse cells are reparative stem cells, and thus their clinical application is warranted.

S. Minatoguchi (✉)
Department of Circulatory and Respiratory Advanced Medicine, Gifu University Graduate School of Medicine, Gifu, Japan

Heart Failure Center, Gifu Municipal Hospital, Gifu, Japan
e-mail: minatos@gifu-u.ac.jp

A. Mikami
Department of Circulatory and Respiratory Advanced Medicine, Gifu University Graduate School of Medicine, Gifu, Japan

T. Tanaka · S. Minatoguchi · Y. Yamada
Department of Cardiology, Gifu University Graduate School of Medicine, Gifu, Japan

M. Dezawa (ed.), *Muse Cells*, Advances in Experimental Medicine and Biology 1103, https://doi.org/10.1007/978-4-431-56847-6_8

Keywords Muse cells · Cardiomyocyte regeneration · Acute myocardial infarction · Cardiac function · Cardiac remodeling

8.1 Introduction

Acute myocardial infarction (AMI) is one of the leading causes of morbidity and mortality [1]. AMI occurs due to the total occlusion of the coronary artery mainly caused by plaque rupture. From the onset of coronary occlusion, the death of myocardial tissues spreads from the endocardial site to epicardial site over time. This is called the wave front phenomenon [2]. Therefore, if the occluded coronary artery is reperfused during the very early phase of coronary occlusion, the death of myocardial tissue will be limited to the endocardial site, the myocardial infarct size will be reduced, and the cardiac function will not be deteriorated. In the clinical setting, it has been clearly established that there is a close relationship between the time to treatment with percutaneous coronary intervention (PCI) and clinical outcome: the shorter the time to reperfusion, the better the clinical outcome [3]. Therefore, the best treatment for AMI is to reperfuse the occluded coronary artery by PCI as soon as possible to salvage the remaining viable cardiomyocytes in the myocardial tissues. However, if treatment fails to reperfuse the occluded coronary artery, or the time to reperfusion is not early enough to salvage the myocardial tissue, the myocardial infarction will become transmural from the endocardial to epicardial sites, and the infarct size will be enlarged. A large myocardial infarction usually results in the loss of a large number of cardiomyocytes, and the necrotic cardiac tissue is eventually replaced by scar formation. The wall thickness of the left ventricle (LV) will be thin and the end-diastolic LV dimension will be enlarged. This change of LV is called "LV remodeling." LV remodeling will be accelerated, and heart failure will progress, leading to a poor prognosis. Although the optimal method to treat the AMI is to reperfuse the occluded coronary artery as soon as possible, as mentioned above, the number of AMI patients whose occluded coronary artery is reperfused by PCI within 90 min from the onset time of AMI, the time duration considered to minimize cardiac damage, is quite low. Therefore, because the time from the onset of AMI to reperfusion is generally longer than 90 min in most cases, treatment in addition to PCI to protect, repair, and regenerate the heart is essential. Realistically, additional treatment will be novel and advanced therapies such as pharmacological intervention and cytokine and/or stem cell therapies.

In this review, several therapies for AMI other than PCI are described, focusing mainly on Muse cell therapy. In particular, the important role of Muse cells in the acute phase of AMI and the use of bone marrow-derived Muse cells for the treatment of AMI are described.

8.2 Pharmacological Intervention

In 1986, Murry CE et al. reported that four repetitions of a short-period coronary ischemia and reperfusion promoted resistance to subsequent prolonged coronary ischemia and markedly reduced the myocardial infarct size in dogs [4]. They named this phenomenon ischemic preconditioning (PC). Thereafter, the PC phenomenon has been noted in the rat [5], rabbit [6], and pig [7] and even in humans [8]. Patients with AMI preceded by anginal attack showed smaller infarct sizes and a better cardiac function in the chronic phase when compared with those without pre-infarct ischemia [9, 10]. Several possible mechanisms by which PC reduces the infarct size have been described. The mechanisms by which PC reduces the infarct size involve adenosine [6, 11], bradykinin [12], opioid [13], noradrenalin [14–16], free radical [17], activation of protein kinase C [18], and the opening of sarcolemmal and mitochondrial KATP channels [19, 20]. Among the abovementioned mechanisms of PC, only the KATP channel opener nicorandil is clinically available. The IONA study demonstrated that the use of KATP channel opener nicorandil significantly decreases the rate of coronary heart disease death, nonfatal MI, or unplanned hospitalization for cardiac chest pain in high-risk patients with angina pectoris [21]. However, nicorandil is effective only when prescribed before the onset of AMI.

Besides PC, Zhao et al. demonstrated that repetitive 5-min ischemia and 5-min reperfusion applied during reperfusion immediately after 60 min of coronary occlusion significantly reduced myocardial infarction [22]. This phenomenon was termed ischemic postconditioning. Ischemic postconditioning was effective in reducing the myocardial infarct size. The mechanisms by which ischemic postconditioning reduces the myocardial infarct size have been reported as activation of the reperfusion injury salvage kinase (RISK) pathway, an active effect via activation of PI3K-Akt or ERK1/2, and phosphorylation of downstream targets such as eNOS producing NO, which inhibits the opening of the mitochondrial permeability transition pore (mPTP) [23]. However, pharmacological intervention involving the mechanism of ischemic postconditioning has not yet been used clinically.

8.3 Cytokine Therapy

The granulocyte colony-stimulating factor (G-CSF), which can mobilize multipotential progenitor cells from the bone marrow (BM) into peripheral blood, has been reported to reduce the myocardial infarct size and improve postinfarction left ventricular (LV) remodeling and function [24–26]. At present, proposed mechanisms by which G-CSF causes cardioprotection are transdifferentiation of subpopulation of BM progenitor cells into cardiac tissues such as cardiomyocytes, vascular endothelial cells, vascular α-smooth muscle actin, and myofibroblasts [25, 27], acceleration of the healing process [25], and the prevention of apoptotic cardiomyocytes [26]. On the basis of animal experiments on G-CSF in acute myocardial infarction, many clinical trials have been performed [28–34]. The safety of using G-CSF for patients with acute myocardial infarction has been confirmed in these clinical trials. Some groups reported that G-CSF is beneficial for treating patients with acute myocardial infarction, but other groups reported no beneficial effects. These differences in the efficacy of G-CSF might have been caused by differences in the doses, timing, and duration of G-CSF use and patient selection. Despite numerous clinical trials, G-CSF has not yet become a standard therapy for acute myocardial infarction.

Erythropoietin (EPO) stimulates the proliferation of early erythroid precursors and the differentiation of later precursors of the erythroid lineage [35]. Recombinant human EPO is currently used frequently in the treatment of anemia associated with end-stage renal disease [36]. Recent studies have suggested that EPO also exerts a cardioprotective effect after acute myocardial infarction (MI) [37, 38]. Although many clinical trials on EPO in acute myocardial infarction have been performed [39], EPO is considered to have no clinical benefit for heart function, reducing infarct size, cardiovascular events, or all-cause mortality. EPO has not yet become a standard therapy for acute myocardial infarction.

8.4 Human AMI and Behavior of Muse Cells in the Peripheral Blood

It has been reported that there is a baseline level of Muse cells circulating in the peripheral blood, and their number increases in stroke patients in the acute phase [40]. However, the dynamics of Muse cells in the peripheral blood in patients with AMI had not been clarified until recently [41]. We examined whether endogenous Muse cells are mobilized after AMI and whether the increase of Muse cells in the peripheral blood correlate with improvement of LV function and attenuation of LV remodeling in the chronic phase at 6 months after AMI.

We defined peripheral blood-Muse cells as $SSEA3^+$ and $CD105^+$ double-positive cells. In 79 patients with AMI, 44 patients with coronary artery disease (CAD), and 64 normal subjects (control), we measured the number of Muse cells in the peripheral blood by FACS. Muse cells were measured on days 0, 1, 7, 14, and 21 after AMI. Plasma sphingosine-1-phosphate (S1P) levels were also measured. Cardiac echocardiography was performed in the acute (within 7 days) and chronic (6 months) phases of AMI. The Muse cell number was significantly higher in the AMI patients at the acute phase within 14 days after the onset of AMI than in the CAD patients and control subjects. While day 0, the date of admission, did not show difference between AMI and CAD/control groups, the number of Muse cells peaked on day 1 and gradually decreased on days 14 and 21, returning to the baseline level. The number of Muse cells positively correlated with plasma S1P levels, and S1P elevation in the blood proceeded the increase of Muse cell number, suggesting that S1P mobilizes endogenous Muse cells into the peripheral blood.

Patients with a greater increase in the number of Muse cells in the peripheral blood in the acute phase showed an improvement of the LV function, represented by recovery of ejection fraction, and attenuation of LV remodeling, represented by left

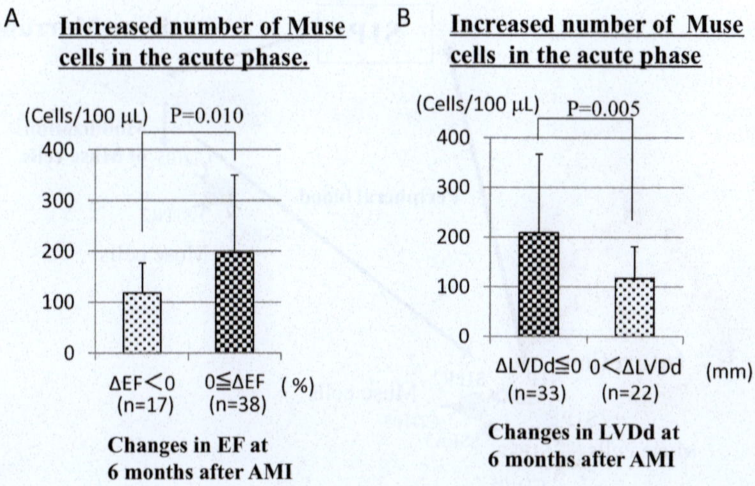

Fig. 8.1 (**A**) Increases in Muse cell numbers in the acute phase in response to AMI in patients with improved EF (ΔEF ≥ 0) or deteriorated EF (ΔEF < 0) and (**B**) attenuated LV remodeling (ΔLVDd < 0) or accelerated LV remodeling (ΔLVDd ≥0) at 6 months after AMI (Modified from Ref. [41]). AMI patients with a higher number of Muse cells in the peripheral blood showed an improvement of EF and attenuation of LV remodeling, while those with a lower number of Muse cells showed a deterioration of EF and worsening of LV remodeling

ventricle end-systolic dimension, in the chronic phase at 6 months after AMI (Fig. 8.1A). However, those with a lower increase level in the number of Muse cells in the peripheral blood showed a deterioration of the LV function and acceleration of LV remodeling in the chronic phase (Fig. 8.1B) [41]. Our study demonstrated that (1) endogenous Muse cells are mobilized into the peripheral blood, following to the elevation of blood S1P level after AMI, and (2) since the increase of peripheral blood-Muse cell number in the acute phase positively correlated with functional recovery and avoidance of heart failure, the number of peripheral blood-Muse cells could be a prognostic indicator in patients with AMI. These results also suggest that Muse cells in the peripheral blood function as reparative stem cells, and Muse cells are mobilized in response to an emergency such as AMI and repair the infarcted cardiac tissue in patients with AMI. A conceptual figure of the behavior of endogenous Muse cells in AMI is shown in Fig. 8.2.

8.5 Muse Cells as a Promising Source for Stem Cell Therapy

In large AMI, extensive tissue damage and loss of functional cardiomyocytes lead to heart failure. Therefore, stem/progenitor cell therapy to replenish cardiac tissue component such as cardiomyocytes and vessels is a fundamental medical treatment for AMI. Many stem cell types have been intensively studied for this purpose. Although

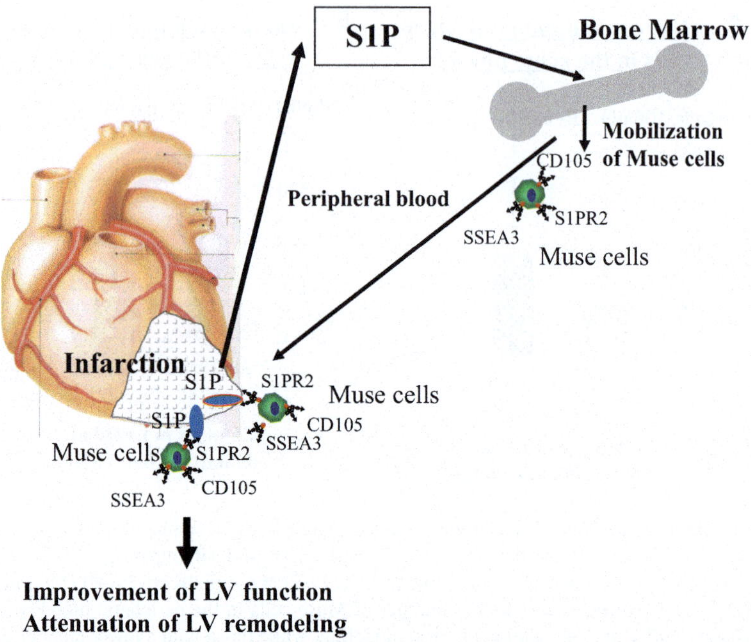

Fig. 8.2 Conceptual figure of endogenous Muse cells behavior in AMI

bone marrow (BM)-mesenchymal stem cells (MSCs) and BM-mononucleated cells (MNCs) have been successfully applied in clinical studies and their safety has been demonstrated, these cells are not clinically relevant [42, 43].

Muse cells can be harvested from the BM, connective tissue of various organs, and peripheral blood as pluripotent surface marker SSEA-3-positive [44–47]. Muse cells make up approximately 0.03% of the mononucleated fraction of the BM, and thus ~30 mL of fresh human BM aspirate yields ~0.15 million Muse cells, which expand to ~one million cells after 3 days in culture [44, 47]. Muse cells express factors related to stress tolerance and pluripotency, are self-renewable, and are able to differentiate into cells of all three germ layers from a single cell in vitro [44, 48].

We recently reported that intravenously administered Muse cells reduce the myocardial infarct size, improve left ventricular function, and attenuate LV remodeling in a rabbit AMI model [49]. AMI model was made in rabbits because rabbits have minimal collateral circulation. Japanese white rabbits underwent 30 min of coronary artery occlusion and reperfusion under anesthesia. Twenty-four hours after the onset of AMI, the rabbits were injected with autologous 3×10^5 of Muse cells/2 mL of saline (Muse group), 3×10^5 of non-Muse cells/2 mL of saline (cells other than Muse cells in MSCs; non-Muse group), or 3×10^5 of BM-MSCs (MSC group) into an ear vein and then followed up for 2 weeks or 2 months. The 2 mL of saline was intravenously injected in the vehicle group. Allograft Muse cells and human xenograft Muse cells at 3×10^5 of cells were also injected and followed up for 2 weeks without using immunosuppressive drugs. The effect of an S1PR2 antagonist on the integration of allograft Muse cells was also evaluated. For the

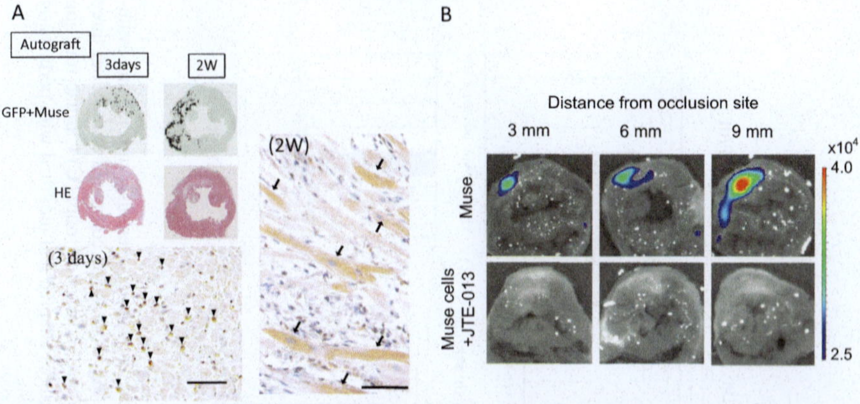

Fig. 8.3 Engraftment of Muse cells into the post-infarct heart. (**A**) Muse cells are labeled with GFP. Engraftment of Muse cells was detected by anti-GFP immunostaining at day 3 and 2 weeks. (**B**) Muse cells are labeled with Nano-lantern. Note that at 2 weeks, Muse cells selectively homed to the infarct area in the AMI heart, while their integration was substantially abrogated when JTE-013, specific antagonist of S1P receptor 2, was co-injected with Muse cells. The suppression of Muse cell homing by JTE-013 suggested that Muse cell migration and homing is mainly controlled by S1P-S1P receptor 2 axis. (Pictures adapted and modified with permission from Yamada et al. (2018), *Cir Res* [49])

6-month experiment, 3×10^5 of allograft Muse cells and vehicle (saline) were intravenously injected and evaluated at 6 months after AMI.

One of the marked characteristics of Muse cells is the high engraftment rate (~14%) of the injected cells to the infarct and infarct border areas of the heart (Fig. 8.3A). Different from the behavior of Muse cells, the other cell types such as MSCs have shown a low or zero engraftment rate when intravenously administered in papers reported previously [50, 51]. The high rate of engraftment of Muse cells to the infarct and infarct border areas is mediated by the S1P-S1PR2 axis; an interaction between S1P produced in the damaged heart and S1P receptor 2 (S1PR2) located on Muse cells [49]. Nano-lantern-labeled Muse cells demonstrated engraftment into the infarct and infarct border areas. This engraftment was completely abolished by a specific S1PR2 antagonist JTE-013 (Fig. 8.3B), suggesting that the engraftment of Muse cells is mediated mainly through S1P-S1PR2 axis.

The intravenous administration of autograft Muse cells after AMI strikingly reduced the myocardial infarct size, improved the left ventricular (LV) function, and attenuated LV remodeling at 2 months after AMI (Fig. 8.4) [49]. The myocardial infarct size was ~52% smaller as compared to the vehicle group.

Autograft, allograft, and human GFP-labeled Muse cells homed to the infarct and infarct border areas of the myocardium. These cells expressed the cardiac markers ANP, troponin I, and α-actinin and expressed vascular endothelial marker CD31 and vascular smooth muscle marker α-smooth muscle actin, suggesting that Muse cells transdifferentiated into cardiomyocytes and vessels spontaneously after engraftment (Fig. 8.5A, B, C, D, E). GCaMP-Muse cell-derived cardiomyocytes also exhibited increased GCaMP3 fluorescence during systole and decreased fluo-

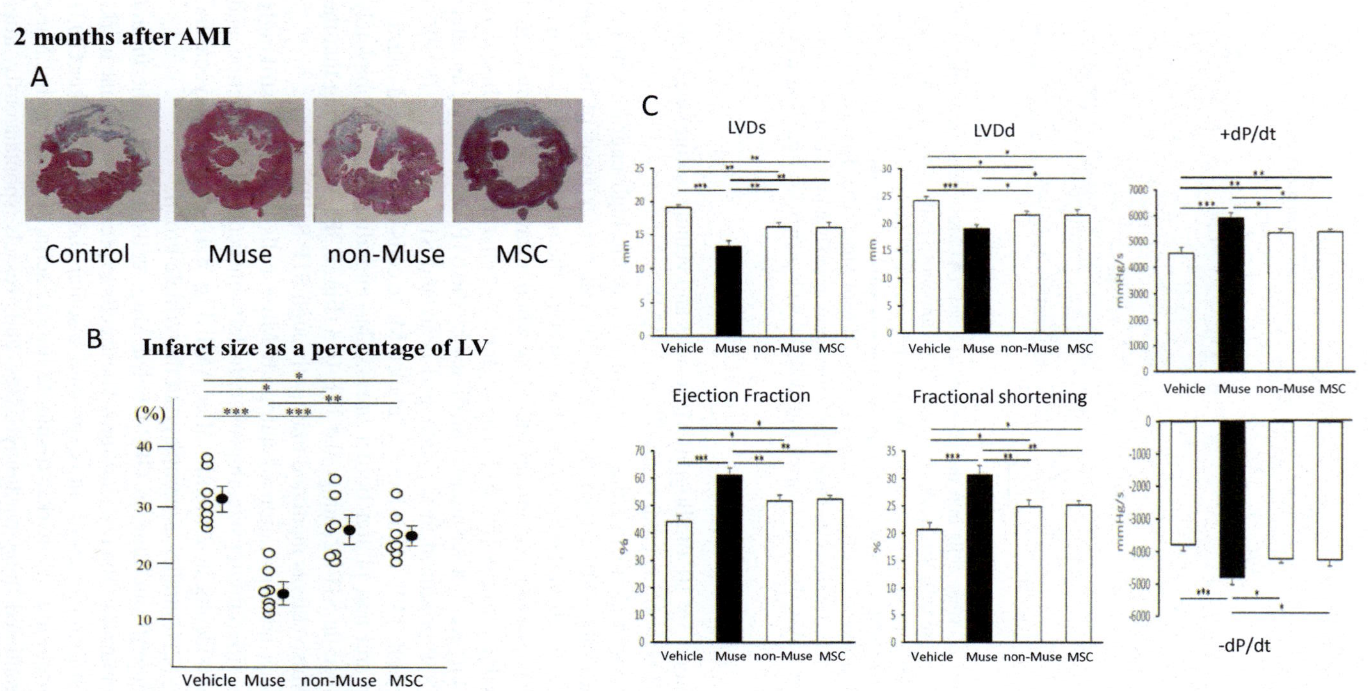

Fig. 8.4 Infarct size reduction and cardiac function recovery by Muse cell injection at 2 months. Autograft Muse cells were intravenously injected and evaluated at 2 months. (**A**) Masson-trichrome staining of LV showed the substantial reduction of infarct area (blue) in the Muse group compared to the other three groups. (**B**) The measurement of infarct size. The Muse group showed ~51% reduction in infarct size compared to the vehicle group. (**C**) Cardiac function demonstrated the substantial recovery of ejection fraction and LVDd with statistical significant differences. (Pictures adapted and modified with permission from Yamada et al. (2018), *Cir Res* [49])

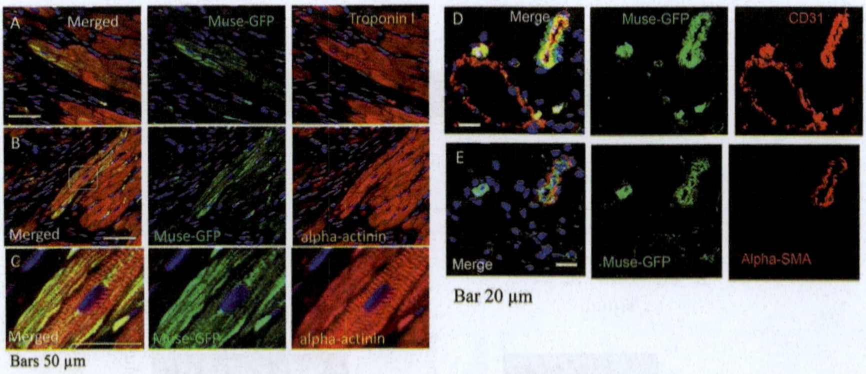

Fig. 8.5 Immunohistochemistry for cardiac troponin-I (**A**) and sarcomeric α-actinin (**B**) at 2 months. Striation-like arrangement of α-actinin (**B**) (box) is enlarged (**C**). Muse cells also spontaneously differentiated into vascular cells at 2 weeks. They expressed CD31 (**D**) and alpha-smooth muscle actin (SMA) (**E**). Bars; A–C = 50 μm; D, E = 20 μm. (Pictures adapted and modified with permission from Yamada et al. (2018), *Cir Res* [49])

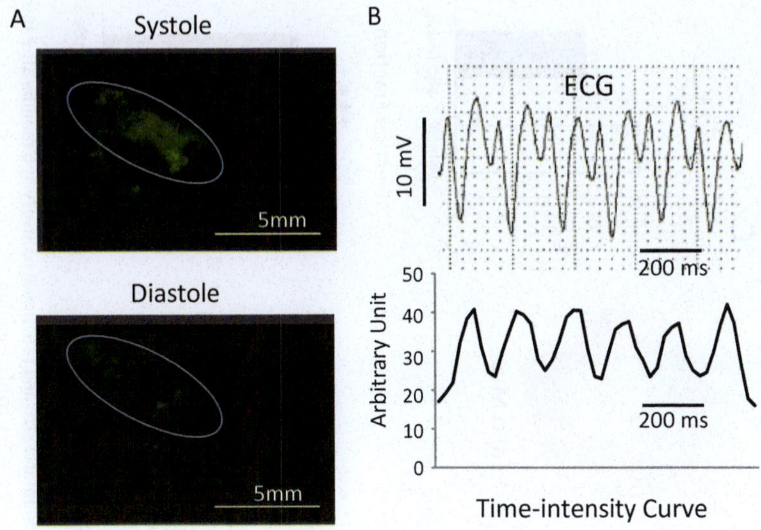

Fig. 8.6 GCaMP3-Muse cell activity in vivo. (**A**) In vivo image of GCaMP3 fluorescence in systole and diastole. (**B**) Electrocardiogram and time-intensity curve of GCaMP3 fluorescence (Pictures adapted and modified with permission from Yamada et al. (2018), *Cir Res* [49])

rescence during diastole, synchronous to electrocardiogram. This suggested that Muse cells differentiated into working cardiomyocytes with physiologic activity (Fig. 8.6A, B).

The beneficial effects of Muse cells, namely, recovery of cardiac function, reduction of infarct size, attenuation of cardiac remodeling, and suppression of fibrosis, lasted for up to 6 months after AMI in the allograft Muse cell experiment (Fig. 8.7).

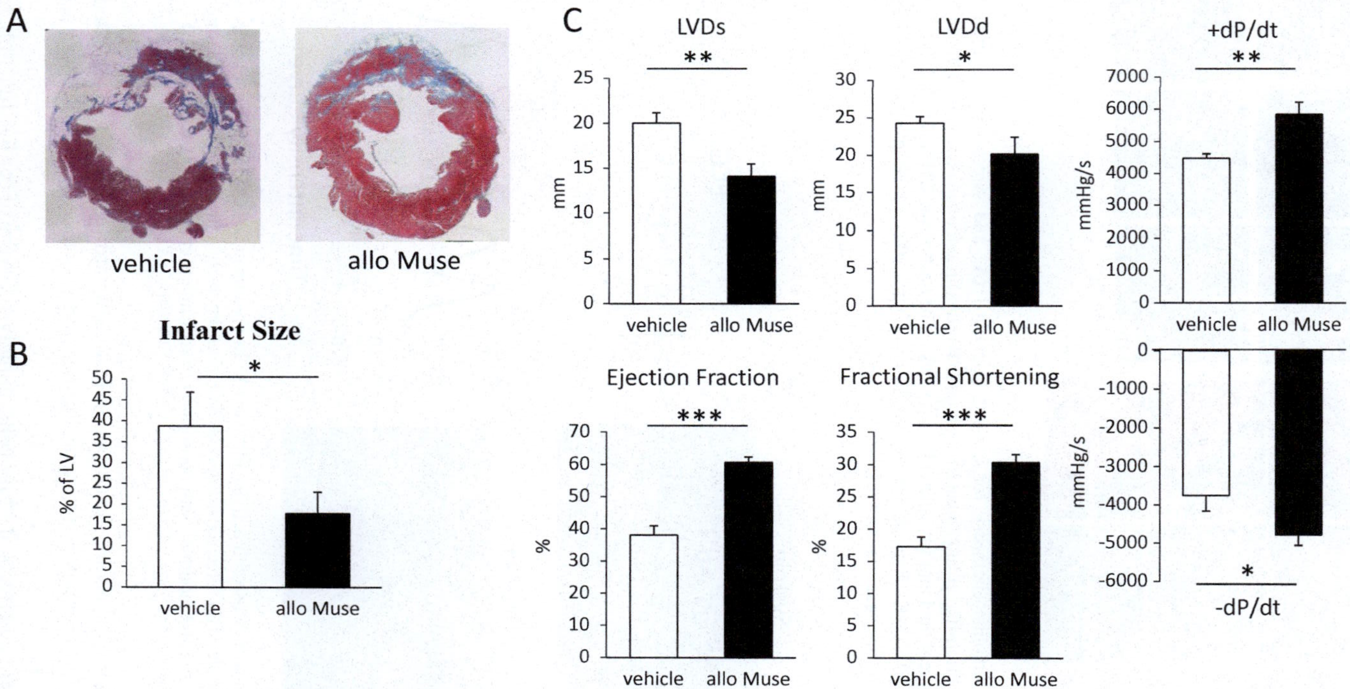

Fig. 8.7 Maintenance of the reduced infarct size and functional recovery at 6 months after intravenous injection of allogenic Muse cells. (**A**) Masson-trichrome staining and (**B**) infarct size at 6 months. (**C**) Cardiac function

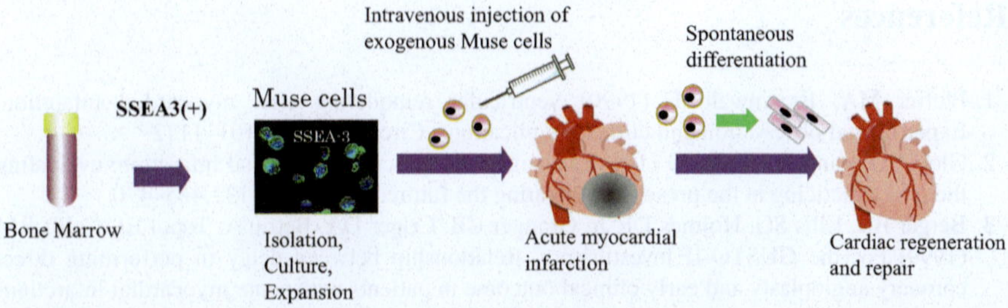

Fig. 8.8 Conceptual figure of stem cell therapy using exogenous Muse cells for the treatment of AMI

Muse cells show paracrine effects such as the production of antifibrosis/fibrinolysis factors MMP-2 and MMP-9 and trophic factors HGF and VEGF, which might have contributed to the reduction in the infarct size, through the degradation of fibrosis, anti-apoptosis, stimulation of endogenous cardiac progenitors, and neovascularization. Importantly, Muse cells not only demonstrated immunomodulatory effect similar to MSCs, i.e., conversion of naïve T cells to regulatory T cells, suppression of the differentiation of monocytes into monocyte-dendritic cell progenitors and into monocyte-dendritic cells, but ~90% of Muse cells expressed HLA-G, an immunotolerance factor expressed in the placenta during pregnancy [49]. GFP-labeled allograft Muse cells that had engrafted to the infarct border area expressed HLA-G on day 3 after AMI. These results suggest the immunotolerance and immunomodulatory effect of Muse cells [49].

All these multiple pleiotropic effects of Muse cells might have contributed to the structural and functional recovery of the heart after AMI.

A conceptual figure of cell therapy using exogenous Muse cells for the treatment of AMI is shown in Fig. 8.8.

8.6 Conclusion

After the onset of AMI, coronary reperfusion therapy by PCI is the first-line therapy to rescue the remaining viable cardiomyocytes. However, if treatment fails to reperfuse the occluded coronary artery or in long time passes until reperfusion, and consequently the infarct size is large and cardiac function is deteriorated, one of the best therapies after treatment with PCI would be regenerative therapy to reconstruct the infarcted cardiac tissue. Muse cell therapy may be a promising fundamental and epoch-making stem cell therapy for the treatment of AMI.

References

1. Pfeffer MA, Brawnwald E (1990) Ventricular remodeling after myocardial infarction. Experimental observation and clinical implications. Circulation 81:1161–1172
2. Glogar D, Yang P, Steurer G (1996) Management of acute myocardial infarction: evaluating the past, practicing in the present, elaborating the future. Am Heart J 132:465–470
3. Berger BB, Ellis SG, Holmes DR Jr, Granger CB, Criger DA, Betriu A, Topol EJ, Califf RM (1999) For the GUSTO-II investigators. Relationship between delay in performing direct coronary angioplasty and early clinical outcome in patients with acute myocardial infarction-results from the global use of strategies to open occluded arteries in acute coronary syndromes (GUSTO-IIb) trial. Circulation 100:14–20
4. Murry CE, Jennings RB, Reimer KA (1986) Preconditioning with ischemia: a delay of lethal cell injury in ischemic myocardium. Circulation 74:1124–1136
5. Yellon DM, Alkhulaifi AM, Browne EE, Pugsley WB (1992) Ischaemic preconditioning limits infarct size in the rat heart. Cardiovasc Res 26:983–987
6. Liu GS, Thornton J, Van Winkle DM, Stanley AWH, Olsson RA, Downey JM (1991) Protection against infarction afforded by preconditioning is mediated by A1 adenosine receptors in rabbit heart. Circulation 84:350–356
7. Scott RJ, Rohmann S, Braun ER, Schaper W (1990) Ischemic preconditioning reduces infarct size in swine myocardium. Circ Res 66:1133–1142
8. Noda T, Minatoguchi S, Fujii K, Hori M, Ito T, Kanmatsuse K, Matsuzaki M, Miura T, Nonogi H, Tada M, Tanaka M, Fujiwara H (1999) Evidence for the delayed effect in human ischemic preconditioning. J Am Coll Cardiol 34:1966–1974
9. Ottani F, Galvani M, Ferrini D, Sorbello F, Limonetti P, Pantoli D, Rusticali F (1995) Prodromal angina limit infarct size. A role for ischemic preconditioning. Circulation 91:291–297
10. Nakagawa Y, Ito H, Kitakaze M, Kusuoka H, Hori M, Kuzuya T, Higashino Y, Fujii K, Minamino T (1995) Effect of angina pectoris on myocardial protection in patients with reperfused anterior wall myocardial infarction: retrospective clinical evidence of "preconditioning.". J Am Coll Cardiol 25:1076–1083
11. Thornton JD, Liu GS, Olsson RA, Downey JM (1992) Intravenous pretreatment with A1-selective adenosine analogues protects the heart against infarction. Circulation 85:659–665
12. Mahiko G, Liu Y, Yang XM, Ardell JL, Cohen MV, Downey JM (1995) Role of bradykinin in protection of ischemic preconditioning in rabbit hearts. Circ Res 77:611–621
13. Wang GY, Wu S, Pei JM, Yu XC, Wong TM (2001) Kappa- but not delta-opioid receptors mediate effects of ischemic preconditioning on both infarct and arrhythmia in rats. Am J Physiol Heart Circ Physiol 280:H384–H391
14. Bankwara Z, Hale SL, Kloner RA (1994) α-Adrenoceptor stimulation with exogenous norepinephrine or release of endogenous catecholamines mimics ischemic preconditioning. Circulation 90:1023–1028
15. Tsuchida A, Liu Y, Liu GS, Choen MV, Downey JM (1994) α1-adrenergic agonist precondition rabbit ischemic myocardium independent of adenosine by direct activation of protein kinase C. Circ Res 75:576–585
16. Kariya T, Minatoguchi S, Ohno T, Yamashita K, Uno Y, Arai M, Koshiji M, Fujiwara T (1997) Hisayoshi Fujiwara infarct size-reducing effect to ischemic preconditioning is related to α1b-adrenoceptors but not to α1a-adrenoceptors in rabbits. J Cardiovasc Pharmacol 30:437–445
17. Tanaka M, Fujiwara H, Yamasaki K, Sasayama S (1994) Superoxide dismutase and N-2-mercaptopropionyl glycine attenuate infarct size limitation effect of ischaemic preconditioning in the rabbit. Cardiovasc Res 28:980–986
18. Ytrehus K, Liu Y, Downey J (1994) Preconditioning protects ischemic rabbit heart by protein kinase C activation. Am J Phys 266:H1145–H1152
19. Gross GJ, Auchampach J (1992) Blockade of ATP-sensitive potassium channels prevent myocardial preconditioning in dogs. Circ Res 70:223–233

20. Hide EJ, Thiemermann C (1996) Limitation of myocardial infarct size in the rabbit by ischaemic preconditioning is abolished by sodium 5-hydroxydecanoate. Cardiovasc Res 31:941–946
21. IONA Study Group (2002) Effect of nicorandil on coronary events in patients with stable angina: the impact of Nicorandil in Angina (IONA) randomized trial. Lancet 359:1269–1275
22. Zhao ZQ, Corvera JS, Halkos ME, Kerendi F, Wang NP, Guyton RA, Vinten-Johansen J (2003) Inhibition of myocardial injury by ischemic postconditioning during reperfusion: comparison with ischemic preconditioning. Am J Physiol Heart Circ Physiol 285:H579–H588
23. Tsang A, Hausenloy DJ, Mocanu MM, Yellon DM (2004) Postconditioning: a form of "modified reperfusion" protects the myocardium by activating the phosphatidylinositol 3-kinase-Akt pathway. Circ Res 95:230–232
24. Orlic D, Kajstura J, Chimenti S et al (2001) Mobilized bone marrow cells repair the infarcted heart, improving function and survival. Proc Natl Acad Sci USA 98:10344–10349
25. Minatoguchi S, Takemura G, Chen XH, Wang N, Uno Y, Koda M, Arai M, Misao Y, Lu C, Suzuki K, Goto K, Komada A, Takahashi T, Kosai K, Fujiwara T, Fujiwara H (2004) Acceleration of the healing process and myocardial regeneration may be important as a mechanism of improvement of cardiac function and remodeling by postinfarction granulocyte colony-stimulating factor treatment. Circulation 109:2572–2580
26. Harada M, Qin Y, Takano H, Sano M, Nishi J, Iwanaga K, Akazawa H, Kunieda T, Zhu W, Hasegawa H, Kunisada K, Nagai T, Nakaya H, Takihara K, Komuro I (2005) G-CSF prevents cardiac remodeling after myocardial infarction by activating the Jak-Stat pathway in cardiomyocytes. Nat Med 11:305–311
27. Kawada H, Fujita J, Kinjo K, Matsuzaki Y, Tsuma M, Miyatake H, Muguruma Y, Tsuboi K, Itabashi Y, Ikeda Y, Ogawa S, Okano H, Hotta T, Ando K, Fukuda K (2004) Nonhematopoietic mesenchymal stem cells can be mobilized and differentiate into cardiomyocytes after myocardial infarction. Blood 104:3581–3587
28. Valgimigli M, Rigolin GM, Cittanti C, Malagutti P, Curello S, Percoco G, Bugli AM, Della Porta M, Bragotti LZ, Ansani L, Mauro E, Lanfranchi A, Giganti M, Feggi L, Castoldi G, Ferrari R (2005) Use of granulocyte-colony stimulating factor during acute myocardial infarction to enhance bone marrow stem cell mobilization in humans: clinical and angiographic safety profile. Eur Heart J 26:1838–1845
29. Kuethe F, Figulla HR, Herzau M, Voth M, Fritzenwanger M, Opfermann T, Pachmann K, Krack A, Sayer HG, Gottschild D, Werner GS (2005) Treatment with granulocyte-colony stimulating factor for mobilization of bone marrow cells in patients with acute myocardial infarction. Am Heart J 150:115 e1–115 e7
30. Ince H, Petzsch M, Kleine HD, Schmidt H, Rehders T, Korber T, Schmichen C, Freund M, Nienaber CA (2005) Preservation from ventricular remodeling by front-integrated revascularization and stem cell liberation in evolving acute myocardial infarction by use of granulocyte-colony-stimulating factor (FIRSTLINE-AMI). Circulation 112:3097–3106
31. Ince H, Petzsch M, Kleine HD, Eckard H, Rehders T, Burska D, Kische S, Freund M, Nienaber CA (2005) Prevention of left ventricular remodeling with granulocyte colony-stimulating factor after acute myocardial infarction: final 1-year results of the front-integrated revascularization in evolving acute myocardial infarction by granulucyte colony-stimulating factor (FIRSTLINE-AMI) trial. Circulation 112:173–180
32. Ripa RS, Jorgensen E, Wang Y, Thune JJ, Nilsson JC, Sondergaard L, Johnsen HE, Kober L, Grande P, Kastrup J (2006) Stem cell mobilization induced by subcutaneous granulocyte-colony stimulating factor to improve cardiac regeneration after acute ST elevation myocardial infarction. Result of the double-blind, randomized, placebo-controlled stem cells in myocardial infarction (STEMMI) trial. Circulation 113:1983–1992
33. Zohlnhofer D, Ott I, Mehilli J, Schomig K, Michalk F, Ibrahim T, Meisetschlager G, von Wedel J, Bollwhein H, Seyfarth M, Dirschinger J, Schmitt C, Schwaiger M, Kastrati A, Schomig A, REVIVAL-2 Investigators (2006) Stem cell mobilization by granulocyte colony-stimulating factor in patients with acute myocardial infarction: a randomized controlled trial. JAMA 295:1003–1010

34. Suzuki K, Nagashima K, Arai M, Uno Y, Misao Y, Takemura G, Nishigaki K, Minatoguchi S, Watanabe S, Tei C, Fujiwara H (2006) Effect of granulocyte colony-stimulating factor treatment at a low dose but a long duration in patients with coronary heart disease. Circ J 70:430–437
35. Kranz SB (1991) Erythropoietin. Blood 77:419–434
36. Eschbach JW (1989) The anemia of chronic renal failure: pathophysiology and the effects of recombinant erythropoietin. Kidney Int 35:134–148
37. Calvillo L, Latini R, Kajstura J, Leri A, Anversa P, Ghezzi P et al (2003) Recombinant human erythropoietin protects the myocardium from ischemia-reperfusion injury and promotes beneficial remodeling. PNAS 100:4802–4806
38. Parsa CJ, Matsumori A, Kim J, Riel RU, Pascal LS, Walton GB et al (2003) A novel protective effect of erythropoietin in the infarcted heart. J Clin Invest 112:999–1007
39. Gao D, Ning N, Niu X, Dang Y, Dong X, Wei J, Zhu C (2012) Erythropoietin treatment in patients with acute myocardial infarction: a meta-analysis of randomized controlled trials. Am Heart J 164:715–721
40. Hori E, Hayakawa Y, Hayashi T, Hori S, Okamoto S, Shibata T et al (2016) Mobilization of pluripotent multilineage-differentiating stress-enduring cells in ischemic stroke. J Stroke Cerebrovasc Dis 25:1473–1481
41. Tanaka T, Nishigaki K, Minatoguchi S, Nawa T, Yamada Y, Kanamori H, Mikami A, Ushikoshi H, Kawasaki M, Dezawa M, Minatoguchi S (2018) Mobilized Muse cells after acute myocardial infarction predict cardiac function and remodeling in the chronic phase. Circ J 82:1319–1326
42. George JC (2010) Stem cell therapy in acute myocardial infarction: a review of clinical trials. Transl Res 155:10–19
43. Fisher SA, Zhang H, Doree C, Mathur A, Martin-Rendon E (2015) Stem cell treatment for acute myocardial infarction. Cochrane Database Syst Rev 9:CD006536
44. Kuroda Y, Kitada M, Wakao S, Nishikawa K, Tanimura Y, Makinoshima H, Goda M, Akashi H, Inutsuka A, Niwa A, Shigemoto T, Nabeshima Y, Nakahata T, Nabeshima Y, Fujiyoshi Y, Dezawa M (2010) Unique multipotent cells in adult human mesenchymal cell populations. Proc Natl Acad Sci USA 107:8639–8643
45. Wakao S, Kitada M, Kuroda Y, Shigemoto T, Matsuse D, Akashi H, Tanimura Y, Tsuchiyama K, Kikuchi T, Goda M, Nakahata T, Fujiyoshi Y, Dezawa M (2011) Multilineage-differentiating stress-enduring (Muse) cells are a primary source of induced pluripotent stem cells in human fibroblasts. Proc Natl Acad Sci USA 108:9875–9880
46. Kuroda Y, Wakao S, Kitada M, Murakami T, Nojima M, Dezawa M (2013) Isolation, culture and evaluation of multilineage-differentiating stress-enduring (Muse) cells. Nat Protoc 8:1391–1415
47. Dezawa M (2016) Muse cells provide the pluripotency of mesenchymal stem cells: direct contribution of Muse cells to tissue regeneration. Cell Transplant 25:849–861
48. Alessio N, Ozcan S, Tatsumi K, Murat A, Peluso G, Dezawa M, Galderisi U (2016) The secretome of MUSE cells contains factors that may play a role in regulation of stemness, apoptosis and immunomodulation. Cell Cycle 16:33. https://doi.org/10.1080/15384101.2016.1211215
49. Yamada Y, Wakao S, Kushida Y, Minatoguchi S, Mikami A, Higashi K, Baba S, Shigemoto T, Kuroda Y, Kanamori H, Amin M, Kawasaki M, Nishigaki K, Taoka M, Isobe T, Muramatsu C, Dezawa M, Minatoguchi S (2018) S1P-S1PR2 axis mediates homing of Muse cells into damaged heart for long lasting tissue repair and functional recovery after acute myocardial infarction. Circ Res 122:1069–1083
50. Nagaya N, Fujii T, Iwase T, Ohgushi H, Itoh T, Uematsu M, Yamagishi M, Mori H, Kangawa K, Kitamura S (2004) Intravenous administration of mesenchymal stem cells improves cardiac function in rats with acute myocardial infarction through angiogenesis and myogenesis. Am J Physiol Heart Circ Physiol 287:H2670–H2676
51. Freyman T, Polin G, Osman H, Crary J, Lu M, Cheng L, Palasis M, Wilensky RL (2006) A quantitative, randomized study evaluating three methods of mesenchymal stem cell delivery following myocardial infarction. Eur Heart J 27:1114–1122

Chapter 9
Application of Muse Cell Therapy to Stroke

Kuniyasu Niizuma, Cesar V. Borlongan, and Teiji Tominaga

Abstract Stroke is defined as a sudden onset of neurologic deficits arising from cerebrovascular complications. It is the second common cause of death around the world and the major cause of disability. Because brain is an organ with complicated neural networks and neurons are highly differentiated, it has been traditionally considered to possess a limited potential for regeneration. The number of stroke patients is increasing, and stroke represents a serious problem from the viewpoint of the national medical economy. Even with the current sophisticated treatments, more than half of stroke patient survivors remain disabled. Therefore, it is imperative to develop a new treatment for promoting functional recovery and repair of the lost neurological circuit. Multilineage-differentiating stress-enduring (Muse) cells are endogenous non-tumorigenic stem cells with pluripotency. After transplantation, Muse cells recognize the injured site through their specific receptor for damage signal, home preferentially into these tissues and spontaneously differentiate into tissue-compatible cells to replace the lost cells, and repair the tissue, delivering functional and structural regeneration. These properties are desirable for the treatment of strokes and advantageous compared to other stem cell therapies. Here, we describe the current status of stem cell therapies for stroke and future possibilities of Muse cell therapy.

K. Niizuma (✉)
Department of Neurosurgical Engineering and Translational Neuroscience, Tohoku University Graduate School of Biomedical Engineering, Sendai, Japan

Department of Neurosurgical Engineering and Translational Neuroscience, Tohoku University Graduate School of Medicine, Sendai, Japan

Department of Neurosurgery, Tohoku University Graduate School of Medicine, Sendai, Japan
e-mail: niizuma@nsg.med.tohoku.ac.jp

C. V. Borlongan
Department of Neurosurgery and Brain Repair, University of South Florida College of Medicine, Tampa, FL, USA

T. Tominaga
Department of Neurosurgery, Tohoku University Graduate School of Medicine, Sendai, Japan

M. Dezawa (ed.), *Muse Cells*, Advances in Experimental Medicine and Biology 1103, https://doi.org/10.1007/978-4-431-56847-6_9

Keywords Brain · Brain ischemia · Cell therapies · Cerebral infarction · Intracerebral hemorrhage · Muse cells · Regeneration · Stem cells · Stroke

9.1 Introduction

The brain is the key organ of the central nervous system. The overall total number of human neocortical neurons and glial cells is 49.3 billion in females and 65.2 billion in males, and they form complicated networks [1]. Even though neurogenesis is enhanced at the subventricular zone of the lateral ventricles and the subgranular zone of the hippocampal dentate gyrus after various kinds of brain injury including stroke [2, 3], the nervous system is highly complicated differentiated system and such endogenous neurogenesis is insufficient to deliver robust functional recovery and reconstruction of neuronal circuit. Thus, the brain has been thought to mount a very limited potential for regeneration in the baseline level. Ischemia-induced multipotent stem cells (iSCs) have been recently identified and isolated from the post-stroke human brain [4]. iSCs are localized near blood vessels within poststroke areas where apoptotic/necrotic neurons are located. However, these endogenous proliferating cells are not sufficient to repair stroke, and no robust and stable functional recovery has been achieved. iSCs actually undergo cell death without appropriate support [5]. Thus either appropriate support for endogenous neurogenesis or exogenous administration of stem cells is required to achieve clinically relevant neurogenesis and reconstruction of neuronal circuit.

9.2 Definition and Treatment of Stroke

Stroke is defined as sudden onset of neurologic deficits due to cerebrovascular complications. It is the second common cause of death around the world and major cause of disability [6]. It consists of ischemic and hemorrhagic strokes; hemorrhagic is further divided into subarachnoid hemorrhage (SAH) and intracerebral hemorrhage (ICH) (Fig. 9.1). Incidence of ischemic stroke is higher; around 80% of strokes are ischemic, and 20% of them are hemorrhagic.

Ischemic stroke results from shortage of cerebral blood flow because of the embolism, atherothrombosis, or small vessel disease. Recently, super-acute revascularization therapy, including intravenous administration of recombinant tissue plasminogen activator and mechanical thrombectomy, has been developed, and those who are successfully treated showed dramatic recovery. However, these revascularization therapies are applicable to less than 10% of patients because of the

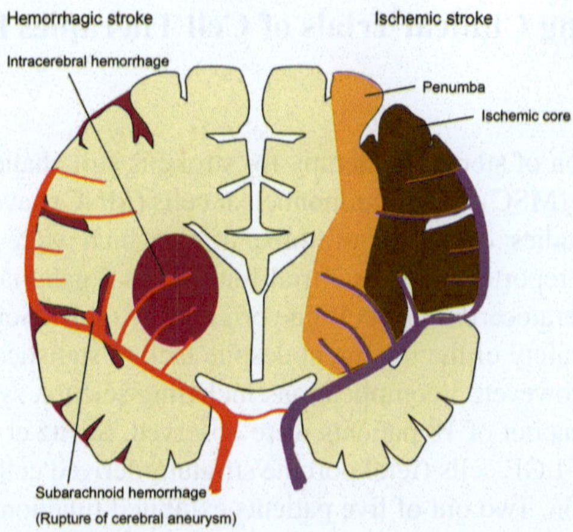

Fig. 9.1 Types of stroke. Stroke can be divided into hemorrhagic (left) and ischemic (right) stroke

short therapeutic time window. Those without revascularization therapy develop brain infarction, and more than half of the patients remain disabled. Thus, it is imperative to develop a new treatment to promote functional recovery and restore the lost neurological circuit. The role of stem cell therapy for ischemic stroke is to salvage of dying brain or regeneration of injured brain, offering a wider therapeutic window after stroke onset.

SAH results from rupture of cerebral aneurysms. Ruptured aneurysms should be treated surgically or endovascularly to avoid re-rupture which cause extremely high mortality. In the subacute phase, the patients develop cerebral vasospasm, which may cause severe brain infarction. Neuronal damage after SAH can be divided into primary damage due to sudden increase of the intracranial pressure and brain infarction due to vasospasm. Possible targets of cell therapy are regeneration of primary brain damage, control of cerebral vasospasm by the protective effect of stem cells to endothelial cells [7], and regeneration of brain infarction due to vasospasm.

ICH results from rupture of weakened blood vessels inside the brain mainly because of hypertension. Bleeding causes neuronal damage because of the direct compression as well as secondary inflammation. Routinely, intracerebral hemorrhage is treated conservatively. Surgery is performed only if the patients have uncontrollable high intracranial pressure or signs of cerebral herniation. Stem cell therapy may be used to control secondary brain injury by regulating neuroinflammation [8] or regeneration of damaged brain.

9.3 Pioneering Clinical Trials of Cell Therapies for Ischemic Stroke

Clinical application of stem cell therapy for stroke is still challenging, and mesenchymal stem cells (MSC) and /or mononuclear cells (MNC) have been mainly used. Initial clinical studies consisted of three trials which were reported in 2005. Kondziolka et al. reported a phase 2 trial for the basal ganglia lesion using LBS-neurons (human teratocarcinoma cell line origin, Layton BioScience, Inc.) [9]. The authors reported safety of the treatment despite lack of statistical meaningful functional recovery. However, 3 complications including seizure, syncope, and chronic subdural hematoma out of 18 patients were observed. Savitz et al. also reported an open-label trial of LGE cells (fetal porcine striatum-derived cells, Genvec, Inc.) for basal ganglia lesion. Two out of five patients exhibited functional improvements in speech, language, and/or motor impairments, but two had complications including temporary deterioration of motor deficits and seizures, and the study was terminated [10]. Bang et al., reported autologous MSC treatment for MCA territory infarction with severe neurological deficits. They demonstrated no adverse effects of MSC treatment, but no functional improvement was observed [11].

Various studies followed the first three trials, and the major ones are summarized in Table 9.1 [12–24]. They all used either MNC or MSC. The administration routes varied: intracerebral, intravenous, and intra-arterial. Most studies used autologous cells and subacute administration of the cells. Two reports monitored the biodistribution of transplanted BM-MNC labeled with technetium-99 m [17, 23]. The transplanted cells accumulated in the infarct area 2 h after the onset but rapidly decreased by 24 h, and then majority of the grafted cell deposition shifted to the liver and spleen.

For safety, all studies in Table 9.1, except the report of Savitz et al. [10] which used porcine cells, reported safety of the stem cell treatment. However, potential risks of cell therapy should be recognized, and the tumorigenesis might occur long after cell transplantation. Amariglio et al. [25] reported a boy with telangiectasia, who was treated by intracerebellar and intrathecal injection of human fetal neural stem cells. Four years after the first treatment, the patient was diagnosed with a multifocal brain tumor. The biopsied tumor was diagnosed as a glioneuronal neoplasm. Human leukocyte antigen typing demonstrated that the tumor was of non-host origin suggesting it was derived from the transplanted neural stem cells.

From these reported clinical trials of cell therapy for ischemic stroke, MNC and MSC seem to be safe with no obvious adverse effects, but their effectiveness was not widely confirmed by randomized studies. Based on previous trials, large clinical trial using MultiStem® (HLM051) which are allogenic cell products, Treatment Evaluation of Acute Stroke for Using in Regenerative Cell Elements (TREASURE) trial [26], has started in Japan in November 2017 targeting moderate to severe stroke patients within 18–36 h of the onset. In parallel with the TREASURE trial, MASTERS-2 trial, targeting patients similar to TREASURE, is also planned in North America and Europe.

Table 9.1 Clinical trials of stem cell therapy for ischemic stroke

Authors	Year	Number of patients/ controls	Route	Cell type	Cell dose	Time window	Result
NSC							
Kondziolka et al. [9]	2005	14/4	IC	Allogenic LBS neurons	5 or 10×10^6	1–6 years	Safe
MSC							
Bang et al. [11]	2005	5/25	IV	Autologous MSC	1 × 10^8	1–2 months	Safe and feasible
Lee et al. [12]	2010	16/36	IV	Autologous MSC	1 × 10^8	1–2 months	Safe and feasible
Honmou et al. [13]	2011	12/0	IV	Autologous MSC	0.5–5 × 10^8	6 months	Safe and feasible
Bhasin et al. [14]	2011	6/6	IV	Autologous MSC	5–6 × 10^7	3–12 months	Safe and feasible
Diez-Tejedor et al. [15]	2014	10/10	IV	Allogenic MSC	NA	2 weeks	Safe and effective
Steinberg et al. [16]	2016	18/0	IC	Allogenic modified MSC (SB632)	$2–10 \times 10^6$	2 years	Safe and effective
MNC							
Barbosa da Fonseca et al. [17]	2009	1/0	IA	Autologous MNC	5×10^8	2 months	Safe
Suárez-Monteagudo et al. [18]	2009	5/0	IC	Autologous MNC	1.4–5.5×10^7	1–10 years	Safe
Savitz et al. [19]	2011	10/0	IV	Autologous MNC	1×10^7/kg	24–72 h	Safe and feasible
Battistela et al. [20]	2011	6/0	IA	Autologous MNC	1.5×10^8	90 days	Safe and feasible
Moniche et al. [21]	2012	10/10	IA	Autologous MNC	1.59×10^8	5–9 days	Safe and feasible
Friedrich et al. [22]	2012	20/0	IA	Autologous MNC	2.3×10^8	3–7 days	Safe and feasible
Rosado-de-Castro et al. [23]	2013	12/0	IA/IV	Autologous MNC	$1–5 \times 10^8$	19–89 days	Safe
Taguchi et al. [24]	2015	12/0	IV	Autologous MNC	2.5–3.4×10^8	7–10 days	Safe and effective
Other cells							
Savitz et al. [10]	2005	5/0	IC	Porcine LGE cells	$3–5 \times 10^6$	1.5–10 years	Terminated

NSC neural stem cells, *IC* intracerebral, *IV* intravenous, *IT* intrathecal, *BM* bone marrow, *MNC* mononuclear cells, *MSC* mesenchymal stem cells

9.4 Properties of Muse Cells and Advantages for Stroke Treatment

For further advancement of stem cell therapy, we have focused on multilineage-differentiating stress-enduring (Muse) cells which are endogenous non-tumorigenic stem cells exhibiting pluripotency, collectable as pluripotent surface marker, SSEA-3, from various kinds of sources such as the bone marrow (BM), adipose tissue, and dermis, as well as from commercially released cultured fibroblasts. Their telomerase activity level is equivalent to that of somatic cells; thus they are suggested to display minimum safety concerns. While their proliferative activity is not exponential, their doubling time is ~1.3 day/cell division, and therefore, clinical scale is available in Muse cells. After transplantation, Muse cells recognize the injured site through their specific receptor for damage signal, home preferentially into these injured tissues by intravenous injection and spontaneously differentiate into tissue-compatible cells to replace the lost cells, and repair the tissue, delivering functional and structural regeneration. It is envisioned that applying this strategy to stroke may involve topical or intravenously administrated Muse cells which are expected to preferentially migrate to and home into the damaged neural tissue without dispersing to unnecessary sites, and spontaneously differentiate to neuronal cells to repair neuronal circuit. If such approach is successful, an efficient and effective treatment protocol to treat strokes may involve collecting Muse cells by SSEA-3, expand them and treating patients by systemic administration, rendering gene introduction and/or induction into purposive cells in cell processing center unnecessary. Muse cells are also expected to engraft in the host tissue for prolonged period because of their immune tolerance [27].

Here, we describe preclinical studies investigating Muse cell therapy for the treatment of ischemic stroke. Two models were used as cerebral infarction models: transient middle cerebral artery occlusion (tMCAO) to mimic large brain infarction in rats and lacunar infarction model which induces small subcortical infarct on the corticospinal tract and pure motor paresis in mice [28]. In both models, Muse cells (normal human dermal fibroblast or NHDF-derived Muse cells for tMCAO, BM-MSC-derived Muse cells for lacunar infarction) were injected into the peri-infarct areas either acute or subacute stage after the onset. Transplanted Muse cells migrated into the injured tissue, which was compatible to the migration and homing capacity of the Muse cells described above. In contrast to MSCs or non-Muse cells (the remainder of the cells after the isolation of Muse cells from NHDFs), Muse cells survived in the hostile stroke environment and differentiated spontaneously into neurons (~60%) and oligodendrocytes (~20%). Muse cells were suggested to differentiate into astrocytes, but such phenotypic commitment was not observed. Our finding that no MSCs or non-Muse cells survived in the stroke brain seems to be compatible to clinical trials which monitored the biodistribution of transplanted BM-MNC labeled with technetium-99 m [17, 23]. In summary, transplanted Muse cells migrated into the injured site, survived for long and spontaneously differentiated into neurons and oligodendrocytes after cerebral ischemia.

9.5 Muse Cell Reconstruct Corticospinal Tract

Intracerebral injection (peri-infarct cortex and basal ganglia for tMCAO model, and peri-infarct basal ganglia for lacunar infarction model) of human Muse cells was reported to reconstruct corticospinal tract after tMCAO [29] and lacunar infarction [30]. In both models, Muse cell treatment significantly improved the motor function at 8–12 weeks follow-up. To investigate detailed mechanisms, retrograde tracer, Fluorogold, was injected into the contralateral dorsal funiculus of C1 level spinal cord 84 days after Muse cell transplantation in the mice tMCAO model. Seven days after Fluorogold injection, green fluorescent protein (GFP)-labeled Muse cells labeled with Fluorogold were detected in the ipsilateral peri-infarct motor cortex, suggesting that Muse cells extended neurites into contralateral spinal cord through pyramidal decussation (Fig. 9.2). Moreover, interruption of nerve axons in the pyramidal tract was confirmed after lacunar infarction (Fig. 9.3A), where Muse cells formed new synapses connecting with motor cortex neurons and participated in the restoration of pyramidal tract, confirmed by anterograde labeling of motor cortical neurons using dextran (Fig. 9.3B) [30]. With anterograde labeling of the Muse cells, dextran-labeled axons were detected at level of midbrain, medulla in the ipsilateral side, and in the cervical spinal cord in the contralateral side, clearly demonstrated the formation of pyramidal decussation (Fig. 9.3C) [30]. Muse cell-derived neurites were positive for glutamatergic neuronal marker, suggesting differentiation of Muse cells into glutamatergic neuron. Human Muse cells that integrated into the motor cortex extended neurites into the corticospinal tract, at least to C1 level of the spinal cord (Fig. 9.4).

9.6 Muse Cells Reconstruct the Sensory Cortex

Muse cells were also reported to contribute to the regeneration of sensory cortex after tMCAO [29]. Integration of GFP-labeled human Muse cells locally injected to peri-infarct area was recognized in the ipsilateral sensory cortex at 84 days (Fig. 9.5A). Furthermore, synaptophysin was detected adjacent to dendrite-like processes belonging to GFP-labeled human Muse cells in the sensory cortex (Fig. 9.5B). The recovery of sensory system was further confirmed by electrophysiological approach, sensory evoked potential (SEP). Bilateral hind limb SEPs were recorded at the primary sensory cortex 84 days after transplantation (Fig. 9.5C). After Muse cell injection, amplitude of SEP was significantly high compared to the vehicle-treated rats (Fig. 9.5C–E). Thus, Muse cells injected after tMCAO exerted the capacity to reconstruct the sensory circuits in both histological and electrophysiological evaluations (Fig. 9.6).

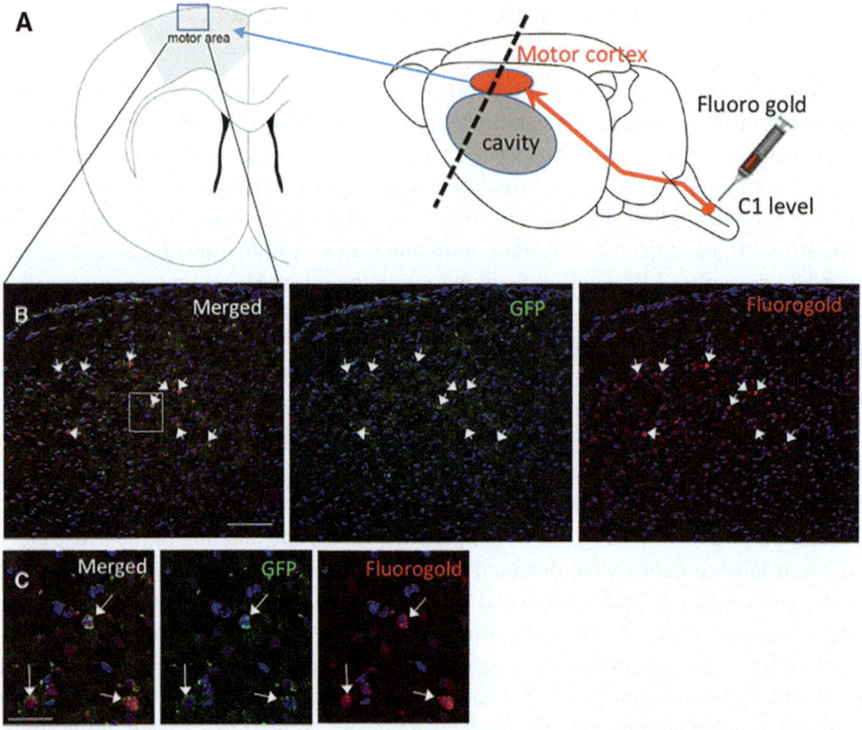

Fig. 9.2 Reconstruction of the corticospinal tract by Muse cell transplantation after transient middle cerebral artery occlusion in rats. Retrograde labeling of Muse cells that integrated into the motor cortex. (**A**) Schematic diagram illustrating the injection site of the Fluorogold at contralateral dorsal funiculus of cervical spinal cord (C1 level) at 84 days after transplantation and the site of detection at the ipsilateral motor cortex at day 91. (**B**) Fluorescent staining with Fluorogold (red) and green fluorescent protein (GFP) (green) revealed that GFP-positive-Muse cells in the ipsilateral motor cortex were retrogradely labeled by Fluorogold (arrows). Scale bar, 100 μm. (**C**) Enlarged image of the white rectangular box in (**B**). Arrows indicate double-positive cells. Scale bar, 50 μm. (Cited from Ref. [29])

9.7 Integrated Muse Cells Directly Regenerate Neural Tissue

Double immunofluorescence of GFP-labeled human Muse cells and neural markers revealed that transplanted Muse cells expressed neuronal markers (NeuN, 62.2%; MAPs 2, 30.6%) and oligodendrocyte marker (GST-pi, 12.1%) after transplantation into lacunar infarction (Fig. 9.7) [30]. Differentiation into astrocytes, however, was not observed (Fig. 9.7).

The time course of functional recovery after Muse cell treatment basically differs from that of bystander effects of MSC or MNC. Behavioral recovery produced by MSC or MNC treatment is considered to involve graft persistence, whereas the bystander effects recapitulate endogenous neurogenesis and vascular regeneration [31–33]. Figure 9.8 demonstrates behavioral analysis after Muse cell transplantation

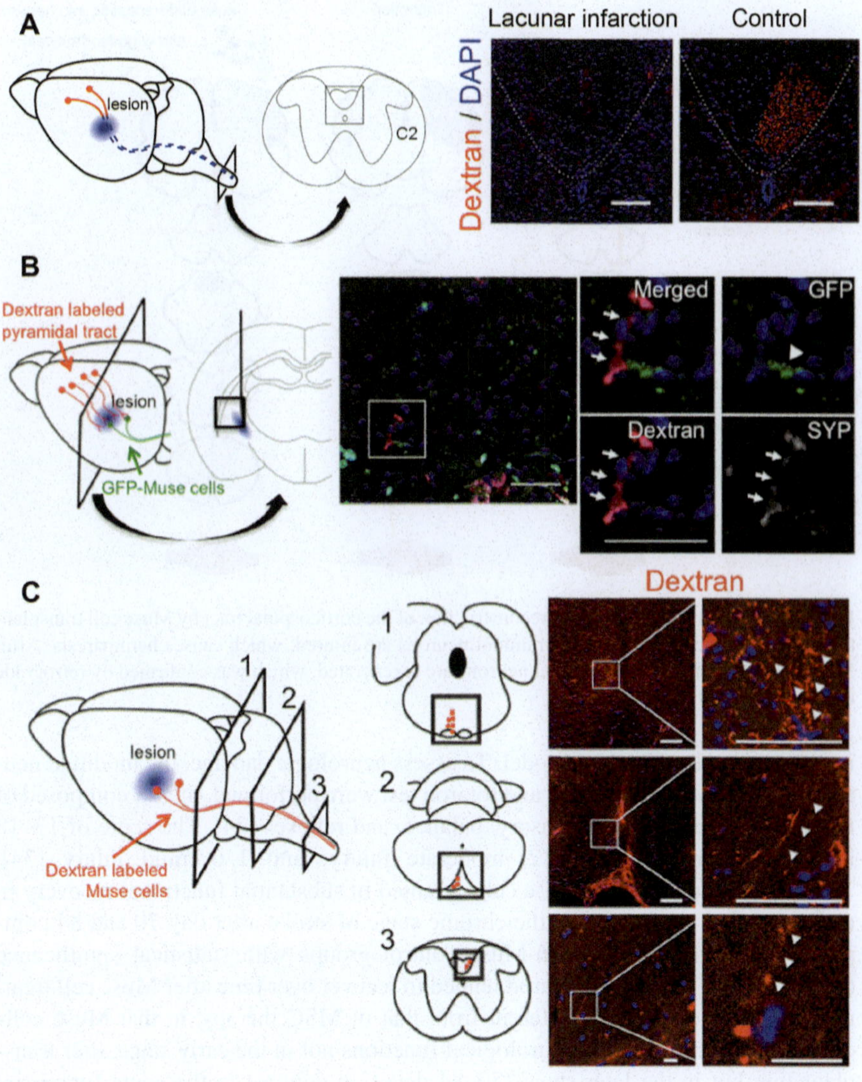

Fig. 9.3 Reconstruction of the corticospinal tract by Muse cell transplantation after lacunar infarction in mice. (**A**) Confirmation of lacunar infarction model and interruption of the motor nerve neurons. Lacunar infarction was made by administration of 2 vasoconstrictive peptides (endothelin-1 and N(omega)-nitro-L-arginine methyl ester). Eight weeks after administration of 2 peptides, the axonal interruption was confirmed by dextran tracing at cervical spinal cord C1–2 level (red signal). Scale bars, 100 μm. (**B**) Motor cortical neurons were anterogradely labeled with dextran. Green fluorescent protein (GFP)-positive Muse cells (arrowhead) transplanted nearby the lesion was observed to connect to dextran-labeled motor neuron axons (red, arrows), which merged with synaptophysin (SYP, arrows), suggesting that Muse cells connected with motor cortex neurons and participated in the restoration of pyramidal tract 8 weeks after Muse cell transplantation. Scale bars, 100 μm. (**C**) Muse cells were anterogradely labeled with dextran. Dextran-labeled axons (red, arrowheads) were detected at level 1: midbrain and level 2: medulla in the ipsilateral side and in cervical spinal cord at level 3 in the contralateral side. Scale bars, 50 μm (levels 1 and 2); and 10 μm (level 3). (Modified from Ref. [30])

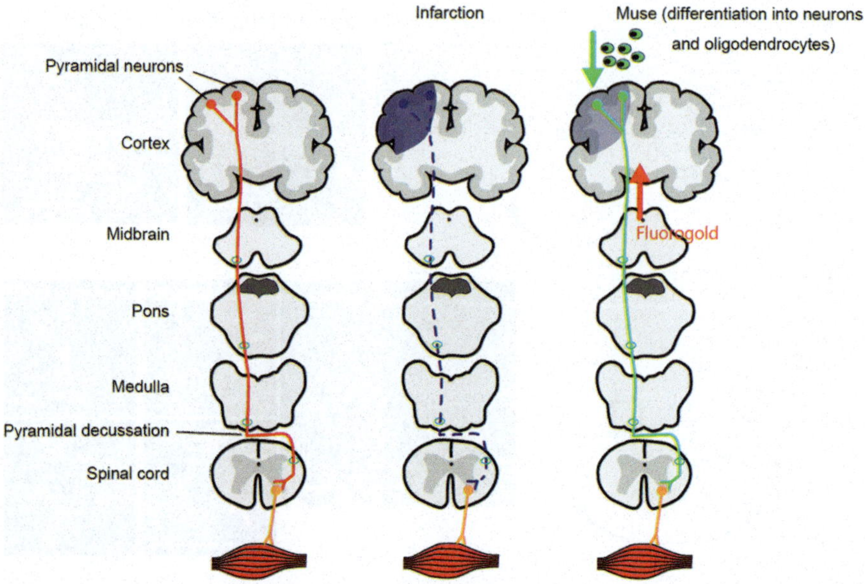

Fig. 9.4 Schematic drawing of the reconstruction of the corticospinal tract by Muse cell transplantation. After cerebral infarction, pyramidal neurons are injured, which causes hemiparesis. After Muse cell transplantation, pyramidal neurons are regenerated, which was confirmed by retrograde tracer, Fluorogold

in rats subjected to tMCAO model. To assess neurologic damage, the modified neurologic severity score (mNSS) and rotarod test were performed. mNSS composed of five tests including motor, sensory, balance, and reflexes [34]. The score of 13–18 indicates severe injury; 7–12, moderate injury; and 1–6, mild injury [34]. Transplantation of human Muse cells resulted in substantial functional recovery in the mNSS and rotarod test at the chronic stage of stroke after day 70 and 84 compared to the vehicle and non-Muse control groups with statistical significance (Fig. 9.8) [29]. mNSS and rotarod tended to recover over time after Muse cell treatment. This time course is different from that of MSC therapy, in that Muse cells dramatically improved the neurological functions not in the early stage after transplantation but in the later stage (70–84 days), in contrast to the acute functional recovery seen with the bystander effect of MSC. Because the process of engraftment, differentiation, axonal regeneration, synapse formation, and reconstruction of the neural network take time, mechanisms and time course of Muse cell treatment represent therapeutic features distinct from those of MSC.

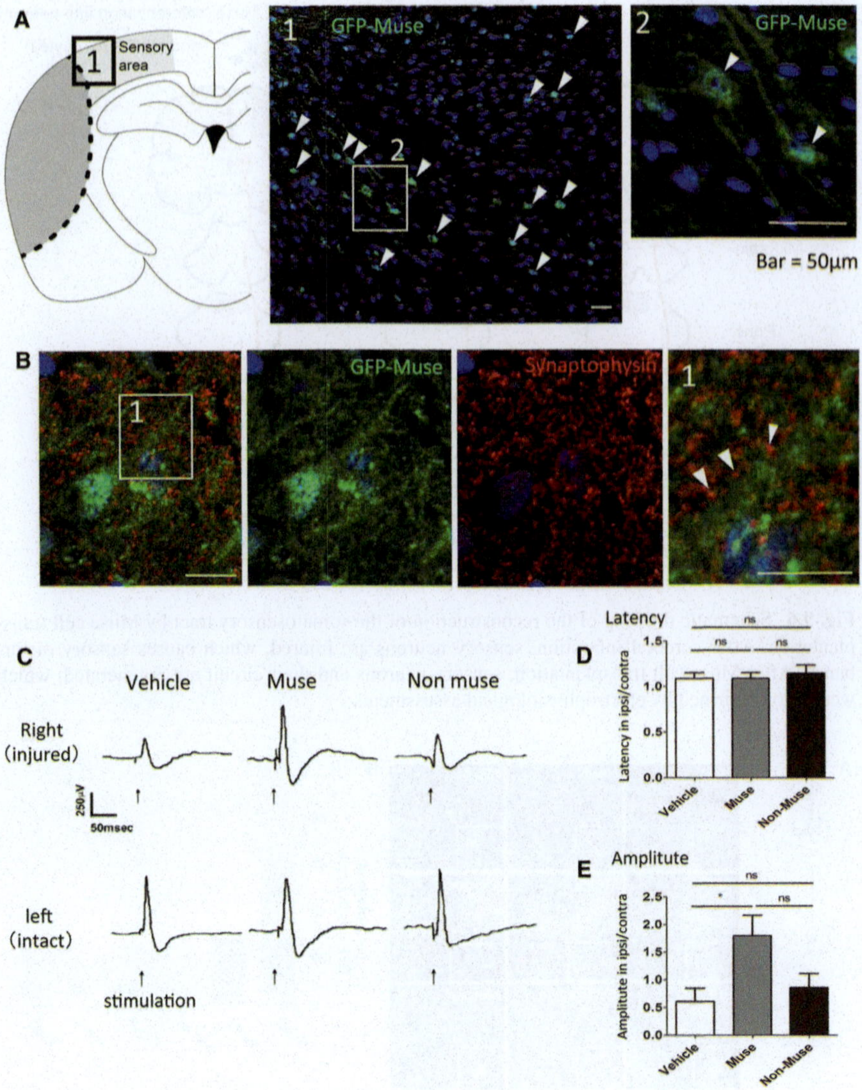

Fig. 9.5 Quantitative analysis of electrophysiologic data 84 days after transplantation. (**A**) Green fluorescent protein (GFP)-positive Muse cells detected at the ipsilateral sensory cortex at day 84 (arrowheads). 1 and 2 are enlarged images of each corresponding box areas. Scale bars, 50 μm. (**B**) Synaptophysin (red) was detected adjacent to dendrite-like structure of GFP-positive Muse cells (green) in the sensory cortex (arrowheads). Scale bars, 50 μm. (**C**) Representative somatosensory evoked potentials (SEP) in the vehicle, Muse, and non-Muse groups. (**D**) Ratio of SEP latency on ipsilateral side to that on contralateral side. (**E**) Ratio of SEP amplitude on ipsilateral side to that on contralateral side. The amplitude of the injured side was significantly high in the Muse groups compared with vehicle groups. *$p < 0.05$. (Modified from Ref. [29])

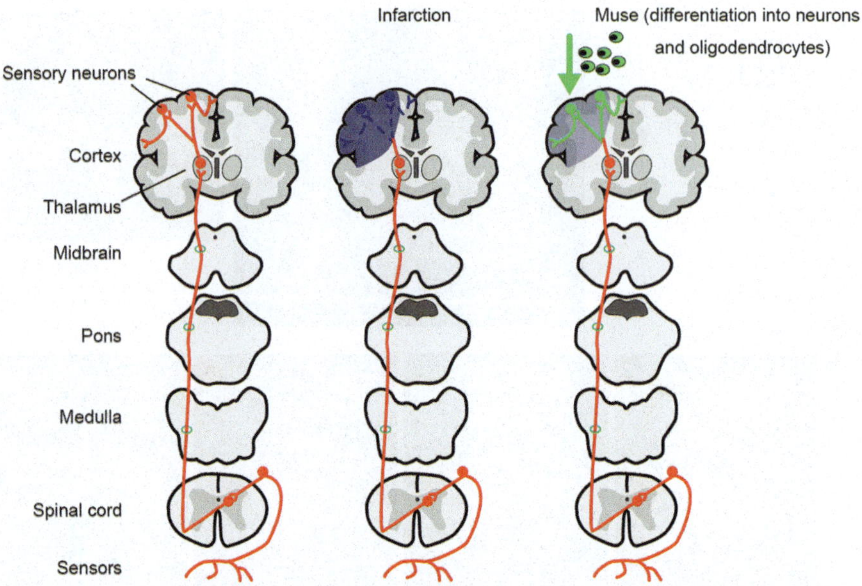

Fig. 9.6 Schematic drawing of the reconstruction of the somatosensory tract by Muse cell transplantation. After cerebral infarction, sensory neurons are injured, which causes sensory disturbance. After Muse cell transplantation, sensory neurons and their circuit are regenerated, which were also confirmed by electrophysiological assessments

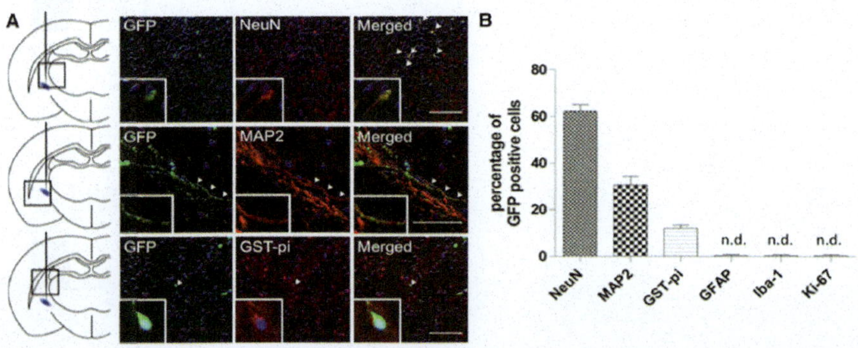

Fig. 9.7 Differentiation of Muse cells 8 weeks after engraftment of lacunar infarction. (**A**) Green fluorescent protein (GFP)-labeled Muse cells, which were positive for NeuN (red), MAP 2 (red), and GST-pi (red), were detected near the lesion site (arrowheads). Scale bars, 100 μm. (**B**) The percentage of each markers in GFP-positive cells. (Modified from Ref. [30])

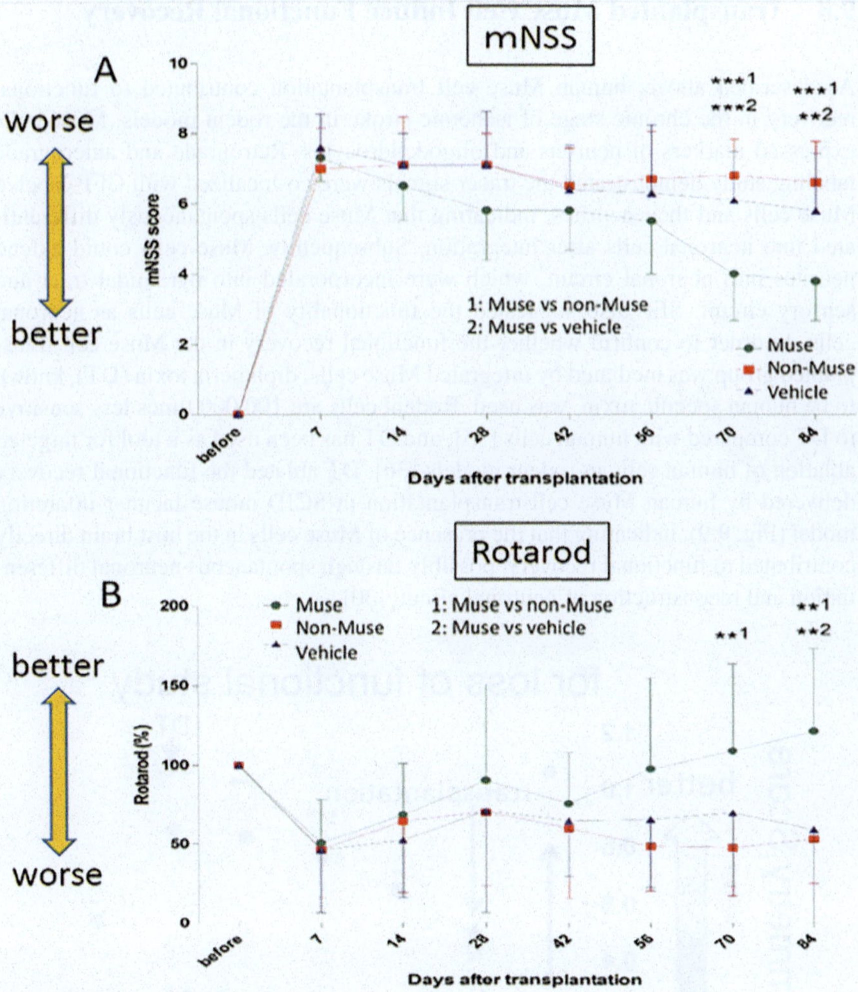

Fig. 9.8 Behavioral analysis after Muse cell transplantation in rats tMCAO model
Transplantation of human Muse cells resulted in substantial functional recovery in the modified neurologic severity score (mNSS, **A**) and rotarod test (**B**) at the chronic stage of stroke after day 70 and 84. $**p < 0.01$, $***p < 0.001$. (Modified from Ref. [29])

9.8 Transplanted Muse Cell Induce Functional Recovery

As described above, human Muse cell transplantation contributed to functional recovery in the chronic stage of ischemic stroke in the rodent models. Muse cells expressed markers of neurons and oligodendrocytes. Retrograde and anterograde labeling study demonstrated the tracer signals were co-localized with GFP-labeled Muse cells and their neurites, indicating that Muse cells spontaneously differentiated into neuronal cells after integration. Subsequently, Muse cells could extend neurites into neuronal circuit, which were incorporated into pyramidal tract and sensory circuit. SEP also suggested the functionality of Muse cells as neuronal cells. In order to confirm whether the functional recovery in the Muse cell transplanted group was mediated by integrated Muse cells, diphtheria toxin (DT), known to be human specific toxin, was used. Rodent cells are 100,000 times less sensitive to DT compared with human cells [35], and DT has been used as a tool for targeted ablation of human cells in rodent models [36]. DT ablated the functional recovery delivered by human Muse cell transplantation in SCID mouse lacunar infarction model (Fig. 9.9), indicating that the presence of Muse cells in the host brain directly contributed to functional recovery, possibly through spontaneous neuronal differentiation and reconstruction of neuronal circuit [30].

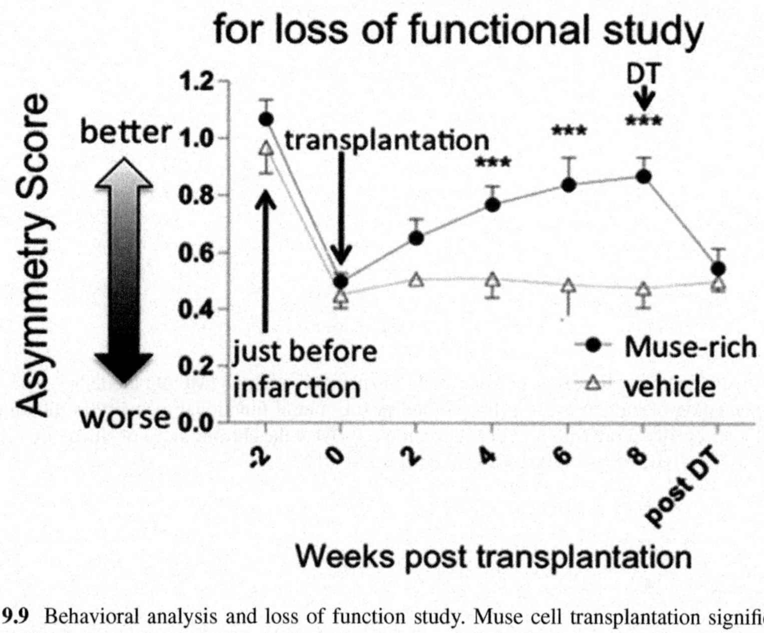

Fig. 9.9 Behavioral analysis and loss of function study. Muse cell transplantation significantly resulted in functional recovery in cylinder test (1: no deficit, 0: severe deficit). The functional recovery was ablated by the administration of diphtheria toxin (DT). ***$p < 0.001$. (Modified from Ref. [30])

9.9 Other Preclinical Studies of Muse Cell Treatment for Ischemic Stroke

Muse cell treatment of ischemic stroke was reported by another research group. Yamauchi et al. [37] showed that local injection of BM-derived Muse cells induced functional recovery, but the treatment effects were inferior to MSC at 42 days after transplantation. Combined with our data, Muse cells may induce functional recovery in the later time period.

Until now, all preclinical Muse cell therapy was done via local injection. Our preliminary data suggested intravenous injection of Muse cells was effective comparable to local Muse cell injection. Similar to local injection described above, intravenously administrated Muse cells migrated to the peri-infarct lesion and differentiated spontaneously into neural cells in the host tissue, leading to functional recovery (unpublished data). The minimally invasive intravenous injection of Muse cells may be more practical and feasible for clinical trials compared to local injections.

9.10 Pioneer Clinical Trials of Cell Therapies for Hemorrhagic Stroke

Clinical trials for hemorrhagic stroke are limited in number compared with ischemic stroke. There are no clinical trials of stem cell therapy for SAH. For ICH, five clinical studies have been reported. All trials used autologous BM-MSC and/or BM-MNC [14, 18, 38–40]. The results are summarized in Table 9.2. The main route of administration of stem cells is intracerebral or intrathecal [18, 38–40]. Only one trial used intravenous administration of BM-MSC [14]. All trials except the study of Li et al. administered stem cell in the chronic stage. All studies found that there are no study-specific adverse effects for BM-MSC or BM MNC therapies. The outcome also showed functional improvements. Li et al. [38] demonstrated that MNC treatment significantly improved the National Institute Stroke Scale (NIHSS) scores and Barthel index compared to the control group, and the improvement achieved 86.7% of the study-group patients. Sharma et al. [39] reported functional independence measure was significantly improved in the MNC-treated group. Zhu et al. [40] also presented that NIHSS scores, Rankin scale, and Barthel index in the MSC-treated group were significantly improved compared to the control group. However, they were not randomized studies, and follow-up period was mainly 6–12 months. In summary, randomized studies for hemorrhagic stroke with longer follow-up period may be needed to determine the roles of stem cell therapy for ICH. Furthermore, optimal dose and timing should also be determined.

Table 9.2 Clinical trials of stem cell therapy for intracerebral hemorrhage

Authors	Year	ICH location	Number of patients/controls	Route	Cell type	Cell dose	Time window	Result
Suarez-Monteagudo et al. [18]	2009	C, S	2/0	IC	Autologous BM-MNC	$1.7–5.5 \times 10^7$	3–8 years	Safe and effective
Bhasin et al. [14]	2011	MCA area	2/1	IV	Autologous BM-MSC	$5–6 \times 10^7$	8–12 months	Safe and feasible
Li et al. [38]	2013	BG	60/40	IC	Autologous BM-MNC	$2.5 \times 10^8–2.3 \times 10^9$	5–7 days	Safe and effective
Sharma et al. [39]	2014	NS	10/0	IT	Autologous BM-MNC	10^6/kg	4–144 months	Safe and feasible
Zhu et al. [40]	2015	BG with surgery	114/96	IC/IT	Autologous BM-MNC (IC) and BM-MSC (IT)	Mean 2×10^9 BM-MNCs/mean 8.4×10^7 BM-MSC	IC injection (BM-MNC): 3.01–6.89 days / IT injection (BM-MSC) after 4 weeks	Safe and effective

ICH intracerebral hemorrhage, *C* cortex, *S* striatum, *MCA* middle cerebral artery, *BG* basal ganglia, *NS* not specified, *IC* intracerebral, *IV* intravenous, *IT* intrathecal, *BM* bone marrow, *MNC* mononuclear cells, *MSC* mesenchymal stem cells

9.11 Preclinical Study of Muse Cell Treatment for Hemorrhagic Stroke

There are no reports to apply Muse cells for the treatment of SAH or cerebral aneurysms. One report described local injection of the Muse cells for the treatment of intracerebral hematoma [41]. Injection of the human bone marrow MSC-derived Muse cells into the hematoma cavity significantly improved functional outcome compared to the non-Muse group. The survival rate of the engrafted cells in the Muse group was significantly higher than in the non-Muse group at day 69, and those cells showed positivity for NeuN (57%) and MAP-2 (41.6%).

9.12 Conclusions

Stroke represents a major health issue worldwide. Because of the limited therapeutic time window and severe morbidity and mortality outcomes, development of novel treatment is an urgent clinical need. Stem cell therapy has emerged as a promising treatment for stroke. While clinical trials of MCS and MNC treatment have demonstrated their safety, the outcomes for functional recovery remain unsatisfactory. Likely, the poor outcomes may be due to the mediocre engraftment of MSC and NSC to the poststroke area, and the treatment effects primarily depend on paracrine or immunomodulation. On the other hand, embryonic stem (ES) cells and neural stem cells (NSC) have ethical concerns related to the origin and source of them. Induced pluripotent stem cells, ES cells, and NSC also have a technical problem related to its potent tumorigenesis. Even if they were differentiated into the target cells before transplantation, the percentage of differentiation is not 100%; thus undifferentiated or partially differentiated cells still pose the risk of tumorigenesis. However, to date no technology is available to guarantee the contamination of undifferentiated cells.

Compared with other cell sources, Muse cells are advantageous. Muse cells have low risks for tumorigenesis, as well as novel properties of robust engraftment and differentiation. Because the ultimate treatment of the stroke may require reconstruction of the lost neurons and neuronal circuits, these unique properties of Muse cells such as migration into damaged area, survival in the poststroke condition, and spontaneous differentiation into various cell types altogether recapitulate important regenerative mechanisms for inducing robust recovery after stroke. Muse cells stand as promising transplantable cell source for stroke treatment.

References

1. Pelvig DP, Pakkenberg H, Stark AK, Pakkenberg B (2008) Neocortical glial cell numbers in human brains. Neurobiol Aging 29:1754–1762
2. Jin K, Minami M, Lan JQ, Mao XO, Batteur S, Simon RP, Greenberg DA (2001) Neurogenesis in dentate subgranular zone and rostral subventricular zone after focal cerebral ischemia in the rat. Proc Natl Acad Sci USA 98:4710–4715
3. Jin K, Wang X, Xie L, Mao XO, Zhu W, Wang Y, Shen J, Mao Y, Banwait S, Greenberg DA (2006) Evidence for stroke-induced neurogenesis in the human brain. Proc Natl Acad Sci USA 103:13198–13202
4. Tatebayashi K, Tanaka Y, Nakano-Doi A, Sakuma R, Kamachi S, Shirakawa M, Uchida K, Kageyama H, Takagi T, Yoshimura S, Matsuyama T, Nakagomi T (2017) Identification of multipotent stem cells in human brain tissue following stroke. Stem Cells Dev 26:787–797
5. Takata M, Nakagomi T, Kashiwamura S, Nakano-Doi A, Saino O, Nakagomi N, Okamura H, Mimura O, Taguchi A, Matsuyama T (2012) Glucocorticoid-induced TNF receptor-triggered T cells are key modulators for survival/death of neural stem/progenitor cells induced by ischemic stroke. Cell Death Differ 19:756–767
6. Donnan GA, Fisher M, Macleod M, Davis SM (2008) Stroke. Lancet 371:1612–1623
7. Khalili MA, Sadeghian-Nodoushan F, Fesahat F, Mir-Esmaeili SM, Anvari M, Hekmati-Moghadam SH (2014) Mesenchymal stem cells improved the ultrastructural morphology of cerebral tissues after subarachnoid hemorrhage in rats. Exp Neurobiol 23:77–85
8. Stonesifer C, Corey S, Ghanekar S, Diamandis Z, Acosta SA, Borlongan CV (2017) Stem cell therapy for abrogating stroke-induced neuroinflammation and relevant secondary cell death mechanisms. Prog Neurobiol 158:94–131
9. Kondziolka D, Steinberg GK, Wechsler L, Meltzer CC, Elder E, Gebel J, Decesare S, Jovin T, Zafonte R, Lebowitz J, Flickinger JC, Tong D, Marks MP, Jamieson C, Luu D, Bell-Stephens T, Teraoka J (2005) Neurotransplantation for patients with subcortical motor stroke: a phase 2 randomized trial. J Neurosurg 103:38–45
10. Savitz SI, Dinsmore J, Wu J, Henderson GV, Stieg P, Caplan LR (2005) Neurotransplantation of fetal porcine cells in patients with basal ganglia infarcts: a preliminary safety and feasibility study. Cerebrovasc Dis 20:101–107
11. Bang OY, Lee JS, Lee PH, Lee G (2005) Autologous mesenchymal stem cell transplantation in stroke patients. Ann Neurol 57:874–882
12. Lee JS, Hong JM, Moon GJ, Lee PH, Ahn YH, Bang OY (2010) A long-term follow-up study of intravenous autologous mesenchymal stem cell transplantation in patients with ischemic stroke. Stem Cells 28:1099–1106
13. Honmou O, Houkin K, Matsunaga T, Niitsu Y, Ishiai S, Onodera R, Waxman SG, Kocsis JD (2011) Intravenous administration of auto serum-expanded autologous mesenchymal stem cells in stroke. Brain 134:1790–1807
14. Bhasin A, Srivastava MV, Kumaran SS, Mohanty S, Bhatia R, Bose S, Gaikwad S, Garg A, Airan B (2011) Autologous mesenchymal stem cells in chronic stroke. Cerebrovasc Dis Extra 1:93–104
15. Diez-Tejedor E, Gutierrez-Fernandez M, Martinez-Sanchez P, Rodriguez-Frutos B, Ruiz-Ares G, Lara ML, Gimeno BF (2014) Reparative therapy for acute ischemic stroke with allogeneic mesenchymal stem cells from adipose tissue: a safety assessment: a phase II randomized, double-blind, placebo-controlled, single-center, pilot clinical trial. J Stroke Cerebrovasc Dis 23:2694–2700
16. Steinberg GK, Kondziolka D, Wechsler LR, Lunsford LD, Coburn ML, Billigen JB, Kim AS, Johnson JN, Bates D, King B, Case C, McGrogan M, Yankee EW, Schwartz NE (2016) Clinical outcomes of transplanted modified bone marrow-derived mesenchymal stem cells in stroke: a phase 1/2a study. Stroke 47:1817–1824
17. Barbosa Da Fonseca LM, Battistella V, de Freitas GR, Gutfilen B, Dos Santos Goldenberg RC, Maiolino A, Wajnberg E, Rosado De Castro PH, Mendez-Otero R, Andre C (2009) Early

tissue distribution of bone marrow mononuclear cells after intra-arterial delivery in a patient with chronic stroke. Circulation 120:539–541
18. Suarez-Monteagudo C, Hernandez-Ramirez P, Alvarez-Gonzalez L, Garcia-Maeso I, de la Cuetara-Bernal K, Castillo-Diaz L, Bringas-Vega ML, Martinez-Aching G, Morales-Chacon LM, Baez-Martin MM, Sanchez-Catasus C, Carballo-Barreda M, Rodriguez-Rojas R, Gomez-Fernandez L, Alberti-Amador E, Macias-Abraham C, Balea ED, Rosales LC, Del Valle Perez L, Ferrer BB, Gonzalez RM, Bergado JA (2009) Autologous bone marrow stem cell neurotransplantation in stroke patients. An open study. Restor Neurol Neurosci 27:151–161
19. Savitz SI, Misra V, Kasam M, Juneja H, Cox CS Jr, Alderman S, Aisiku I, Kar S, Gee A, Grotta JC (2011) Intravenous autologous bone marrow mononuclear cells for ischemic stroke. Ann Neurol 70:59–69
20. Battistella V, de Freitas GR, da Fonseca LM, Mercante D, Gutfilen B, Goldenberg RC, Dias JV, Kasai-Brunswick TH, Wajnberg E, Rosado-De-Castro PH, Alves-Leon SV, Mendez-Otero R, Andre C (2011) Safety of autologous bone marrow mononuclear cell transplantation in patients with nonacute ischemic stroke. Regen Med 6:45–52
21. Moniche F, Gonzalez A, Gonzalez-Marcos JR, Carmona M, Pinero P, Espigado I, Garcia-Solis D, Cayuela A, Montaner J, Boada C, Rosell A, Jimenez MD, Mayol A, Gil-Peralta A (2012) Intra-arterial bone marrow mononuclear cells in ischemic stroke: a pilot clinical trial. Stroke 43:2242–2244
22. Friedrich MA, Martins MP, Araujo MD, Klamt C, Vedolin L, Garicochea B, Raupp EF, Sartori El Ammar J, Machado DC, Costa JC, Nogueira RG, Rosado-De-Castro PH, Mendez-Otero R, Freitas GR (2012) Intra-arterial infusion of autologous bone marrow mononuclear cells in patients with moderate to severe middle cerebral artery acute ischemic stroke. Cell Transplant 21(Suppl 1):S13–S21
23. Rosado-De-Castro PH, Schmidt Fda R, Battistella V, Lopes De Souza SA, Gutfilen B, Goldenberg RC, Kasai-Brunswick TH, Vairo L, Silva RM, Wajnberg E, Alvarenga Americano Do Brasil PE, Gasparetto EL, Maiolino A, Alves-Leon SV, Andre C, Mendez-Otero R, Rodriguez De Freitas G, Barbosa Da Fonseca LM (2013) Biodistribution of bone marrow mononuclear cells after intra-arterial or intravenous transplantation in subacute stroke patients. Regen Med 8:145–155
24. Taguchi A, Sakai C, Soma T, Kasahara Y, Stern DM, Kajimoto K, Ihara M, Daimon T, Yamahara K, Doi K, Kohara N, Nishimura H, Matsuyama T, Naritomi H, Sakai N, Nagatsuka K (2015) Intravenous autologous bone marrow mononuclear cell transplantation for stroke: phase1/2a clinical trial in a homogeneous group of stroke patients. Stem Cells Dev 24:2207–2218
25. Amariglio N, Hirshberg A, Scheithauer BW, Cohen Y, Loewenthal R, Trakhtenbrot L, Paz N, Koren-Michowitz M, Waldman D, Leider-Trejo L, Toren A, Constantini S, Rechavi G (2009) Donor-derived brain tumor following neural stem cell transplantation in an ataxia telangiectasia patient. PLoS Med 6:e1000029
26. Osanai T, Houkin K, Uchiyama S, Minematsu K, Taguchi A, Terasaka S (2017) Treatment evaluation of acute stroke for using in regenerative cell elements (TREASURE) trial: rationale and design. Int J Stroke 1747493017743057
27. Uchida N, Kushida Y, Kitada M, Wakao S, Kumagai N, Kuroda Y, Kondo Y, Hirohara Y, Kure S, Chazenbalk G, Dezawa M (2017) Beneficial effects of systemically administered human Muse cells in Adriamycin nephropathy. J Am Soc Nephrol 28:2946–2960
28. Uchida H, Sakata H, Fujimura M, Niizuma K, Kushida Y, Dezawa M, Tominaga T (2015) Experimental model of small subcortical infarcts in mice with long-lasting functional disabilities. Brain Res 1629:318–328
29. Uchida H, Morita T, Niizuma K, Kushida Y, Kuroda Y, Wakao S, Sakata H, Matsuzaka Y, Mushiake H, Tominaga T, Borlongan CV, Dezawa M (2016) Transplantation of unique subpopulation of fibroblasts, Muse cells, ameliorates experimental stroke possibly via robust neuronal differentiation. Stem Cells 34:160–173

30. Uchida H, Niizuma K, Kushida Y, Wakao S, Tominaga T, Borlongan CV, Dezawa M (2017) Human Muse cells reconstruct neuronal circuitry in subacute lacunar stroke model. Stroke 48:428–435
31. Yasuhara T, Matsukawa N, Hara K, Maki M, Ali MM, Yu SJ, Bae E, Yu G, Xu L, McGrogan M, Bankiewicz K, Case C, Borlongan CV (2009) Notch-induced rat and human bone marrow stromal cell grafts reduce ischemic cell loss and ameliorate behavioral deficits in chronic stroke animals. Stem Cells Dev 18:1501–1514
32. Shen LH, Li Y, Chen J, Zacharek A, Gao Q, Kapke A, Lu M, Raginski K, Vanguri P, Smith A, Chopp M (2007) Therapeutic benefit of bone marrow stromal cells administered 1 month after stroke. J Cereb Blood Flow Metab 27:6–13
33. Chen J, Zhang ZG, Li Y, Wang L, Xu YX, Gautam SC, Lu M, Zhu Z, Chopp M (2003) Intravenous administration of human bone marrow stromal cells induces angiogenesis in the ischemic boundary zone after stroke in rats. Circ Res 92:692–699
34. Chen J, Li Y, Wang L, Zhang Z, Lu D, Lu M, Chopp M (2001) Therapeutic benefit of intravenous administration of bone marrow stromal cells after cerebral ischemia in rats. Stroke 32:1005–1011
35. Pappenheimer AM Jr, Harper AA, Moynihan M, Brockes JP (1982) Diphtheria toxin and related proteins: effect of route of injection on toxicity and the determination of cytotoxicity for various cultured cells. J Infect Dis 145:94–102
36. Cummings BJ, Uchida N, Tamaki SJ, Salazar DL, Hooshmand M, Summers R, Gage FH, Anderson AJ (2005) Human neural stem cells differentiate and promote locomotor recovery in spinal cord-injured mice. Proc Natl Acad Sci USA 102:14069–14074
37. Yamauchi T, Kuroda Y, Morita T, Shichinohe H, Houkin K, Dezawa M, Kuroda S (2015) Therapeutic effects of human multilineage-differentiating stress enduring (MUSE) cell transplantation into infarct brain of mice. PLoS One 10:e0116009
38. Li ZM, Zhang ZT, Guo CJ, Geng FY, Qiang F, Wang LX (2013) Autologous bone marrow mononuclear cell implantation for intracerebral hemorrhage-a prospective clinical observation. Clin Neurol Neurosurg 115:72–76
39. Sharma A, Sane H, Gokulchandran N, Khopkar D, Paranjape A, Sundaram J, Gandhi S, Badhe P (2014) Autologous bone marrow mononuclear cells intrathecal transplantation in chronic stroke. Stroke Res Treat 2014:234095
40. Zhu J, Xiao Y, Li Z, Han F, Xiao T, Zhang Z, Geng F (2015) Efficacy of surgery combined with autologous bone marrow stromal cell transplantation for treatment of intracerebral hemorrhage. Stem Cells Int 2015:318269
41. Shimamura N, Kakuta K, Wang L, Naraoka M, Uchida H, Wakao S, Dezawa M, Ohkuma H (2017) Neuro-regeneration therapy using human Muse cells is highly effective in a mouse intracerebral hemorrhage model. Exp Brain Res 235:565–572

Chapter 10
Muse Cell: A New Paradigm for Cell Therapy and Regenerative Homeostasis in Ischemic Stroke

Satoshi Kuroda, Masaki Koh, Emiko Hori, Yumiko Hayakawa, and Takuya Akai

Abstract Multilineage-differentiating stress enduring (Muse) cells are one of the most promising donor cells for cell therapy against ischemic stroke, because they can differentiate into any type of cells constructing the central nervous system (CNS), including the neurons. They can easily be isolated from the bone marrow stromal cells (BMSCs), which may also contribute to functional recovery after ischemic stroke as donor cells. In this chapter, we concisely review their biological features and then future perspective of Muse cell transplantation for ischemic stroke. In addition, we briefly refer to the surprising role of Muse cells to maintain the homeostasis in the living body under both physiological and pathological conditions.

Keywords Bone marrow · Stem cell · Transplantation · Ischemic stroke · Clinical trial

10.1 Introduction

For these three decades, cell therapy has been expected to promote functional recovery after central nervous system (CNS) disorders, including ischemic stroke. A variety of cells have been studied as the candidate donor cells for this purpose. These include embryonic stem (ES) cells, neural stem cells, inducible pluripotent stem (iPS) cells, bone marrow stromal cells (BMSCs), and so on. Of these, the BMSCs may be the most promising cells among them when considering its clinical application, because they can be obtained from the patients themselves and easily expanded without any ethical and immunological problems [1–3]. However, Dezawa and coworkers have successfully isolated stress-tolerant adult human stem cells from

S. Kuroda (✉) · M. Koh · E. Hori · Y. Hayakawa · T. Akai
Department of Neurosurgery, Graduate School of Medicine and Pharmaceutical Science, University of Toyama, Toyama, Japan
e-mail: skuroda@med.u-toyama.ac.jp

M. Dezawa (ed.), *Muse Cells*, Advances in Experimental Medicine and Biology 1103, https://doi.org/10.1007/978-4-431-56847-6_10

cultured skin fibroblasts and BMSCs. These cells can self-renew, express a set of genes associated with pluripotency, and differentiate into endodermal, ectodermal, and mesodermal cells both in vitro and in vivo. They are named as multilineage-differentiating stress enduring (Muse) cells [4]. In this chapter, we review the biological features of both BMSCs and Muse cells and discuss their possible capacities for cell therapy against ischemic stroke.

10.2 Bone Marrow Stromal Cells (BMSCs)

The BMSCs are non-hematopoietic cells and one of the major mesenchymal stem cells in humans. Originally, they were isolated from the bone marrow as adherent, fibroblast-like-shaped cells. In the bone marrow, the BMSCs regulate the proliferation and differentiation of hematopoietic cells under physiological condition (Fig. 10.1). In addition, the BMSCs are now known to have the ability for differentiation into multiple cell types, immunomodulation, paracrine effect, and hematopoietic support. Based on these unique biological features, the BMSCs have been widely expected as a transplantable resource for cell therapy [5].

Recent studies have rapidly clarified the mechanisms through which the transplanted BMSCs enhance functional recovery after ischemic stroke. It is well known that the BMSCs aggressively migrate into the injured tissues in vitro and in vivo. Chemokine system may be involved in their migration capacity into cerebral infarct.

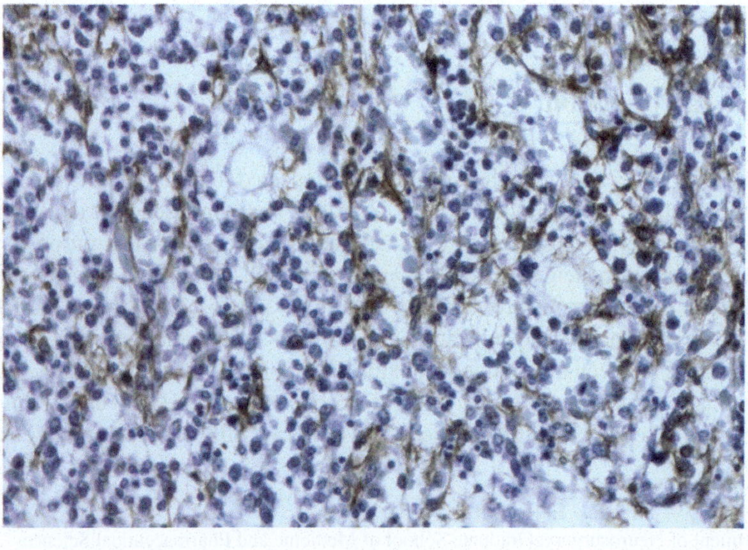

Fig. 10.1 Distribution of CD90-positive cells in the human bone marrow. CD90 is an immunohistochemical marker for the non-hematopoietic cells. The CD90-positive cells morphologically simulate the fibroblasts. Original magnification: ×200

Of these, a specific receptor for stromal cell-derived factor (SDF)-1α, CXCR4 is reported to play an important role in their migration in the CNS [6]. There are few studies whether the engrafted BMSCs retain their proliferative activity in the host brain or not. Yano et al. [7] labeled the GFP-expressing BMSCs with a superparamagnetic iron oxide (SPIO) agent and transplanted into the ipsilateral striatum of the mice infarct brain ("double labeling" technique) and found that the BMSCs actively proliferate, migrate toward the lesion, and partially express the neuronal phenotype in the host brain after transplantation [7].

Nowadays, the BMSCs are known to produce some neuroprotective or neurotrophic factors and support the survival of the host neural cells [8]. This hypothesis is readily reasonable because the BMSC per se support the homing, proliferation, and differentiation of the hematopoietic cells in the bone marrow by producing a variety of cytokines [9]. Interestingly, the conditioned medium of BMSCs significantly promote neurite outgrowth from the dorsal root ganglion [10]. They also release soluble neuroprotective factors, including nerve growth factor (NGF), hepatocyte growth factor (HGF), and brain-derived neurotrophic factor (BDNF), and significantly ameliorate glutamate-induced damage of neurons [11]. The BMSCs markedly promote the neurite extension from the neurons in the organotypic slice of the brain and spinal cord [12, 13]. The BMSCs also protect the neurovascular integrity between basement membrane and astrocyte end-feet and ameliorate brain damage in stroke-prone spontaneous hypertensive rats [14]. Shichinohe et al. demonstrated that the BMSCs serve the "nursing effect" to the damaged neurons and activate the neural stem cells in the host brain by producing neurotrophic factors. Of these, BDNF may be the most powerful factor to protect and repair the damaged neurons [15]. Therefore, the transplanted BMSCs most likely trigger endogenous signaling pathways of survival and repair in neurons by secreting soluble neurotrophic factors (Fig. 10.2).

Both neutrophils and macrophages are well known to play an important role in the early inflammation after cerebral infarct [16]. Indeed, their inflammatory response may be an essential process to clear cellular debris and initiate the healing pathways. Simultaneously, however, these inflammatory reactions may also give rise to cytotoxic damage to the surviving neurons, astrocytes, and endothelial cells in the peri-infarct area [16]. The BMSCs have currently been investigated as donor cells for novel cell therapy to prevent and to treat clinical disease associated with aberrant immune response. In the host, the BMSCs may attenuate pro-inflammatory cytokine and chemokine induction and reduce pro-inflammatory cell migration into sites of injury and infection [17]. Furthermore, intravenous injection of the multipotent adult stem cell (MAPC), an adherent, human multipotent adult stem cell derived from bone marrow, restores the expression of multiple genes and pathways involved in immune and inflammatory responses after stroke, which indicates that immunomodulation of the splenic response by the intravenous administration of MAPC may create a more favorable environment for brain repair after stroke [18]. Therefore, the transplanted BMSCs may prevent excessive inflammatory response and prevent further tissue damage in the peri-infarct area through their immuno-modulatory capacity (Fig. 10.2).

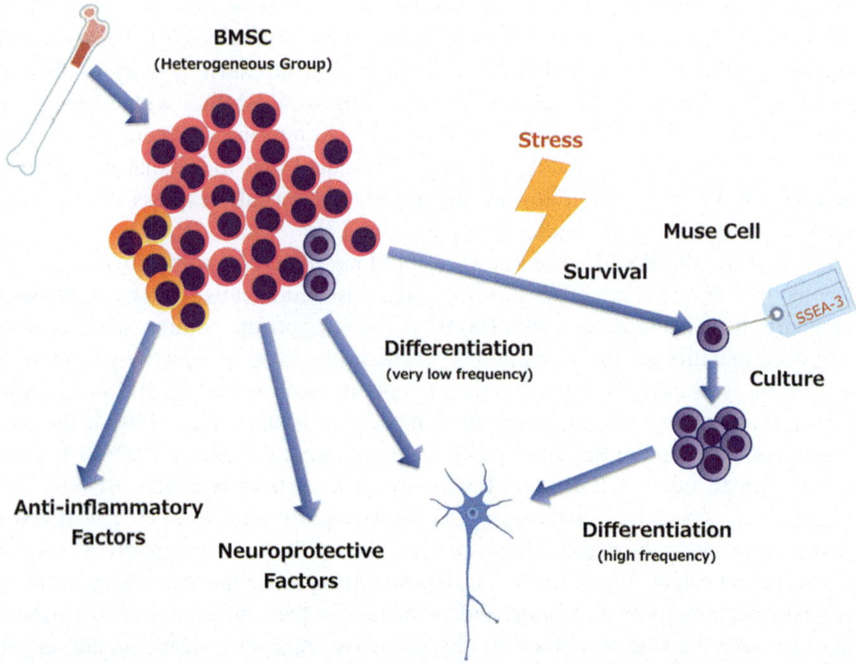

Fig. 10.2 A diagram shows the biological differences between the BMSCs and Muse cells. The BMSCs can be harvested from the bone marrow and consist of at least three "heterogeneous" cell populations, which include the cells secreting anti-inflammatory factors, those producing neuroprotective factors, and those with the potential to differentiate into multilineage cells, including neurons. Of these BMSCs, Muse cells are a certain subpopulation that can survive under extremely serious conditions. They are positive for SSEA-3 that is originally a specific marker for ES cells. Muse cells are biologically very specific cell populations that can renew themselves and differentiate into multilineage cells, including the neurons

The BMSCs are believed to differentiate into neural cells in the host's brain, while with very low frequency. This theory is based on the findings that BMSC simulate neuronal morphology and express the proteins specific for neurons in vitro [19] or in vivo [20, 21]. It may sound strange that the BMSC have the ability to differentiate into the neural cells. However, the BMSC per se express the genes related to neuronal and glial cells [22]. Recent studies also show that the BMSCs can alter their gene expression profile in response to exogenous stimuli and increase the genes related to the neural cells [22–24]. Wislet-Gendebien et al. [25] co-cultured the BMSCs with cerebellar granule cells and assessed their fates. They found that the nestin-expressing BMSCs express other neuronal markers and that BMSC-derived neuron-like cells fire single-action potentials in response to neurotransmitters such as glutamate [25]. Hokari et al. [11] also demonstrated that a certain subpopulation of the BMSCs morphologically simulated the neuron and expressed the neuron-specific proteins without any evidence of cell fusion, when co-cultured with the neurons [11]. These findings strongly suggest that at least a certain subpopulation of the BMSCs have the potential to alter their gene expression profile and to differentiate into the neural cells in response to the surrounding environment. More importantly, the findings indicate that only the subgroup of BMSCs with potential of neural differentiation can survive in the host brain for a long time (>4 weeks). According to recent work by Liu et al., the BMSC may enhance the axonal sprouting from the survived cortical neurons in the peri-infarct area [26]. Chiba et al. have also found that the BMSC are integrated into the neural circuits of the host spinal cord and promote functional recovery [27]. These biological properties of BMSC may play a key role to enhance functional recovery after ischemic stroke (Fig. 10.2).

Based on these observations, the exogenous transplantation of BMSCs is now believed to enhance functional recovery through multiple mechanisms, including nursing effect, anti-inflammatory action, and neural cell differentiation, in patients with ischemic stroke. This speculation seems quite natural, because the BMSCs are isolated only from their adhesive characteristics and contain heterogeneous cell populations. Therefore, several cell subgroups of BMSCs "independently" contribute to functional recovery through each biological features. According to previous observations, the cells with neuroprotective or anti-inflammatory actions occupy the majority of BMSCs and may play a key role in protecting the surviving neural cells during the first 2–4 weeks after BMSC transplantation. Subsequently, the remaining small subgroup of BMSCs has the potential to differentiate into the neural cells and restore neuronal circuits in the damaged brain (Fig. 10.3) [3]. Cell fusion, one of their biological characteristics, may also play some important role in promoting functional recovery after the insult [15, 28, 29].

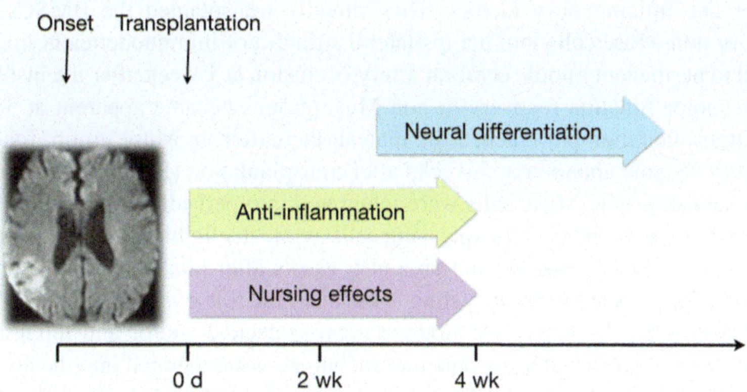

Fig. 10.3 A diagram shows the concept of multifunctional involvement in infarct brain by the BMSC. The BMSCs are heterogeneous cell populations and contribute to functional recovery through multiple mechanisms at different timing. (Cited from Kuroda et al. (2016) with permission [3])

10.3 Muse Cells

As aforementioned, Muse cells were found by Dezawa and coworkers [4]. When transplanted into immunodeficient mice by local or intravenous injection, Muse cells are integrated into various kinds of damaged organs, including the skin, muscle, liver, and brain, and differentiate into the host cell types in the respective tissues. They can efficiently be isolated as SSEA-3-positive cells. Unlike authentic ES cells, their proliferation activity is not very high, and they do not form teratoma in immunodeficient mouse testes, which would be a positive feature for clinical use (Fig. 10.2). These findings are quite attractive, because non-tumorigenic stem cells with the ability to generate the multiple cell types of the three germ layers can be obtained through easily accessible adult human mesenchymal cells without introducing exogenous genes [4]. Interestingly, they have proven that Muse cells are a primary source of induced pluripotent stem (iPS) cells in human fibroblasts [30]. Therefore, Muse cells are the responsible subpopulation of BMSCs that differentiate into the multilineage cells in response to the microenvironment of each organ. Furthermore, Muse cells promptly committed to neural/neuronal-lineage cells when co-cultured with stroke brain slices in vitro and significantly improve motor function when directly injected into the rat brain subjected to middle cerebral artery occlusion at 2 days after the insult [31]. Therefore, Muse cells, this unique and promising adult stem cell, are expected available for application into clinical situation for ischemic stroke through further studies.

In fact, recent studies strongly suggest the possibility of Muse cells as biologically powerful stem cells for patients with ischemic stroke. Thus, Yamauchi et al. [29] isolated Muse cells from the cultured human BMSCs, using SSEA-3 as a marker specific for Muse cells. In this study, he BMSC-derived cells negative for SSEA-3 were defined as non-Muse cells, which would have no potential to differentiate into any lineage cells and support the host cells by secreting neuroprotective and/or anti-inflammatory factors. They directly transplanted the BMSCs, Muse cells, or non-Muse cells into the ipsilateral striatum of immunodeficient mice subjected to permanent middle cerebral artery occlusion at 1 week after the insult. As a result, motor function recovery in non-Muse groups became apparent at 3 weeks after transplantation but reached the plateau thereafter. In Muse group, functional recovery became apparent at 5 weeks after transplantation (Fig. 10.4). In immunohistochemistry, only Muse cells were integrated into peri-infarct cortex and differentiate into Tuj-1- and NeuN-expressing cells, while negligible number of non-Muse cells remained in the peri-infarct area at 6 weeks after transplantation (Fig. 10.5) [29]. The findings are very interesting to consider the role of each subset of BMSCs, i.e., non-Muse cells may only produce neuroprotective and/or anti-inflammatory factors for 3–4 weeks after transplantation, but are not integrated into the host brain thereafter. On the other hand, Muse cells survived in the host brain and enhanced functional recovery by differentiating into the neurons, although they require longer time to yield therapeutic effects than non-Muse cells. More interestingly, direct injection of BMSCs that contains both Muse cells and non-Muse cells started to

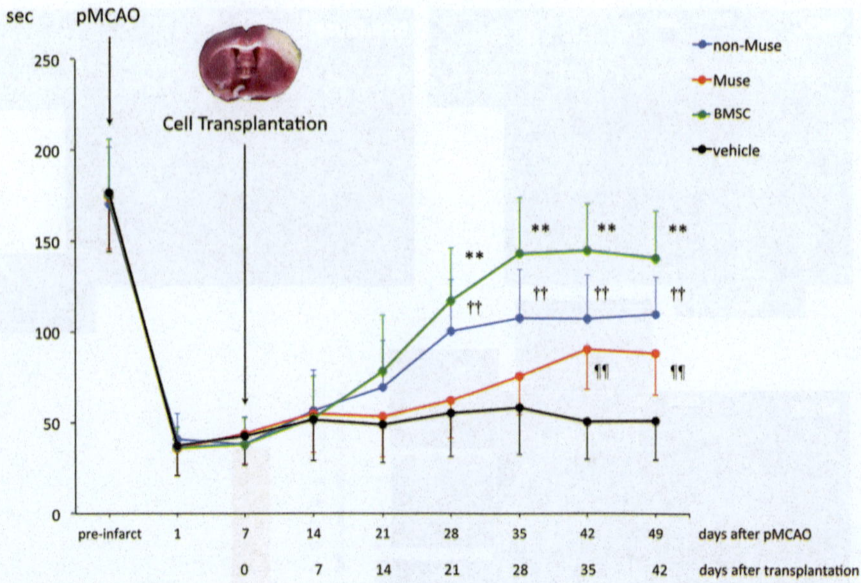

Fig. 10.4 A line graph shows the temporal profile of functional recovery in vehicle-, BMSC-, non-Muse cell-, and Muse cell-treated mice subjected to permanent middle cerebral artery occlusion. **, ††, ¶¶ $P < 0.01$ vs. vehicle-treated mice. (Modified from Yamauchi et al. (2015) with permission [29])

promote functions recovery at 3 weeks after transplantation, and their therapeutic effects on motor function was maximal, when compared with those in Muse group and non-Muse group (Fig. 10.4). Therefore, "multiple" actions by "heterogeneous" subsets of donor cells may be powerful to obtain maximal functional recovery after ischemic stroke (Fig. 10.6). However, it is well known that Muse cells occupy only small fraction in the BMSCs. Therefore, it would be very promising to expand Muse cells in vitro and transplant them with the non-Muse cells at the proper ration in order to further advance therapeutic effects of BMSC transplantation for ischemic stroke.

Very recently, an interesting hypothesis has been posed that Muse cells as well as BMSCs are playing a key role in the homeostasis/turnover of peripheral tissues and, if needed, could be mobilized from the bone marrow into the circulating blood during tissue injury and stress, facilitating the regeneration of damaged organs in situ [5, 32]. Therefore, we have recently conducted clinical testing to serially quantify the number of the circulating Muse cells in the peripheral blood of 29 patients with acute ischemic stroke for 1 month. Peripheral blood was obtained from all patients on admission and at day 7 and 30. As a result, the number of Muse cells robustly increased within 24 h after the onset, compared with the controls, but their baseline number and temporal profile widely varied among patients. No clinical data predicted the baseline number of Muse cells at the onset. Multivariate analysis revealed that smoking and alcohol intake significantly affect the increase in circulating Muse

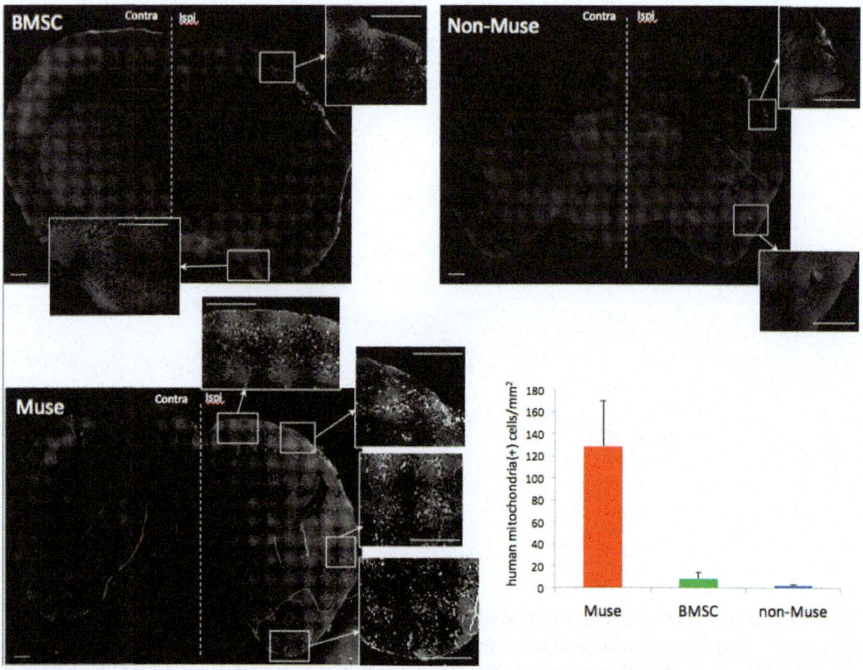

Fig. 10.5 Low-power photomicrographs of fluorescence immunohistochemistry using antihuman mitochondria antibody in BMSC-, non-Muse cell-, and Muse cell-treated mice at 42 days after transplantation. A large number of human mitochondria-positive cells are engrafted in the peri-infarct area in Muse cell group, but not in both BMSC and non-Muse cell groups. *Graph shows the number of human mitochondria-positive cells/mm² in ipsilateral cortex of each group*. Scale bars = 500 μm. (Modified from Yamauchi et al. (2015) with permission [29])

cells. The odds ratio was 0.0027 (P = 0.0336) and 1688 (P = 0.0220) for smoking and alcohol intake, respectively. The findings strongly suggested that pluripotent Muse cells are mobilized from the bone marrow into peripheral blood in acute stage of ischemic stroke. Smoking and alcohol intake significantly affect their temporal profile [33]. Therapeutic interventions that increase endogenous Muse cells may improve functional outcome after ischemic stroke. Based on these clinical results, we have very recently evaluated temporal profile of the circulating Muse cells in the mice subjected to permanent middle cerebral artery occlusion. Interestingly, the circulating Muse cells significantly decreased at 7 and 28 days after the insult, which was a little bit different from the finding in humans. Histological analysis revealed that Muse cells migrated toward the peri-infarct area and partly expressed the neuron-specific marker. On the other hand, non-Muse cells were found in the lung and spleen, but not in the brain (unpublished data). These findings strongly suggest that the BMSCs are mobilized from the bone marrow in acute phase of ischemic stroke and then non-Muse cells are migrated into the spleen and lung.

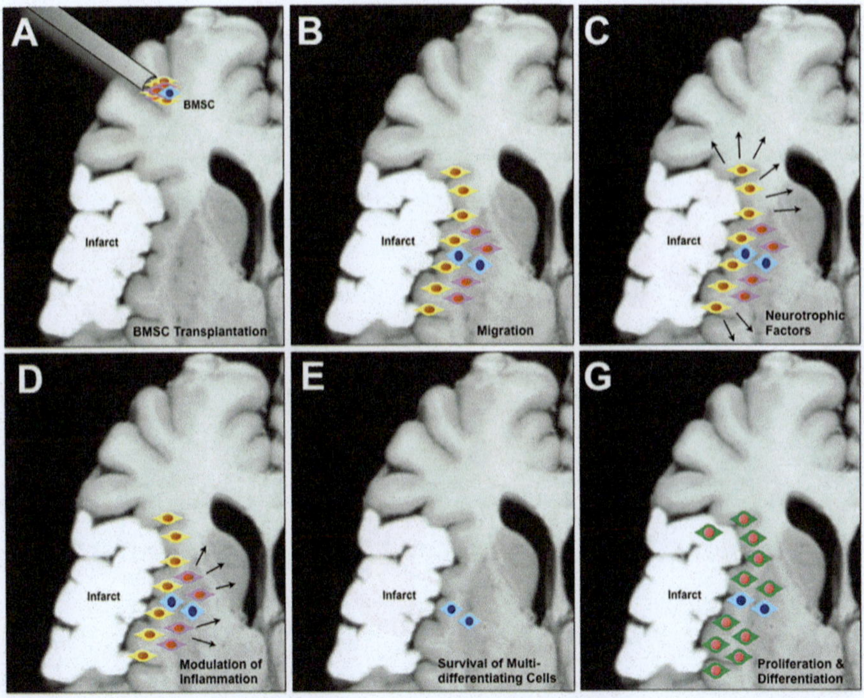

Fig. 10.6 Possible mechanism of functional recovery after ischemic stroke by BMSC transplantation. (Cited from Kuroda et al. (2013) with permission [2])

They may control the spleen and suppress excessive inflammatory reactions in the whole body, which in turn protect the ischemic brain. Muse cells, but not non-Muse cells, can pass through the lung and migrate toward the ischemic brain. They are integrated into the brain around cerebral infarct and differentiate into the neural cells, including the neurons (Fig. 10.7).

10.4 Conclusion

Basic and translational studies on the BMSCs and Muse cells have markedly increased our knowledge on cell therapy for ischemic stroke. Especially, the circulating Muse cells may play an important role in maintaining homeostasis in the whole body under both physiological and pathological conditions. Furthermore, Muse cells would also be one of the most promising donor cells to enhance functional recovery after ischemic stroke in clinical situation in very near future.

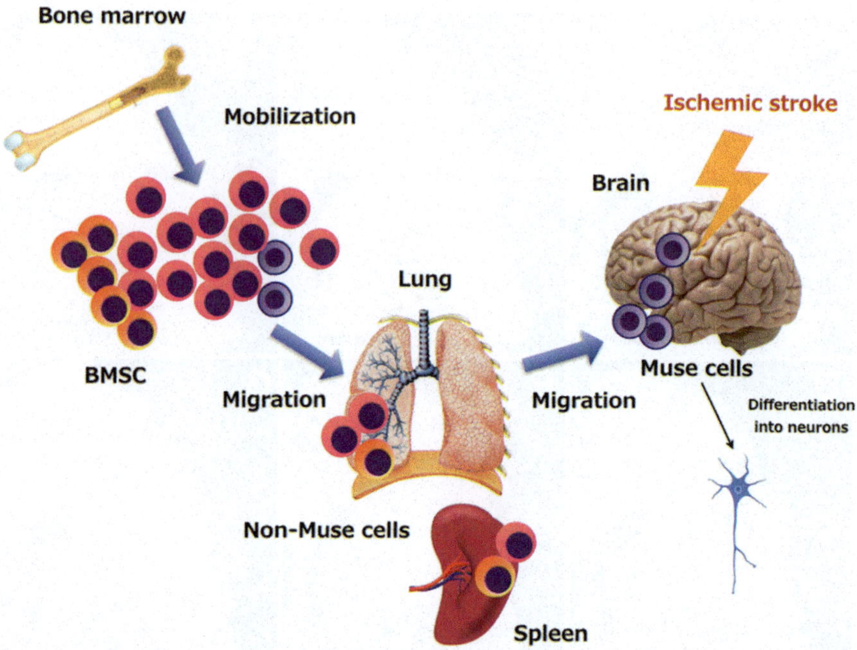

Fig. 10.7 A diagram shows that the BMSCs are mobilized from the bone marrow in acute phase of ischemic stroke. Non-Muse cells migrate into the lung and spleen. They control the spleen and suppress excessive inflammatory reactions against ischemic stroke. On the other hand, Muse cells, but not non-Muse cells, can migrate into the peri-infarct area and differentiate into the neural cells

References

1. Kuroda S (2008) How should we bridge the missing steps in translational research for stroke therapy? – a critical review. Jpn J Stroke 30:875–880
2. Kuroda S (2013) Bone marrow stromal cell transplantation for ischemic stroke – its multifunctional feature. Acta Neurobiol Exp (Wars) 73:57–65
3. Kuroda S (2016) Current opinion of bone marrow stromal cell transplantation for Ischemic stroke. Neurol Med Chir (Tokyo) 56:293–301
4. Kuroda Y, Kitada M, Wakao S et al (2010) Unique multipotent cells in adult human mesenchymal cell populations. Proc Natl Acad Sci USA 107:8639–8643
5. Miura Y (2015) Human bone marrow mesenchymal stromal/stem cells: current clinical applications and potential for hematology. Int J Hematol 103(2):122–128
6. Shichinohe H, Kuroda S, Yano S, Hida K, Iwasaki Y (2007) Role of SDF-1/CXCR4 system in survival and migration of bone marrow stromal cells after transplantation into mice cerebral infarct. Brain Res 1183:138–147
7. Yano S, Kuroda S, Shichinohe H, Hida K, Iwasaki Y (2005) Do bone marrow stromal cells proliferate after transplantation into mice cerebral infarct? – a double labeling study. Brain Res 1065:60–67
8. Zhong C, Qin Z, Zhong CJ, Wang Y, Shen XY (2003) Neuroprotective effects of bone marrow stromal cells on rat organotypic hippocampal slice culture model of cerebral ischemia. Neurosci Lett 342:93–96

9. Kortesidis A, Zannettino A, Isenmann S, Shi S, Lapidot T, Gronthos S (2005) Stromal-derived factor-1 promotes the growth, survival, and development of human bone marrow stromal stem cells. Blood 105:3793–3801
10. Neuhuber B, Timothy Himes B, Shumsky JS, Gallo G, Fischer I (2005) Axon growth and recovery of function supported by human bone marrow stromal cells in the injured spinal cord exhibit donor variations. Brain Res 1035:73–85
11. Hokari M, Kuroda S, Shichinohe H, Yano S, Hida K, Iwasaki Y (2008) Bone marrow stromal cells protect and repair damaged neurons through multiple mechanisms. J Neurosci Res 86:1024–1035
12. Kamei N, Tanaka N, Oishi Y et al (2007) Bone marrow stromal cells promoting corticospinal axon growth through the release of humoral factors in organotypic cocultures in neonatal rats. J Neurosurg Spine 6:412–419
13. Shichinohe H, Kuroda S, Tsuji S et al (2008) Bone marrow stromal cells promote neurite extension in organotypic spinal cord slice: significance for cell transplantation therapy. Neurorehabil Neural Repair 22:447–457
14. Ito M, Kuroda S, Sugiyama T et al (2012) Transplanted bone marrow stromal cells protect neurovascular units and ameliorate brain damage in stroke-prone spontaneously hypertensive rats. Neuropathology 32:522–533
15. Shichinohe H, Ishihara T, Takahashi K et al (2015) Bone marrow stromal cells rescue ischemic brain by trophic effects and phenotypic change toward neural cells. Neurorehabil Neural Repair 29:80–89
16. Barone FC, Feuerstein GZ (1999) Inflammatory mediators and stroke: new opportunities for novel therapeutics. J Cereb Blood Flow Metab 19:819–834
17. Auletta JJ, Deans RJ, Bartholomew AM (2012) Emerging roles for multipotent, bone marrow-derived stromal cells in host defense. Blood 119:1801–1809
18. Yang B, Hamilton JA, Valenzuela KS et al (2017) Multipotent adult progenitor cells enhance recovery after stroke by modulating the immune response from the spleen. Stem Cells 35:1290–1302
19. Woodbury D, Schwarz EJ, Prockop DJ, Black IB (2000) Adult rat and human bone marrow stromal cells differentiate into neurons. J Neurosci Res 61:364–370
20. Azizi SA, Stokes D, Augelli BJ, DiGirolamo C, Prockop DJ (1998) Engraftment and migration of human bone marrow stromal cells implanted in the brains of albino rats – similarities to astrocyte grafts. Proc Natl Acad Sci USA 95:3908–3913
21. Kopen GC, Prockop DJ, Phinney DG (1999) Marrow stromal cells migrate throughout forebrain and cerebellum, and they differentiate into astrocytes after injection into neonatal mouse brains. Proc Natl Acad Sci USA 96:10711–10716
22. Yamaguchi S, Kuroda S, Kobayashi H et al (2006) The effects of neuronal induction on gene expression profile in bone marrow stromal cells (BMSC)-a preliminary study using microarray analysis. Brain Res 1087:15–27
23. Bossolasco P, Cova L, Calzarossa C et al (2005) Neuro-glial differentiation of human bone marrow stem cells in vitro. Exp Neurol 193:312–325
24. Hermann A, Liebau S, Gastl R et al (2006) Comparative analysis of neuroectodermal differentiation capacity of human bone marrow stromal cells using various conversion protocols. J Neurosci Res 83:1502–1514
25. Wislet-Gendebien S, Hans G, Leprince P, Rigo JM, Moonen G, Rogister B (2005) Plasticity of cultured mesenchymal stem cells: switch from nestin-positive to excitable neuron-like phenotype. Stem Cells 23:392–402
26. Liu Z, Li Y, Zhang ZG et al (2010) Bone marrow stromal cells enhance inter- and intracortical axonal connections after ischemic stroke in adult rats. J Cereb Blood Flow Metab 30:1288–1295
27. Chiba Y, Kuroda S, Maruichi K et al (2009) Transplanted bone marrow stromal cells promote axonal regeneration and improve motor function in a rat spinal cord injury model. Neurosurgery 64:991–999 discussion 999-1000

28. Ito M, Kuroda S, Sugiyama T et al (2011) Validity of bone marrow stromal cell expansion by animal serum-free medium for cell transplantation therapy of cerebral infarct in rats – a serial MRI study. Transl Stroke Res 2:294–306
29. Yamauchi T, Kuroda Y, Morita T et al (2015) Therapeutic effects of human multilineage-differentiating stress enduring (MUSE) cell transplantation into infarct brain of mice. PLoS One 10:e0116009
30. Wakao S, Kitada M, Kuroda Y et al (2011) Multilineage-differentiating stress-enduring (Muse) cells are a primary source of induced pluripotent stem cells in human fibroblasts. Proc Natl Acad Sci USA 108:9875–9880
31. Uchida H, Morita T, Niizuma K et al (2015) Transplantation of unique subpopulation of fibroblasts, Muse cells, ameliorates experimental stroke possibly via robust neuronal differentiation. Stem Cells 34(1):160–173
32. Dezawa M (2016) Muse cells provide the pluripotency of mesenchymal stem cells: direct contribution of Muse cells to tissue regeneration. Cell Transplant 25:849–861
33. Hori E, Hayakawa Y, Hayashi T et al (2016) Mobilization of pluripotent multilineage-differentiating stress-enduring cells in ischemic stroke. J Stroke Cerebrovasc Dis 25:1473–1481

Chapter 11
Application of Muse Cell Therapy for Kidney Diseases

Nao Uchida, Naonori Kumagai, and Yoshiaki Kondo

Abstract The kidney plays an essential role in the maintenance of homeostasis in healthy individuals, e.g., by regulating the amount of water and concentration of electrolyte in the body. Owing to the structural complexity, renal dysfunction is caused by a myriad of diseases and conditions, and in severe cases, it progresses to end-stage renal disease in which patients require renal replacement therapy, i.e., maintenance dialysis or kidney transplantation. The currently available therapeutic modalities, with the exception of renal transplantation, cannot recover severely deteriorated renal function. Thus, regenerative medicine holds considerable promise as a potential means for developing next-generation renal therapeutics. Mesenchymal stem cell (MSC) transplantation has been investigated in acute kidney injury and chronic kidney disease models, and clinical studies have already been started for some kinds of kidney diseases. However, most of these studies concluded that the main underlying mechanism of therapeutic effect of MSC transplantation was paracrine. Recently, we reported that Muse cell therapy in a murine model of chronic kidney disease resulted in differentiation of intravenously injected Muse cells into glomerular cells after preferential homing to damaged glomerulus and improvement in renal function. The result suggested the potentiality of Muse cell therapy for glomerular regeneration. Muse cells are a promising cell source for regenerative therapy for kidney diseases.

Keywords FSGS · Kidney disease · Muse cell · MSC · Xenotransplantation · Regenerative therapy · Podocyte regeneration

---

N. Uchida (✉) · N. Kumagai
Departments of Pediatrics, Tohoku University Graduate School of Medicine, Sendai, Japan
e-mail: n-uchida@med.tohoku.ac.jp

Y. Kondo
Department of Healthcare Services Management, Nihon University School of Medicine, Tokyo, Japan

M. Dezawa (ed.), *Muse Cells*, Advances in Experimental Medicine and Biology 1103, https://doi.org/10.1007/978-4-431-56847-6_11

11.1 Introduction

The kidney plays an essential role in the maintenance of homeostasis in healthy individuals by continuously adjusting urinary excretion of water and electrolytes to match net intake and endogenous production by excretion of waste products of metabolism; maintenance of acid–base balance; and secretion of hormones that regulate systemic blood pressure, red blood cell production, and bone metabolism. Renal dysfunction leads to life-threatening conditions, such as edema, hypertension, azotemia, and abnormal electrolyte concentration, which might cause cardiac arrest. Renal dysfunction is associated with high morbidity and mortality. Patients with end-stage renal disease (ESRD) require renal replacement therapy, i.e., maintenance dialysis or kidney transplantation. Maintenance dialysis is expensive and time-intensive and adversely impacts the patient's quality of life. Kidney transplantation enables the patient to regain own renal function; however, limited availability of transplantable organ is a major constraint. Moreover, it also creates a potential health hazard for live donors who may face health problems in later life owing to reduced renal function. Therefore, search for new and better methods for treating kidney diseases is a key research imperative. Regenerative medicine holds considerable promise as a potential means for developing next-generation renal therapeutics. In this part, we provide an overview of the development and structure of the kidney, renal diseases, and review the currently available evidences from studies on cell therapy for kidney diseases. Finally, we report on the results of Muse cell therapy in a murine model of chronic kidney disease (CKD) and discuss the prospects of its potential application in the treatment of various renal diseases.

11.2 Development and Regeneration of Kidneys

The kidneys are two bean-shaped organs located on the right and left sides of the retroperitoneal space. Each kidney contains approximately one million nephrons, which represent the functional and structural unit of the kidney. Twenty-five percent of the cardiac output flows into the kidneys via renal arteries. Broadly, the kidney is roughly composed of four parts, i.e., the glomerulus, renal tubules, interstitial component, and intrarenal vessels. Each part comprises various types of highly differentiated cells which perform specialized roles. The circulating blood which flows into the kidney is filtered at the glomerulus; the glomerular filtrate which contains both waste products and useful substances passes through the renal tubules where water, electrolytes, and other necessary substances are reabsorbed (Fig. 11.1a). The glomerulus consists of three types of cells: podocytes, mesangial cells, and endothelial cells (Fig. 11.1b). These cells form a fine filter that allows the passage of substances according to their molecular size and electrical charge and regulates the glomerular filtration rate in coordination with renal tubules. Renal tubular cells reabsorb and actively secrete substances to regulate the net amount of water and electrolytes and

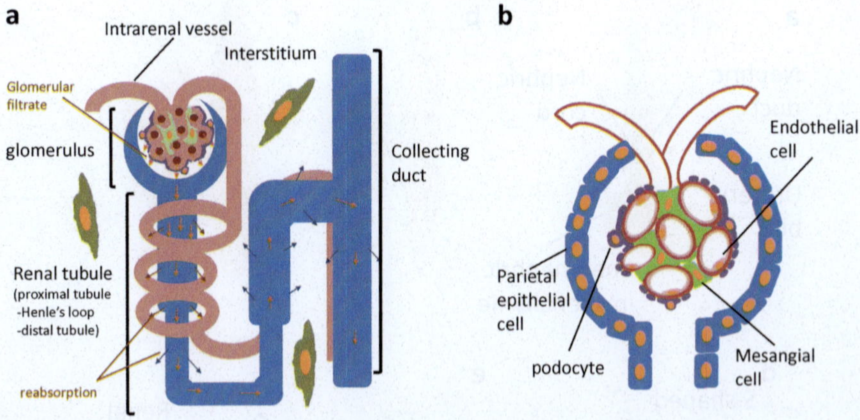

Fig. 11.1 The structure of the kidney. (**A**) The nephron and the interstitium. Blood flows into the glomerulus via the intrarenal vessel and is filtered through the endothelial cell, glomerular basement membrane, and slit diaphragm of the podocyte. The glomerular filtrate passes through the renal tubules where necessary substances are reabsorbed. (**B**) The glomerulus. The glomerulus consists of three types of cells: podocyte, mesangial cell, and endothelial cell

the acid–base balance in the body. Generally, mesangial, endothelial, and renal tubular cells have the capacity to self-regenerate after injury; however, podocytes lack the ability to proliferate and self-regenerate because they are highly differentiated cells that have a complex structure. Thus, the glomeruli are more vulnerable to injury than the renal tubules. Once the glomerulus is severely injured and becomes sclerotic, the loss of function is difficult to recover. Therefore, regeneration of podocytes is one of the key therapeutic targets in the realm of regenerative medicine for kidney dysfunction.

The mammalian kidney develops from the intermediate mesoderm (IM), a region of the embryonic mesoderm that lies between the paraxial mesoderm and the lateral plate mesoderm [1]. The IM generates two components, the mesenchymal "nephrogenic cord" and the nephric duct. The permanent kidney starts to develop when the nephric duct gives rise to the ureteric bud (UB) at weeks 4–5 of gestation in humans. The UB grows toward and penetrates the metanephric mesenchyme (MM) generated from the caudal nephric cord (Fig. 11.2a). Reciprocal induction between the UB and MM initiates nephrogenesis. The UB bifurcates within the MM and continues to branch repeatedly (Fig. 11.2b–c). Endothelial progenitors migrate into the S-shaped bodies that are generated from the MM (Fig. 11.2d). The UB gives rise to the entire epithelium of the renal collecting system, including that of the collecting ducts, calyces, renal pelvis, and ureter. The MM, including the nephron progenitor cells, gives rise to the nephron epithelia via mesenchymal-to-epithelial transition (Fig. 11.2e). The sequential inductive interactions between the UB and MM lead to the development of the kidney [2], which consists of more than 20 types of specialized cells.

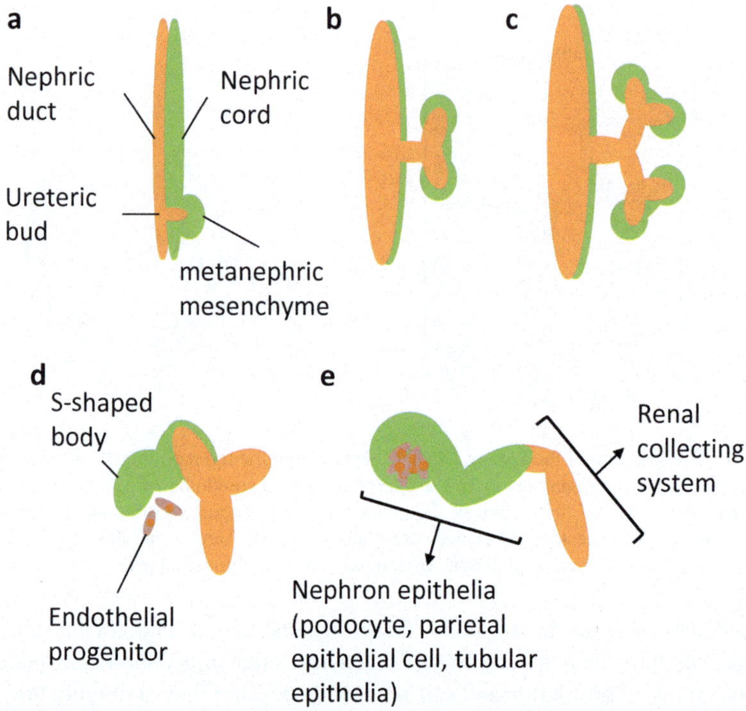

Fig. 11.2 Development of the kidney. (**A**) The nephric duct gives rise to the ureteric bud (UB), which penetrates the metanephric mesenchyme (MM) that generated from the nephric cord. (**B–C**) The UB repeatedly bifurcates within the MM with reciprocal induction to initiate nephrogenesis. (**D**) The MM forms the S-shaped body. Endothelial progenitors move into the S-shaped body to generate glomerular capillary. (**E**) The proximal part of the S-shaped body develops into the glomerulus, and the distal part generates the proximal tubule, Henle's loop, and distal tubule. The UB develops into the collecting system

Although the origin of nephron epithelia is the same, the MM, the regenerative capacity of the various types of cells in adult kidneys, varies. Proximal tubular cells have an excellent capacity to regenerate after kidney injury, while podocytes or visceral epithelial cells have limited regenerative capacity. As simple undifferentiated epithelial cells mature into podocytes, they establish their characteristic complex cell architecture and subsequently lose their mitotic activity [3]. A specific arrangement of the actin cytoskeleton is required for normal podocyte function, which is incompatible with the formation of the actin contractile ring required for cytokinesis. Although under certain stressed conditions, podocytes show aneuploidy and express replication markers, they cannot undergo successful cytokinesis [4]. The limited regenerative capacity of podocytes defines the particular vulnerability of the glomerulus. The glomerulus is the first segment of the nephron; therefore, disruption of the glomerulus results in functional impairment of the entire nephron.

Injured podocytes lack the ability of self-regeneration; however recent studies suggest the presence of podocyte precursor in intrarenal [5] and extrarenal tissues [6]. Recent publications mentioned podocyte regeneration or turnover by parietal epithelial cells [5]. Becker et al. examined renal biopsy specimens from six male recipients of kidneys transplanted from female donors for receiver-derived podocytes [6]. Small number of glomerular cells expressed Y chromosome and, the podocyte marker, Wilms tumor-1 antigen (WT-1). The importance of these phenomena is not clearly understood; however, it represents a positive development for renal regeneration therapy as it demonstrates that podocyte precursors can immigrate into the damaged site of the glomerulus even from outside the kidneys and differentiate into mature podocytes even in adults.

11.3 Renal Diseases and Conventional Treatment

Owing to the structural complexity, a myriad of diseases and conditions may lead to renal dysfunction. We can group them into acute kidney injury (AKI) and CKD (Table 11.1). However, recent data suggest that the two are closely interrelated. AKI may increase the risk of CKD and also directly cause ESRD. Conversely, CKD is a strong risk factor for development of AKI.

Table 11.1 Causes and treatments of AKI and CKD

			Causes	Treatments	
AKI	Prerenal		Hypovolemia	Fluid therapy	Sometimes hemodialysis is required
			Decreased cardio output	Therapy for heart failure	
	Renal	Glomerulus	Acute glomerular nephritis	Steroid, Immunosuppressants, PE	
		Tubular and interstitium	Allergy for drugs, nephrotoxic drugs	Discontinuation of causative drugs	
		Vessels	TTP, HUS	PE, control of blood pressure and electrolytes	
	Postrenal		Posterior uretheral valve	Surgical treatment	
			Prostate cancer, stone		
CKD			Hypertension, diabetes mellitus, chronic glomerular nephritis, inherited kidney diseases (Alport syndrome, etc.)	Control blood pressure, correction of acidosis, optimal salt and protein intake	In ESRD, maintenance hemodialysis or kidney transplantation is required

AKI acute kidney injurym, *PE* plasma exchange, *CKD* chronic kidney disease, *ESRD* end-stage renal disease, *TTP* thrombotic thrombocytopenic purpura, *HUS* hemolytic uremic syndrome

11.3.1 Acute Kidney Injury (AKI)

AKI is defined as a rapid decline in the glomerular filtration rate (GFR) resulting in retention of nitrogenous wastes and water, as indicated by elevation of creatinine levels and decrease in urine output [7]. AKI is responsible for approximately two million deaths annually worldwide [8]. Clinically, AKI is classified into three groups based on the primary etiology: prerenal, renal, and postrenal. Prerenal AKI results from a decrease in renal perfusion including hypovolemia, impaired cardiac output, and decreased vascular resistance. Renal AKI occurs from various causes, such as drugs, and infections that injure any part of the kidney, i.e., the glomerulus, tubules, interstitial portion, or intrarenal vessels. Postrenal AKI is caused by obstruction of urinary tract from any causes such as posterior urethral valve, cancer, and intratubular precipitation [8].

Treatment for AKI depends on the cause, e.g., appropriate fluid therapy for hypovolemia, elimination of causative drugs for interstitial nephritis, and surgical approach for urinary tract obstruction. While kidney cells are severely damaged, dialysis sometimes would be required to support the depleted renal function. The injury of kidney cells can be lethal or sublethal. AKI was believed to be reversible; however, recent data suggests that reversibility of renal function depends on the degree to which sublethally injured cells can restore normal function and promote regeneration [8]. Thus, AKI often results in CKD in surviving patients. Despite substantial improvement in supportive therapy, AKI is associated with a high mortality and morbidity. Therefore, exploration of novel therapeutic modalities for AKI is a key imperative. Clinical trials have demonstrated the potential of cell therapy using mesenchymal stem cells (MSCs) as a novel therapeutic option for AKI [9].

11.3.1.1 Chronic Kidney Disease (CKD)

CKD is defined as abnormalities of kidney structure or function, which persists for >3 months by the Kidney Disease: Improving Global Outcomes (KDIGO) [10]. Globally, diabetes mellitus and hypertension are the leading causes of CKD [11]. Other causes include chronic glomerulonephritis, renovascular disease, and inherited diseases [11]. While the definition of CKD tends to vary across different countries, the prevalence of CKD in adults is estimated to vary from 8 to 20% [12]. While the rate of progression of CKD depends on the underlying disease, presence of comorbid conditions, and treatments, these patients are at a high risk of progression to ESRD. The prevalence of ESRD continues to increase [13] and has become a major global public health issue. In addition, CKD is associated with high mortality, high risk of hospitalization, and cardiovascular diseases [12, 14].

Effective therapies for control of blood pressure and hyperlipidemia, correction of acidemia, and optimal salt and protein intake may slow the progression of CKD. However, the currently available therapeutic modalities, with the exception of renal transplantation, are not able to recover severely deteriorated renal function.

Some clinical trials are currently underway to evaluate the efficacy of MSC therapy in patients with CKD mainly with expectation of immunoregulation [15].

11.4 Stem Cell Therapy: A New Era in the Treatment of Kidney Diseases

Could stem cell therapy replace lethal kidney cells and recover renal function? In this section, we reviewed evidences from preclinical and clinical studies on cell therapy for kidney diseases (Table 11.2) [16–23].

11.4.1 Preclinical Studies

In the early 2000s, transplantation of whole bone marrow cells (BMCs), including hematopoietic and mesenchymal lineage, was investigated in various animal models of kidney disease. Imasawa et al. first reported the potential of BMCs to differentiate into mesangial cells. They transplanted whole BMCs from GFP-transgenic mice into lethally irradiated B6 mice and observed GFP-positive mesangial cells in the glomeruli of the recipients [24]. Ito et al. demonstrated sequential increase in transplanted-BMC-derived mesangial cells in Thy-1 glomerulonephritis [25]. Rookmaaker et al. reported the differentiation of BMCs into endothelial cells and mesangial cells [26]. In a study by Prodromidi et al., a small number of BMC-derived podocytes were shown to have integrated into the glomeruli in COL4A3$^{-/-}$ mice (Alport syndrome, inherited kidney disease model) [27]. Several studies have shown the ability of BMCs to differentiate into various types of renal cells [26–33], which suggests that BMCs include renal progenitor cells. BMCs represent a heterogeneous cell population which largely consists of hematopoietic stem cells (HSCs) and MSCs. In 2004, Morigi et al. examined the therapeutic effects of HSCs and MSCs in a cisplatin-induced AKI model [34]. MSCs attenuated kidney injury owing to their migration and differentiation into renal tubular epithelial cells. HSCs did not alleviate kidney injury and were less frequently observed in renal tubules.

Now, among the various types of stem cells, MSCs are the most extensively investigated stem cells in relation to cell therapy for kidney diseases [35]. These studies have demonstrated the outstanding regenerative potential of MSCs. They are multipotent cells that have the ability to differentiate not only into mesenchymal cells but also into cells with visceral mesodermal, neuroectodermal, and endodermal characteristics, both in vitro [36] and in vivo [37–39]. They also produce various kinds of cytokines and growth factors [40–42] and induce production of several growth factors in host cells [43]. In addition, MSCs carrying mRNA and micro RNA were shown to release microvesicles that mediate mRNA expression of moieties related to apoptosis and immunoregulation in the resident cells [44]. Processes

Table 11.2 Summary of preclinical reports on stem cell therapies for kidney diseases

Stem cell type	Disease	Model	Species		Induction	Route	Effectiveness	Mechanism of repairment				References
									Cell migration			
			Stem cell	Recipient				Paracrine	Glomeruli	Tubules	Differentiation	
MSCs	AKI	Ischemia/ reperfusion	Mouse (BM)	Mouse	–	iv	Effective	+	–	–	–	Hu et al. [16]
		Ischemia/ reperfusion	Rat (FM)	Rat	–	iv	Effective	+	–	–	–	Tsuda et al. [17]
		Cisplatin-treated	Human (BM)	Mouse (NOD/ SCID)	–	iv	Effective	+	–	–	–	Morigi et al. [18]
	CKD	Alport syndrome	Mouse (BM)	Mouse (COL4A3-deficient)	–	iv	Limited	+	–	–	–	Ninichuk et al. [19]
		UUO	Human(AF)	Mouse	–	iv	Effective	n.e.	–	+	–	Sun et al. [20]
iPSCs	AKI	Ischemia/ reperfusion	Human (fibroblast)	Mouse	+	sct	Effective	+	–		–	Toyohara et al. [21]
	CKD	n.r.										
ESCs	AKI	Cisplatin-treated	Human embryo	Mouse	+	iv	Effective	Suspected	n.e.	+	n.e.	Luo et al. [22]
Muse cells	AKI	n.r.										
	CKD	DOX nephropathy	Human(BM)	Mouse	–	iv	Effective	+	+	n.e.	+	Uchida et al. [23]

n.r. no report, *n.e.* not examined, *iv* intravenous injection, *sct* renal subcapsular transplantation, *MSCs* mesenchymal stem cells, *iPSCs* induced pluripotent stem cells, *ESCs* embryonic stem cells, *AKI* acute kidney injury, *CKD* chronic kidney disease, *UUO* unilateral ureteral obstruction, *DOX* doxorubicin, *BM* bone marrow, *FM* fetal membrane, *AF* amniotic fluid, *NOD/SCID* nonobese diabetic/severe combined immunodeficiency involved in renal regeneration mimic the natural process of kidney development [45]. Growth factors have been shown to regulate renal recovery in animal models of kidney injury. These factors include hepatocyte growth factor (HGF) [46], vascular endothelial cell growth factor (VEGF) [47], epithelial growth factor (EGF) [48], and insulin-like growth factor-I (IGF-I) [49]. Specific receptor knocked-out mice showed delay in recovery from AKI which indicates the importance of these factors in renal repair [50, 51]. MSCs are expected to supply these factors.

To date, the therapeutic potential of MSCs has been examined in various models of kidney disease. Their therapeutic effects were initially believed to be exerted by replenishment of damaged renal cells; however, differentiation of MSCs into renal cells has been documented only in a few studies [52, 53]. Even intraperitoneal administration of MSCs and conditioned medium showed renoprotective effects [54]. Therefore, most recent studies suggest that MSCs may prevent tissue injury via paracrine mechanism [18, 53, 55, 56].

11.4.2 Clinical Trials

While preclinical studies in animal models have yielded promising results both in the contexts of acute and chronic kidney injury, clinical trials are currently in the early phase [57]. In a clinical study of 16 patients undergoing on-pump cardiac surgery, allogenic MSC infusion was shown to be safe at all the tested dosages; further, a protective effect of these cells on kidney function was demonstrated, and its use was associated with shorter length of hospital stay [58]. Currently, over a hundred clinical trials of MSCs in patients with various kinds of kidney injury are ongoing, and some trials have been completed. Promising results are expected in the near future (data from www. clinicaltrial.gov).

11.5 Muse Cell Therapy as Real Regeneration Method of Kidneys

As mentioned above, cell therapy with MSCs is a promising novel therapeutic option. The main mechanism of their beneficial effects is believed to be paracrine. MSCs produce various kinds of growth factors and cytokines which promote mitosis of resident cells and angiogenesis and inhibit inflammatory responses and apoptosis. MSCs release microvesicles containing mRNA or microRNAs which serve as mitogens and have an anti-apoptotic effect. Direct differentiation into renal cells has been disregarded, while direct differentiation of MSCs in vivo has been reported in other organs [59, 60]. In addition, BMCs were shown to include podocyte progenitors [6]. However, remarkable paracrine effects of MSCs are likely to have masked the differentiation of even a small number of MSCs into renal cells.

MSCs are usually collected just as adherent cells from the bone marrow and other mesenchymal tissues. Consequently, the isolation of MSCs according to the International Society for Cellular Therapy criteria produces heterogeneous, non-clonal cultures of stromal cells which contain stem cells with different multipotent properties [57]. Owing to the heterogeneity of MSCs, their actions are pleiotropic.

Muse cells are stress-tolerant non-tumorigenic endogenous stem cells. They are collectable as pluripotent surface marker stage-specific embryonic antigen-3 (SSEA-3)-positive cells from connective tissue of various organs as well as from the bone marrow as approximately 0.03% of the mononucleated fraction [61, 62]. Muse cells also express the pluripotency markers Sox2, Oct3/4, and Nanog, and a single cell may potentially differentiate into cells representative of all three germ layers. They show self-renewability of triploblastic differentiation over several generations in vitro, which suggest their pluripotent-like properties [62]. These properties are reproducible in Muse cells that are directly collected from BM aspirates, which indicates that their properties are not newly acquired by in vitro manipulation or modified under culture conditions [62]. Differentiation of Muse cells into triploblastic lineages is also demonstrated by their in vivo reparative effects; intravenously or topically administered naïve Muse cells migrate to and integrate into the damaged tissues with a high selectivity and replenish the lost cells by spontaneous in vivo differentiation into tissue-compatible cells, leading to tissue repair in models of stroke, liver cirrhosis, muscle degeneration, and skin ulcers of diabetes mellitus [62–64]. Muse cells circulate in the peripheral blood in healthy individuals, and endogenous circulating Muse cells are increased in stroke patients [65] and acute myocardial infarction patients [66] possibly owing to mobilization from the BM in response to severe damages. The use of naturally existing reparative stem cells that do not require induction of genes or induction into purposive cells prior to transplantation is of immense clinical relevance as a therapeutic strategy for kidney diseases [67]. We investigated the therapeutic effects of Muse cells in kidney disease using a mouse model of Adriamycin nephropathy (DOX nephropathy).

First, we investigated the ability of Muse cells to differentiate into renal lineage cells in vitro. Muse and non-Muse cells (SSEA-3-negative cell fraction of MSCs) were subjected to renal induction with a cytokine mix [68]. Muse cells expressed much higher levels of the developmental renal markers WT1 and EYA1 as compared to non-Muse cells (Fig. 11.3).

Next, we evaluated the therapeutic effects of Muse and non-Muse cells in a mouse model of kidney disease. We used Adriamycin (doxorubicin hydrochloride) nephropathy, which is a well-established rodent model of chronic proteinuric kidney disease and mimics human focal segmental glomerulosclerosis (FSGS) with podocyte loss, focal segmental and global sclerosis, tubulointerstitial inflammation, and fibrosis [69]. We established FSGS model in two different strains: severe combined immunodeficiency (SCID) and BALB/c mouse. One week after single dose of doxorubicin, 2×10^4 human Muse cells (Muse group), non-Muse cells (non-Muse group), or an equivalent volume of sterile saline (vehicle group) were infused via the tail vein, and the structural and functional recovery evaluated after 5 and 7 weeks. Both Muse and non-Muse cells were naïve, i.e., without any preconditioning

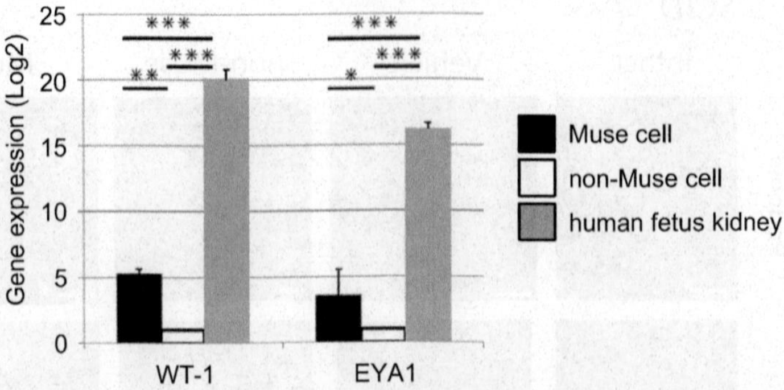

Fig. 11.3 Gene expressions of renal lineage markers WT1 and EYA1, 3 weeks after induction. Expression levels were normalized to those of glyceraldehyde-3-phosphate dehydrogenase. *$P < 0.05$; **$P < 0.01$; ***$P < 0.001$

or induction. In FSGS-SCID mice, integration of human Muse cells was confirmed using a human-specific antibody (the anti-human Golgi complex) and GFP. GFP or human Golgi-positive cells showed positive expression of podocyte markers (podocin and WT1), mesangial cell marker (megsin), and endothelial markers (CD31 and von Willebrand factor, vWF) (Fig. 11.4). Of the GFP-positive cells in the glomerulus, 31.1% ± 1.7% expressed podocin, 12.6% ± 1.7% expressed megsin, and 41.0% ± 4.3% expressed CD31. Cell fusion was ruled out by fluorescence in situ hybridization analysis, suggesting that migrated Muse cells spontaneously differentiated into glomerular cells following the theory of site. Creatinine clearance, a surrogate marker of GFR, at 7 weeks was significantly higher in the Muse group compared to that in the vehicle group. Non-Muse group did not show any improvement when compared with the vehicle group. Surprisingly, even in FSGS-BALB/c mice or normal immune model mice without immunosuppressant, human Muse cells migrated into glomerulus and differentiated into podocytes, mesangial cells, and epithelial cells. The improvement in renal function was more distinct than that in the FSGS-SCID experiment; however, the therapeutic effect of Muse cells was diminished after 5 weeks due to immunorejection (Fig. 11.5). At 5 weeks, GFP- or human Golgi-positive glomerular cells that expressed podocin, WT1, megsin, vWF, or CD31 were only present in the Muse group. While few non-Muse cells were detected in mice kidney, non-Muse group showed slight improvement in renal function. The results were consistent with those of previous studies that used MSCs [56, 70]. The results seem plausible because non-Muse cells, which account for a large proportion of MSCs, exert renoprotective effects via paracrine mechanism. In addition, Muse cells were shown to have a paracrine capacity [64] and a similar immunomodulatory ability [23]. Because renal dysfunction in FSGS model aggravates with chronic inflammation, normal immune mice may more clearly reflect the efficacy of cell therapy mediated via paracrine mechanism and immunomodulation.

In addition to the capacity to engraft and differentiate, migratory potential is another important attribute of Muse cells. We examined the migratory capacity of

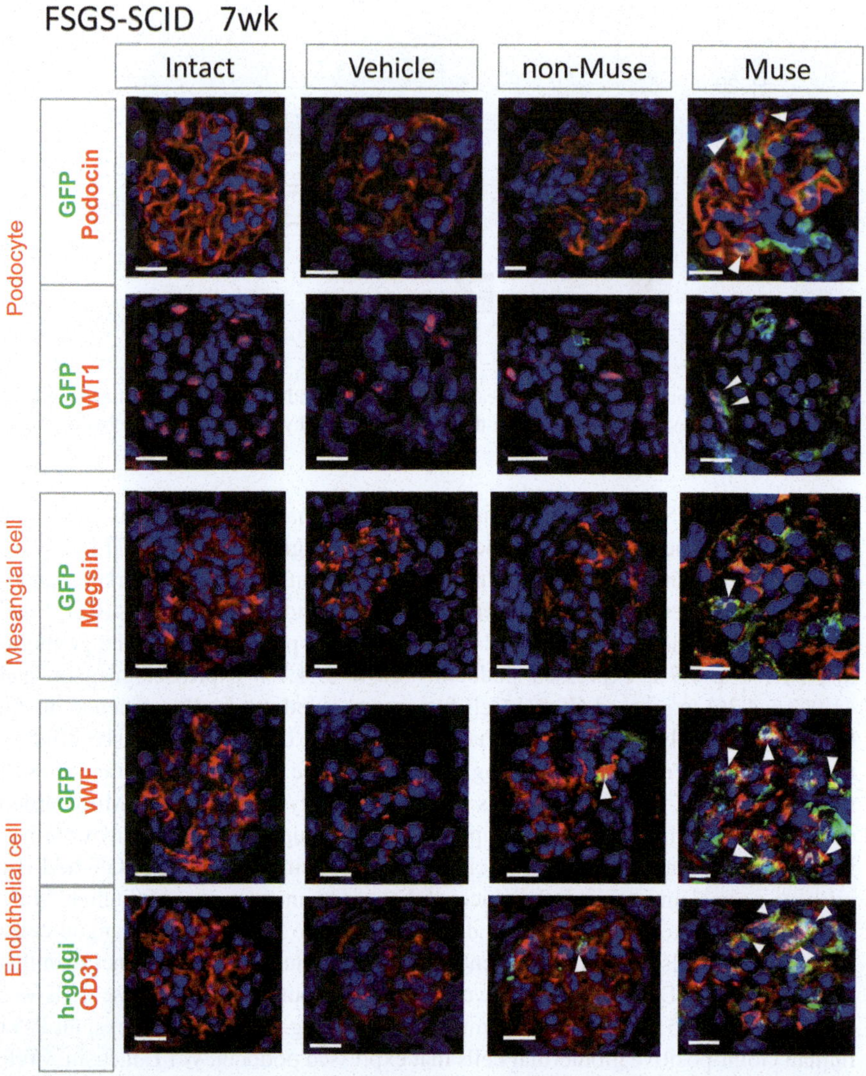

Fig. 11.4 Muse cells spontaneously differentiate into glomerular cells in FSGS-SCID mice (7 weeks old). Expressions of the podocyte markers podocin and WT1, the mesangial cell marker megsin, and the endothelial cell markers vWF and CD31 in the glomerulus of intact SCID and FSGS-SCID of the vehicle, non-Muse, and Muse groups. Human Muse and non-Muse cells were detected as GFP- or human Golgi complex-positive cells. Scale bar, 20 mm

human Muse cells and non-Muse cells toward serum from both normal and FSGS-SCID mice using a Matrigel invasion chamber. Compared with the serum of intact normal mice, the number of Muse cells that migrated to FSGS mouse serum was significantly greater than that of non-Muse cells (Fig. 11.6A). Because of the outstanding capacity for migration and differentiation into injured cells, Muse cells

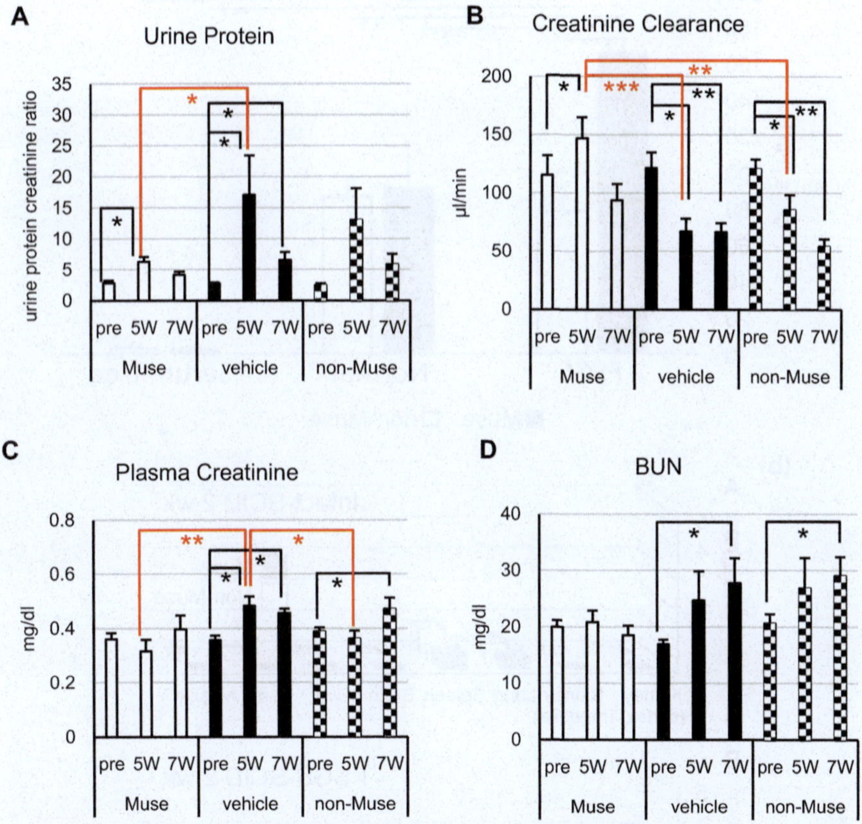

Fig. 11.5 Recovery of renal functions in FSGS-BALB/c mice at 5 weeks post-Muse cell infusion. (**A**) Urine protein-to-creatinine ratio (urine protein), (**B**) creatinine clearance, (**C**) plasma creatinine, and (**D**) BUN. (**A**) At 5 weeks, urine protein levels in the Muse group were significantly attenuated compared with that in the vehicle group ($P < 0.05$). (**B**) Creatinine clearance in the Muse group was significantly higher than that in the vehicle ($P < 0.001$) and non-Muse ($P < 0.01$) groups; the results indicated substantial recovery of renal function in the Muse group at 5 weeks. (**C**) Plasma creatinine levels in the Muse ($P < 0.01$) and non-Muse ($P < 0.05$) groups were significantly lower than that in the vehicle group at 5 weeks. (**D**) BUN was stable in the Muse group, whereas it was significantly increased in the vehicle and non-Muse groups at 7 weeks

were found distributed in the kidney of FSGS-SCID 2–7 weeks after intravenous injection; in contrast, non-Muse cells were distributed mainly in the lungs and spleen at 2 weeks and were scarcely detected in any organs at 7 weeks (Fig. 11.6B). The distribution of non-Muse cells was consistent with the distribution of MSCs reported in previous studies [71]. Considering the heterogeneity of MSCs, our results suggest that Muse cells have higher levels of stemness and reparatory ability as compared to that of other MSC populations. However, both FSGS-SCID and BALB/c mice showed significant improvement in renal function; since the number of integrated Muse cells was considered not to be very high, all of the functional

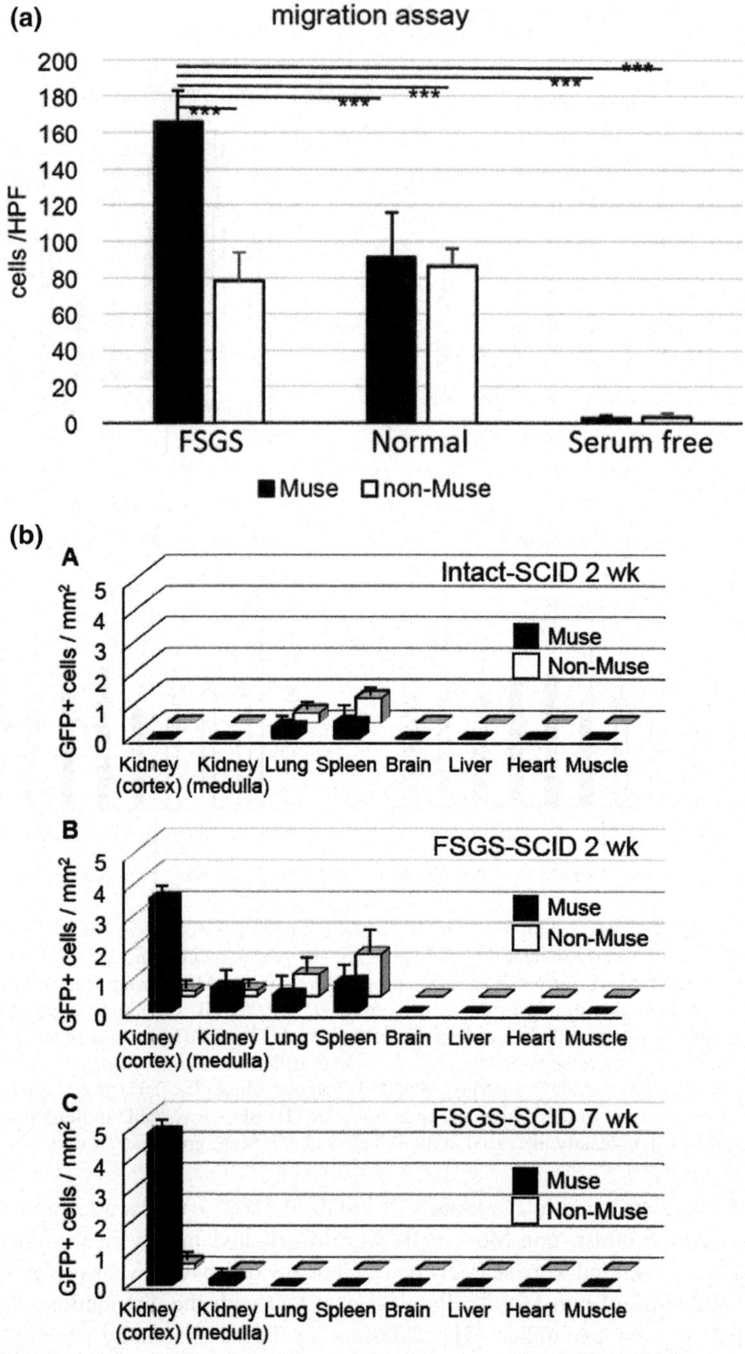

Fig. 11.6 (**a**) Migration assay. Muse cells exhibit migration capacity toward the serum sourced from FSGS-SCID mice. The number of migrated human Muse and non-Muse cells was counted after 22 h of incubation in a Matrigel invasion chamber with serum from normal SCID or FSGS-SCID mice. Serum-free was set as the control. ***$P < 0.001$. (**b**) Distribution of GFP(+) cells per square millimeter in the kidney (cortex and medulla), lung, spleen brain, liver, heart, and muscle tissues of Muse and non-Muse groups. (*A*) The intact SCID (2 weeks); (*B*) FSGS-SCID (2 weeks); and (*C*) FSGS-SCID (7 weeks)

improvement may not be attributed to the replenishment of damaged renal cells by Muse cells. Muse cells also seem to exert a renoprotective effect via paracrine mechanism as previously reported [64]. To summarize, the functional recovery attributable to Muse cells likely reflects the synergistic effect of multiple mechanisms that include cell differentiation, paracrine effects, and immunomodulation.

11.6 Future Prospects of Muse Cell Therapy in Kidney Diseases

Our results are innovative because they demonstrate spontaneous in vivo differentiation of Muse cells, a small subpopulation of MSCs, into three types of glomerular cells, which has scarcely been reported in previous studies that employed MSCs.

The successful demonstration of the beneficial effects of MSCs in preclinical studies paved the way for clinical trials of MSC therapy in patients with kidney diseases. Muse cells are a subpopulation of MSCs; therefore, their clinical use is not a wild tale. More importantly, the distinct migratory and differentiation potential of Muse cells envisages promising prospects of their potential use in kidney diseases.

In clinical settings, one of the great advantages of Muse cells over other pluripotent stem cells, like iPS cells or embryonic stem cells, is their non-tumorigenicity [62, 72, 73] because Muse cells are innate somatic stem cells with non-tumorigenic property. Secondly, simple procedure is available in Muse cell treatment; collection of Muse cells, such as SSEA-3+ cells from accessible sources, expands and administrates to patients by intravenous drip. As shown in our study, intravenously administrated naïve Muse cells migrated to the damaged kidney and spontaneously differentiated into glomerular cells and repair the tissue. All of the above procedures do not require gene transfer or renal cell induction in cell processing center. Thirdly, Muse cells exhibit a remarkable immunosuppressive property. Muse cells demonstrated therapeutic effects in normal immune mice for up to 5 weeks, presumably because of their immunosuppressive property. The results suggest that allotransplantation of human Muse cells may be effective for a longer period.

As mentioned above, progression of kidney diseases occurs over years and, in many cases, is associated with chronic inflammation. This makes it particularly a suitable target for cell therapy. It might be possible to collect auto-MSCs in the early or mid-stage of the disease and culture them ex vivo to recover their therapeutic potential that tends to deteriorate under uremic conditions [74]. This renders auto-Muse cell therapy as a distinctly feasible therapeutic option in clinical settings.

Finally, the most distinguishing feature of Muse cells is their ability to differentiate into damaged cells. Our study used FSGS model that is characterized by glomerular sclerosis. Because the glomerulus is more vulnerable and hard to recover than the renal tubules, Muse cell therapy may demonstrate better therapeutic effect in glomerular diseases such as chronic glomerular nephritis and diabetic nephropathy as compared to that in tubular diseases.

In conclusion, kidney diseases represent a major health issue worldwide. Renal regenerative therapy is one of the potential therapeutic options for kidney diseases, and clinical trials of MSCs have already been started. Muse cells, a small subpopulation of MSCs that exhibit a high level of stemness, the ability to migrate to the damaged site and to replenish the loss of glomerular cells, improved renal function in FSGS model mice. Muse cells are a promising cell source for regenerative therapy for kidney diseases.

References

1. Perantoni AO (2003) Renal development: perspectives on a Wnt-dependent process. Semin Cell Dev Biol 14:201–208
2. Little MH (2015) Kideny development, disease, repair and regeneration. Acdemic press, Boston
3. Schell C, Wanner N, Huber TB (2014) Glomerular development–shaping the multi-cellular filtration unit. Semin Cell Dev Biol 36:39–49
4. Nagata M (2016) Podocyte injury and its consequences. Kidney Int 89:1221–1230
5. Andeen NK, Nguyen TQ, Steegh F, Hudkins KL, Najafian B, Alpers CE (2015) The phenotypes of podocytes and parietal epithelial cells may overlap in diabetic nephropathy. Kidney Int 88:1099–1107
6. Becker JU, Hoerning A, Schmid KW, Hoyer PF (2007) Immigrating progenitor cells contribute to human podocyte turnover. Kidney Int 72:1468–1473
7. Kellum JA, Lameire N (2013) Diagnosis, evaluation, and management of acute kidney injury: a KDIGO summary (Part 1). Crit Care 17:204
8. Basile DP, Anderson MD, Sutton TA (2012) Pathophysiology of acute kidney injury. Compr Physiol 2:1303–1353
9. Westenfelder C, Togel FE (2011) Protective actions of administered mesenchymal stem cells in acute kidney injury: relevance to clinical trials. Kidney int Suppl 1:103–106
10. Levey AS, Eckardt KU, Tsukamoto Y, Levin A, Coresh J, Rossert J, De Zeeuw D, Hostetter TH, Lameire N, Eknoyan G (2005) Definition and classification of chronic kidney disease: a position statement from Kidney Disease: Improving Global Outcomes (KDIGO). Kidney Int 67:2089–2100
11. Jha V, Garcia-Garcia G, Iseki K, Li Z, Naicker S, Plattner B, Saran R, Wang AY, Yang CW (2013) Chronic kidney disease: global dimension and perspectives. Lancet 382:260–272
12. Thomas B, Matsushita K, Abate KH, Al-Aly Z, Arnlov J, Asayama K, Atkins R, Badawi A, Ballew SH, Banerjee A, Barregard L, Barrett-Connor E, Basu S, Bello AK, Bensenor I, Bergstrom J, Bikbov B, Blosser C, Brenner H, Carrero JJ, Chadban S, Cirillo M, Cortinovis M, Courville K, Dandona L, Dandona R, Estep K, Fernandes J, Fischer F, Fox C, Gansevoort RT, Gona PN, Gutierrez OM, Hamidi S, Hanson SW, Himmelfarb J, Jassal SK, Jee SH, Jha V, Jimenez-Corona A, Jonas JB, Kengne AP, Khader Y, Khang YH, Kim YJ, Klein B, Klein R, Kokubo Y, Kolte D, Lee K, Levey AS, Li Y, Lotufo P, HMA ER, Mendoza W, Metoki H, Mok Y, Muraki I, Muntner PM, Noda H, Ohkubo T, Ortiz A, Perico N, Polkinghorne K, Al-Radaddi R, Remuzzi G, Roth G, Rothenbacher D, Satoh M, Saum KU, Sawhney M, Schottker B, Shankar A, Shlipak M, DAS S, Toyoshima H, Ukwaja K, Umesawa M, Vollset SE, Warnock DG, Werdecker A, Yamagishi K, Yano Y, Yonemoto N, MES Z, Naghavi M, Forouzanfar MH, CJL M, Coresh J, Vos T (2017) Global cardiovascular and renal outcomes of reduced GFR. J Am Soc Nephrol 28:2167–2179
13. Sharma S, Sarnak MJ (2017) Epidemiology: the global burden of reduced GFR: ESRD, CVD and mortality. Nat Rev Nephrol 13:447–448
14. Go AS, Chertow GM, Fan D, McCulloch CE, Hsu CY (2004) Chronic kidney disease and the risks of death, cardiovascular events, and hospitalization. N Engl J Med 351:1296–1305

15. Roushandeh AM, Bahadori M, Roudkenar MH (2017) Mesenchymal stem cell-based therapy as a new horizon for kidney injuries. Arch Med Res 48:133–146
16. Hu J, Zhang L, Wang N, Ding R, Cui S, Zhu F, Xie Y, Sun X, Wu D, Hong Q, Li Q, Shi S, Liu X, Chen X (2013) Mesenchymal stem cells attenuate ischemic acute kidney injury by inducing regulatory T cells through splenocyte interactions. Kidney Int 84:521–531
17. Tsuda H, Yamahara K, Otani K, Okumi M, Yazawa K, Kaimori JY, Taguchi A, Kangawa K, Ikeda T, Takahara S, Isaka Y (2014) Transplantation of allogenic fetal membrane-derived mesenchymal stem cells protects against ischemia/reperfusion-induced acute kidney injury. Cell Transplant 23:889–899
18. Morigi M, Introna M, Imberti B, Corna D, Abbate M, Rota C, Rottoli D, Benigni A, Perico N, Zoja C, Rambaldi A, Remuzzi A, Remuzzi G (2008) Human bone marrow mesenchymal stem cells accelerate recovery of acute renal injury and prolong survival in mice. Stem Cells 26:2075–2082
19. Ninichuk V, Gross O, Segerer S, Hoffmann R, Radomska E, Buchstaller A, Huss R, Akis N, Schlondorff D, Anders HJ (2006) Multipotent mesenchymal stem cells reduce interstitial fibrosis but do not delay progression of chronic kidney disease in collagen4A3-deficient mice. Kidney Int 70:121–129
20. Sun D, Bu L, Liu C, Yin Z, Zhou X, Li X, Xiao A (2013) Therapeutic effects of human amniotic fluid-derived stem cells on renal interstitial fibrosis in a murine model of unilateral ureteral obstruction. PLoS One 8:e65042
21. Toyohara T, Mae S, Sueta S, Inoue T, Yamagishi Y, Kawamoto T, Kasahara T, Hoshina A, Toyoda T, Tanaka H, Araoka T, Sato-Otsubo A, Takahashi K, Sato Y, Yamaji N, Ogawa S, Yamanaka S, Osafune K (2015) Cell therapy using human induced pluripotent stem cell-derived renal progenitors ameliorates acute kidney injury in mice. Stem Cells Transl Med 4:980–992
22. Luo J, Zhao X, Tan Z, Su Z, Meng F, Zhang M (2013) Mesenchymal-like progenitors derived from human embryonic stem cells promote recovery from acute kidney injury via paracrine actions. Cytotherapy 15:649–662
23. Uchida N, Kushida Y, Kitada M, Wakao S, Kumagai N, Kuroda Y, Kondo Y, Hirohara Y, Kure S, Chazenbalk G, Dezawa M (2017) Beneficial effects of systemically administered human muse cells in adriamycin nephropathy. J Am Soc Nephrol 28:2946–2960
24. Imasawa T, Utsunomiya Y, Kawamura T, Zhong Y, Nagasawa R, Okabe M, Maruyama N, Hosoya T, Ohno T (2001) The potential of bone marrow-derived cells to differentiate to glomerular mesangial cells. J Am Soc Nephrol 12:1401–1409
25. Ito T, Suzuki A, Imai E, Okabe M, Hori M (2001) Bone marrow is a reservoir of repopulating mesangial cells during glomerular remodeling. J Am Soc Nephrol 12:2625–2635
26. Rookmaaker MB, Smits AM, Tolboom H, Van 't Wout K, Martens AC, Goldschmeding R, Joles JA, Van Zonneveld AJ, Grone HJ, Rabelink TJ, Verhaar MC (2003) Bone-marrow-derived cells contribute to glomerular endothelial repair in experimental glomerulonephritis. Am J Pathol 163:553–562
27. Prodromidi EI, Poulsom R, Jeffery R, Roufosse CA, Pollard PJ, Pusey CD, Cook HT (2006) Bone marrow-derived cells contribute to podocyte regeneration and amelioration of renal disease in a mouse model of Alport syndrome. Stem Cells 24:2448–2455
28. Poulsom R, Forbes SJ, Hodivala-Dilke K, Ryan E, Wyles S, Navaratnarasah S, Jeffery R, Hunt T, Alison M, Cook T, Pusey C, Wright NA (2001) Bone marrow contributes to renal parenchymal turnover and regeneration. J Pathol 195:229–235
29. Kale S, Karihaloo A, Clark PR, Kashgarian M, Krause DS, Cantley LG (2003) Bone marrow stem cells contribute to repair of the ischemically injured renal tubule. J Clin Invest 112:42–49
30. Fang TC, Alison MR, Cook HT, Jeffery R, Wright NA, Poulsom R (2005) Proliferation of bone marrow-derived cells contributes to regeneration after folic acid-induced acute tubular injury. J Am Soc Nephrol 16:1723–1732

31. Ikarashi K, Li B, Suwa M, Kawamura K, Morioka T, Yao J, Khan F, Uchiyama M, Oite T (2005) Bone marrow cells contribute to regeneration of damaged glomerular endothelial cells. Kidney Int 67:1925–1933
32. Li B, Morioka T, Uchiyama M, Oite T (2006) Bone marrow cell infusion ameliorates progressive glomerulosclerosis in an experimental rat model. Kidney Int 69:323–330
33. Sugimoto H, Mundel TM, Sund M, Xie L, Cosgrove D, Kalluri R (2006) Bone-marrow-derived stem cells repair basement membrane collagen defects and reverse genetic kidney disease. Proc Natl Acad Sci USA 103:7321–7326
34. Morigi M, Imberti B, Zoja C, Corna D, Tomasoni S, Abbate M, Rottoli D, Angioletti S, Benigni A, Perico N, Alison M, Remuzzi G (2004) Mesenchymal stem cells are renotropic, helping to repair the kidney and improve function in acute renal failure. J Am Soc Nephrol 15:1794–1804
35. Borges FT, Schor N (2018) Regenerative medicine in kidney disease: where we stand and where to go. Pediatr Nephrol 33:1457–1465
36. Jiang Y, Jahagirdar BN, Reinhardt RL, Schwartz RE, Keene CD, Ortiz-Gonzalez XR, Reyes M, Lenvik T, Lund T, Blackstad M, Du J, Aldrich S, Lisberg A, Low WC, Largaespada DA, Verfaillie CM (2002) Pluripotency of mesenchymal stem cells derived from adult marrow. Nature 418:41–49
37. Toma C, Pittenger MF, Cahill KS, Byrne BJ, Kessler PD (2002) Human mesenchymal stem cells differentiate to a cardiomyocyte phenotype in the adult murine heart. Circulation 105:93–98
38. Luo H, Xu C, Liu Z, Yang L, Hong Y, Liu G, Zhong H, Cai X, Lin X, Chen X, Wang C, Zhang N, Xu W (2017) Neural differentiation of bone marrow mesenchymal stem cells with human brain-derived neurotrophic factor gene-modified in functionalized self-assembling peptide hydrogel in vitro. J Cell Biochem. https://doi.org/10.1002/jcb.26408
39. Sato Y, Araki H, Kato J, Nakamura K, Kawano Y, Kobune M, Sato T, Miyanishi K, Takayama T, Takahashi M, Takimoto R, Iyama S, Matsunaga T, Ohtani S, Matsuura A, Hamada H, Niitsu Y (2005) Human mesenchymal stem cells xenografted directly to rat liver are differentiated into human hepatocytes without fusion. Blood 106:756–763
40. Chen L, Tredget EE, Wu PY, Wu Y (2008) Paracrine factors of mesenchymal stem cells recruit macrophages and endothelial lineage cells and enhance wound healing. PLoS One 3:e1886
41. Baer PC, Geiger H (2010) Mesenchymal stem cell interactions with growth factors on kidney repair. Curr Opin Nephrol Hypertens 19:1–6
42. Cantaluppi V, Biancone L, Quercia A, Deregibus MC, Segoloni G, Camussi G (2013) Rationale of mesenchymal stem cell therapy in kidney injury. Am J Kidney Dis 61:300–309
43. Wakabayashi K, Nagai A, Sheikh AM, Shiota Y, Narantuya D, Watanabe T, Masuda J, Kobayashi S, Kim SU, Yamaguchi S (2010) Transplantation of human mesenchymal stem cells promotes functional improvement and increased expression of neurotrophic factors in a rat focal cerebral ischemia model. J Neurosci Res 88:1017–1025
44. Bruno S, Grange C, Collino F, Deregibus MC, Cantaluppi V, Biancone L, Tetta C, Camussi G (2012) Microvesicles derived from mesenchymal stem cells enhance survival in a lethal model of acute kidney injury. PLoS One 7:e33115
45. Maeshima A, Nakasatomi M, Nojima Y (2014) Regenerative medicine for the kidney: renotropic factors, renal stem/progenitor cells, and stem cell therapy. Biomed Res Int 2014:595493
46. Matsumoto K, Mizuno S, Nakamura T (2000) Hepatocyte growth factor in renal regeneration, renal disease and potential therapeutics. Curr Opin Nephrol Hypertens 9:395–402
47. Togel F, Zhang P, Hu Z, Westenfelder C (2009) VEGF is a mediator of the renoprotective effects of multipotent marrow stromal cells in acute kidney injury. J Cell Mol Med 13:2109–2114
48. Humes HD, Cieslinski DA, Coimbra TM, Messana JM, Galvao C (1989) Epidermal growth factor enhances renal tubule cell regeneration and repair and accelerates the recovery of renal function in postischemic acute renal failure. J Clin Invest 84:1757–1761
49. Miller SB, Martin DR, Kissane J, Hammerman MR (1992) Insulin-like growth factor I accelerates recovery from ischemic acute tubular necrosis in the rat. Proc Natl Acad Sci USA 89:11876–11880

50. Chen J, Chen JK, Harris RC (2012) Deletion of the epidermal growth factor receptor in renal proximal tubule epithelial cells delays recovery from acute kidney injury. Kidney Int 82:45–52
51. Zhou D, Tan RJ, Lin L, Zhou L, Liu Y (2013) Activation of hepatocyte growth factor receptor, c-met, in renal tubules is required for renoprotection after acute kidney injury. Kidney Int 84:509–520
52. Behr L, Hekmati M, Fromont G, Borenstein N, Noel LH, Lelievre-Pegorier M, Laborde K (2007) Intra renal arterial injection of autologous mesenchymal stem cells in an ovine model in the postischemic kidney. Nephron Physiol 107:65–76
53. Togel F, Weiss K, Yang Y, Hu Z, Zhang P, Westenfelder C (2007) Vasculotropic, paracrine actions of infused mesenchymal stem cells are important to the recovery from acute kidney injury. Am J Physiol Renal Physiol 292:F1626–F1635
54. Bi B, Schmitt R, Israilova M, Nishio H, Cantley LG (2007) Stromal cells protect against acute tubular injury via an endocrine effect. J Am Soc Nephrol 18:2486–2496
55. Semedo P, Correa-Costa M, Antonio Cenedeze M, Maria Avancini Costa Malheiros D, Antonia dos Reis M, Shimizu MH, Seguro AC, Pacheco-Silva A, Saraiva Camara NO (2009) Mesenchymal stem cells attenuate renal fibrosis through immune modulation and remodeling properties in a rat remnant kidney model. Stem Cells 27:3063–3073
56. Choi S, Park M, Kim J, Hwang S, Park S, Lee Y (2009) The role of mesenchymal stem cells in the functional improvement of chronic renal failure. Stem Cells Develop 18:521–529
57. Squillaro T, Peluso G, Galderisi U (2016) Clinical trials with mesenchymal stem cells: an update. Cell Transplant 25:829–848
58. Togel FE, Westenfelder C (2012) Kidney protection and regeneration following acute injury: progress through stem cell therapy. Am J Kidney Dis 60:1012–1022
59. Dong BT, Tu GJ, Han YX, Chen Y (2015) Lithium enhanced cell proliferation and differentiation of mesenchymal stem cells to neural cells in rat spinal cord. Int J Clin Exp Pathol 8:2473–2483
60. Theise ND, Badve S, Saxena R, Henegariu O, Sell S, Crawford JM, Krause DS (2000) Derivation of hepatocytes from bone marrow cells in mice after radiation-induced myeloablation. Hepatology 31:235–240
61. Kuroda Y, Wakao S, Kitada M, Murakami T, Nojima M, Dezawa M (2013) Isolation, culture and evaluation of multilineage-differentiating stress-enduring (Muse) cells. Nat Protoc 8:1391–1415
62. Kuroda Y, Kitada M, Wakao S, Nishikawa K, Tanimura Y, Makinoshima H, Goda M, Akashi H, Inutsuka A, Niwa A, Shigemoto T, Nabeshima Y, Nakahata T, Nabeshima Y, Fujiyoshi Y, Dezawa M (2010) Unique multipotent cells in adult human mesenchymal cell populations. Proc Natl Acad Sci USA 107:8639–8643
63. Uchida H, Morita T, Niizuma K, Kushida Y, Kuroda Y, Wakao S, Sakata H, Matsuzaka Y, Mushiake H, Tominaga T, Borlongan CV, Dezawa M (2016) Transplantation of unique subpopulation of fibroblasts, muse cells, ameliorates experimental stroke possibly via robust neuronal differentiation. Stem Cells 34:160–173
64. Kinoshita K, Kuno S, Ishimine H, Aoi N, Mineda K, Kato H, Doi K, Kanayama K, Feng J, Mashiko T, Kurisaki A, Yoshimura K (2015) Therapeutic potential of adipose-derived SSEA-3-positive Muse cells for treating diabetic skin ulcers. Stem Cells Translat Med 4:146–155
65. Hori E, Hayakawa Y, Hayashi T, Hori S, Okamoto S, Shibata T, Kubo M, Horie Y, Sasahara M, Kuroda S (2016) Mobilization of pluripotent multilineage-differentiating stress-enduring cells in ischemic stroke. J Stroke Cerebrovasc Dis 25:1473–1481
66. Tanaka T, Nishigaki K, Minatoguchi S, Nawa T, Yamada Y, Kanamori H, Mikami A, Ushikoshi H, Kawasaki M, Dezawa M, Minatoguchi S (2018) Mobilized muse cells after acute myocardial infarction predict cardiac function and remodeling in the chronic phase. Circ J 82:561–571
67. Dezawa M (2016) Muse cells provide the Pluripotency of mesenchymal stem cells: direct contribution of muse cells to tissue regeneration. Cell Transplant 25:849–861

68. Batchelder CA, Lee CC, Matsell DG, Yoder MC, Tarantal AF (2009) Renal ontogeny in the rhesus monkey (Macaca mulatta) and directed differentiation of human embryonic stem cells towards kidney precursors. Differentiation 78:45–56
69. Lee VW, Harris DC (2011) Adriamycin nephropathy: a model of focal segmental glomerulosclerosis. Nephrology (Carlton) 16:30–38
70. Rampino T, Gregorini M, Bedino G, Piotti G, Gabanti E, Ibatici A, Sessarego N, Piacenza C, Balenzano CT, Esposito P, Bosio F, Soccio G, Frassoni F, Dal Canton A (2011) Mesenchymal stromal cells improve renal injury in anti-Thy 1 nephritis by modulating inflammatory cytokines and scatter factors. Clin Sci (Lond) 120:25–36
71. Kean TJ, Lin P, Caplan AI, Dennis JE (2013) MSCs: delivery routes and engraftment, cell-targeting strategies, and immune modulation. Stem Cells Int 2013:732742
72. Kuroda Y, Dezawa M (2014) Mesenchymal stem cells and their subpopulation, pluripotent muse cells, in basic research and regenerative medicine. Anat Rec (Hoboken) 297:98–110
73. Wakao S, Kitada M, Kuroda Y, Shigemoto T, Matsuse D, Akashi H, Tanimura Y, Tsuchiyama K, Kikuchi T, Goda M, Nakahata T, Fujiyoshi Y, Dezawa M (2011) Multilineage-differentiating stress-enduring (Muse) cells are a primary source of induced pluripotent stem cells in human fibroblasts. Proc Natl Acad Sci USA 108:9875–9880
74. Klinkhammer BM, Kramann R, Mallau M, Makowska A, van Roeyen CR, Rong S, Buecher EB, Boor P, Kovacova K, Zok S, Denecke B, Stuettgen E, Otten S, Floege J, Kunter U (2014) Mesenchymal stem cells from rats with chronic kidney disease exhibit premature senescence and loss of regenerative potential. PLoS One 9:e92115

Chapter 12
Liver Regeneration Supported by Muse Cells

Satoshi S. Nishizuka, Yuji Suzuki, Hirokatsu Katagiri, and Yasuhiro Takikawa

Abstract Cellular compensation from extrahepatic resources is expected to improve the prognosis of liver diseases. Currently, liver dysfunction is treated by a variety of modalities including drugs, cytokines, vascular interventions, energy devices, surgery, and liver transplantation; however, in recent years there have been few significant advancements in treatment efficacy. A next-generation therapeutic strategy for liver disease, cellular compensatory therapy (i.e., cell therapy), is now being considered for clinical practice. Liver dysfunction is attributed to a lack of sufficient functional cells. However, processes involved in recovery of liver function are not fully elucidated, which has complicated the interpretation of treatment effects at the cellular level. Our genotyping study of living donor liver transplantation revealed that a variety of graft liver tissues contained the donor genotype, indicating that extrahepatic cells had differentiated into liver component cells during liver regeneration. Multilineage-differentiating stress-enduring (Muse) cells appear to be a strong candidate for extrahepatic resources that can contribute to liver regeneration. Muse cells are defined as stage-specific embryonic antigen 3-expressing cells that contribute to tissue regeneration and have the potential to differentiate into three germ layers. The significant advantage of Muse cells over other "pluripotent cells" is that Muse cells are present in bone marrow/blood as well as a variety of connective tissues, which provides safety and ethical advantages for clinical applications. Here, we review current therapeutic topics in liver diseases and discuss the potential for cell therapy using Muse cells based on our recent studies of Muse cell administration in a mouse model of physical partial hepatectomy.

S. S. Nishizuka (✉)
Division of Biomedical Research and Development, Institute of Biomedical Sciences, Iwate Medical University, Morioka, Japan
e-mail: snishizu@iwate-med.ac.jp

Y. Suzuki · Y. Takikawa
Division of Hepatology, Department of Internal Medicine, Iwate Medical University School of Medicine, Morioka, Japan

H. Katagiri
Department of Surgery, Iwate Medical University School of Medicine, Morioka, Japan

M. Dezawa (ed.), *Muse Cells*, Advances in Experimental Medicine and Biology 1103, https://doi.org/10.1007/978-4-431-56847-6_12

Keywords Liver regeneration · Hepatectomy · Physical partial hepatectomy · Partial hepatectomy · Living donor liver transplant · Chimerism · Extrahepatic cells · Mesenchymal stem cells · Multilineage-differentiating stress-enduring cells · SSEA-3

12.1 Introduction

The legend of *Prometheus* describes the liver as an organ that has the power to recover from acute loss of physiological mass. Although probable, yet not fully realistic in a practical sense, "organ regeneration" requires differentiation into the multiple cell types that comprise the organ. Liver regeneration after resection (i.e., hepatectomy) generally implies "volume gain" after reduction of liver volume in clinical practice, but how volume loss is compensated is unclear. Notably, hepatocyte hypertrophy is reported to play a central role in "volume gain" in response to liver volume loss, particularly in the acute stages of liver damage [1, 2]. We recently demonstrated that liver regeneration is a multistep process in which the size of hepatocytes increases in the acute phase, followed by cellular proliferation and subsequent differentiation [3]. These observations suggest that "organ regeneration" of the liver requires cells that can differentiate into multiple distinct cell types such as hepatocytes and cholangiocytes. Thus, therapies for liver disease would be predicted to require either fully differentiated cells that can compensate for loss of cellular function and volume or an enriched cell population with proliferative potential to differentiate into multiple cell types.

Current modalities for liver diseases are roughly divided into three major categories: (i) drugs, (ii) cytokines, and (iii) vascular/surgical interventions. Drugs and cytokines generally target the entire liver to treat diffuse types of liver disease, such as hepatitis and nonalcoholic fatty liver disease [4, 5], whereas interventions are mostly performed to treat local lesions such as neoplasms and injury [6]. Although these statuses are distinct, structure reconstruction and volume compensation of the liver are mandatory for both diffuse and local liver diseases. Some of these modalities are established as "standard therapy," but none were based on the concept of tissue-level regeneration.

As one candidate for assisting current therapies for liver disease, cell therapy has been discussed for decades as an alternative for patients for whom no other modalities were effective. Cell therapies for liver diseases can be roughly divided into two groups that involve hepatocytes or mesenchymal stem cells (MSCs). Hepatocyte transplantation was developed for patients with metabolic disorders that have a hepatic basis and acute or chronic liver failure. However, due to difficulties in acquiring good quality "donor" hepatocytes, the number of hepatocyte transplantation cases was approximately 80 worldwide as of 2011 [7]. Although hepatocyte transplantation reportedly can be performed safely and improve disease status [8–10], practical constrains such as a shortage of donors, limited success of hepatocyte engraftment in a severely damaged liver, and difficulties in obtaining high-quality hepatocytes remain before this approach can be applicable in a routine fashion [11].

Because MSCs are fractions generally isolated from the bone marrow or adipose tissue and lack genetic or biochemical modifications, the risk of tumor formation following treatment with these cells is extremely low despite their ability to display multiple germ layer differentiation markers in vitro [12, 13]. Recent observations suggest that paracrine-mediated functions of MSCs attenuate acute liver failure [14]. However, which cell populations present in crude MSCs isolates make significant contributions to biological activities that encourage tissue repair is unclear [12, 15]. Based on the estimation that only a very small fraction of cells in MSCs actually contribute to liver regeneration, we found that Muse cells directly committed to the replacement of multiple liver component cells [3]. Muse cells exist in bone marrow/blood as well as a variety of connective tissues and have demonstrated a differentiating ability in various tissue types in the context of tissue repair and regeneration [16]. We examined whether "extrahepatic" Muse cells were involved in liver regeneration processes that occur after physical partial hepatectomy (PPHx). Using a mouse PPHx as a model of human living donor liver transplantation (LDLT), we demonstrated how Muse cells were integrated into the damaged liver during liver regeneration and participated in the tissue reconstruction [3].

The human LDLT procedure uses a healthy donor graft explanted with a portion of liver that is physically removed from the healthy organ, namely, hepatectomy. This unique feature of LDLT provides an opportunity to trace what types of cells are involved in liver tissue repair and regeneration. Cells in the liver, post-engraftment of liver transplantation, have been shown to possess the recipient genotype, which suggests that extrahepatic cells are indeed integrated in the graft liver [17–21]. Interestingly, among the cells engrafted, cholangiocytes were reported to possess multipotency as evidenced by their expression of both hepatocyte and cholangiocyte lineage markers [20].

In this chapter, we will first review current therapies for liver diseases and then discuss possibilities for cellular therapy, with a special focus on Muse cells. Based on our current findings that suggest a substantial role for Muse cells in post PPHx liver regeneration, we will discuss the biology and potential applications for cell therapy using Muse cells.

12.2 Hepatectomy

Hepatectomy involves surgical resection for localized liver disease and LDLT donors. In humans, hepatectomy has been performed to treat various liver neoplastic diseases including hepatocellular carcinoma [22] and metastatic liver tumors [23, 24], as well as for the acquisition of graft liver for LDLT [25, 26]. At 1 year post-operation, the percentage of liver growth of the remnant left liver of LDLT donors has been reported to be 146% in average, whereas the "simple" right liver graft becomes 152% of the original graft volume in average [27, 28]. However, the detailed mechanism of liver volume recovery in humans remains unclear because of the difficulties in acquiring multiple biopsy samples from the injury front. Additionally, no markers exist to trace the lineage of cells participating in regeneration. Therefore, to study liver regeneration mechanisms at a cellular level, partial hepatectomy (PHx) of rodents has been widely used. Most PHx procedures for rodents use a 2/3 PHx because, in contrast to the human liver, the rodent liver is divided into distinct four lobes [29, 30]. Also in contrast to the human liver, in rodents, a simple ligation of the lobe stem is sufficient to accurately remove the liver and leave an intended volume. In fact, ligation of the base of the left lateral lobe and the median lobes corresponds to a 2/3 volume reduction [31, 32]. However, there is no "injury" region in the remnant liver resulting from the "hepatectomy" mimicked by the ligation technique. PHx in humans is almost always concomitant with an injury caused by transection, where acute inflammatory reactions and wound-healing processes actively occur. Thus, to assess liver regeneration mechanism after an injury at the cellular level, we recently introduced the PPHx technique, which involves a transection line made in the middle of the left lobe of the rodent liver that results in approximately a 30% hepatectomy. The PPHx also retains a long transection (i.e., injured) line, such that the procedure is relatively similar to that of a hepatectomy in humans in terms of the degree of invasiveness and the recovery process from the local liver injury. This model is thus particularly useful for studying host reactions in acute local liver regeneration.

12.3 LDLT

Whether extrahepatic factors (i.e., cytokines or extrahepatic cells) are involved in human liver repair regeneration has long been controversial. Examination of the involvement of such factors has been extremely difficult to investigate directly in humans because it requires cellular labeling in patients undergoing hepatectomy. The only example from which insight can be gained into cellular involvement in graft liver is human LDLT genotyping [33]. For LDLT, a section of liver (i.e., graft) transected out from a healthy donor is transplanted into a patient from whom the entire malfunctioning liver has been removed (Fig. 12.1). Therefore, at a histological level, the graft liver has a transection line equivalent to the liver injury created by discontinuation of the hepatic parenchyma and vessels [34]. The transplanted graft liver appears to undergo three important steps in the recovery process: (a) acute inflammatory reactions, particularly in the area of injury; (b) semi-acute volume recovery via hepatocyte hypertrophy; and (c) liver vascular regeneration.

All three steps require cellular resources for local tissue repair and liver volume compensation. To clarify how the liver acquires volume through these steps, we must first evaluate hepatocytes in LDLT grafts in terms of liver volume rendering and histological examinations. Among 13 available LDLT graft liver cases treated at Iwate Medical University Hospital that were examined at three time points, 8 cases showed liver volume gains (range, 108–169% of initial graft volume), whereas the other 5 cases had reduced volume (range, 82–99%; Fig. 12.2A). However,

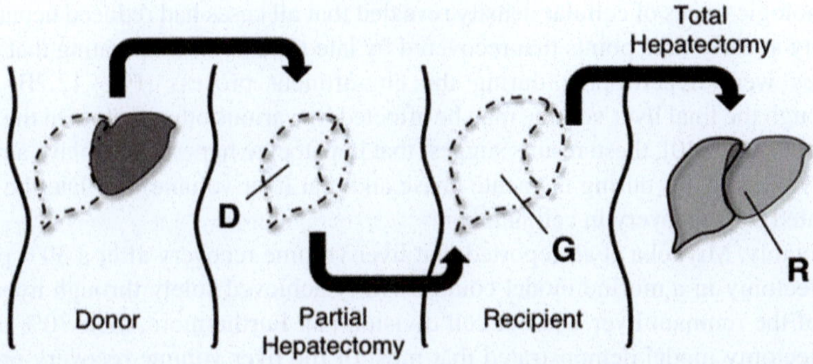

Fig. 12.1 Schematic illustration of LDLT. The right lobe of the donor is removed by hepatectomy (D) and transplanted into the recipient (R) as a graft (G). During liver regeneration, including tissue repair processes in the graft, two genotypes, from donor and recipient, are present in the graft liver. *D* donor, *R* recipient, and *G* graft liver. (The data are reproduced from Katagiri et al. [3])

Fig. 12.2 *Changes in volume and hepatocyte number in graft liver tissue*. (**A**) Change in the LDLT graft liver volume in recipients, relative to the initial "zero-point" graft volume, represented as 100%. A horizontal dashed line indicates 80% relative to the initial volume. (**B**) The number of hepatocytes counted in needle biopsy specimens at three different time points. The view area is 200 × 200 μm². (The data are reproduced from Katagiri et al. [35])

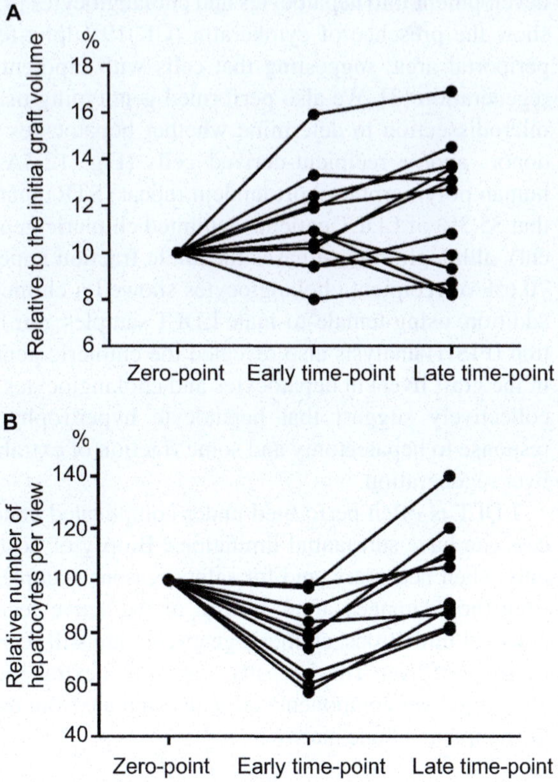

chronologic traces of cellular density revealed that all cases had reduced hepatocyte density at early time points that recovered by late time points, indicating that hepatocytes were hypertrophic during the engraftment process (Fig. 12.2B) [35]. Although the final liver volume may be affected by various other factors in the clinical course [36–40], these results suggest that hepatocyte hypertrophy plays a role in liver volume gains during the acute phase and that liver volume may later be compensated by a recovery in cell number.

Notably, Miyaoka et al. reported that liver volume recovery after a 30% partial hepatectomy in a murine model could be fully achieved solely through hypertrophy of the remnant liver without cell division [2]. Furthermore, their 70% partial hepatectomy model demonstrated that most of the liver volume recovery process is primarily due to hepatocyte hypertrophy and subsequent cell division. However, putative "liver progenitor cells (LPCs)" could emerge and contribute to liver regeneration in circumstances when the liver is severely injured and hepatocyte proliferation process is disrupted [41]. Although the origin of the LPCs remains to be identified, one source for LPCs could be the Canals of Hering, where hepatocytes and cholangiocytes are connected, which implies the potential for structural development into hepatocytes and cholangiocytes [42]. Our observations on LDLT show the presence of cytokeratin (CK)19^{+}/alpha-fetoprotein (AFP)$^{+}$ cells in the periportal area, suggesting that cells with bipotential play certain roles in liver regeneration [3]. We also performed genotyping using extracted DNA from laser microdissection to determine whether hepatocytes and cholangiocytes comprise donor- and/or recipient-derived cells (Fig. 12.3A–C). Genotyping analysis of human polymorphic short tandem repeat (STR) markers for hepatocytes revealed that 35.3% of LDLT patients exhibited chimeric genotypes (i.e., donor and recipient), although in most cases the allele fraction appeared to be small. Conversely, 70.6% of recipient cholangiocytes showed a chimeric genotype (Fig. 12.3D). In addition, using female-to-male LDLT samples, our fluorescence in situ hybridization (FISH) analysis also revealed the chimeric genotype (i.e., XY chromosomes in the graft liver) in hepatocytes and cholangiocytes (Fig. 12.3E–F). These results collectively suggest that hepatocyte hypertrophy plays an important role in response to hepatectomy and some fraction of extrahepatic cells may contribute to liver regeneration.

LDLT is often performed under complicated conditions, and the recovery process can have substantial limitations. Biopsy of postoperative LDLT is performed only when necessary, and for safety reasons, samples must be taken at site distant from the "damaged area" instead of the active regeneration front. Despite these practical limitations, chimeric genotypes are still seen in substantial fractions of the LDLT graft liver. These results suggest that extrahepatic cells capable of differentiating into liver component cells can originate from the blood flow, most likely from the recipient's bone marrow.

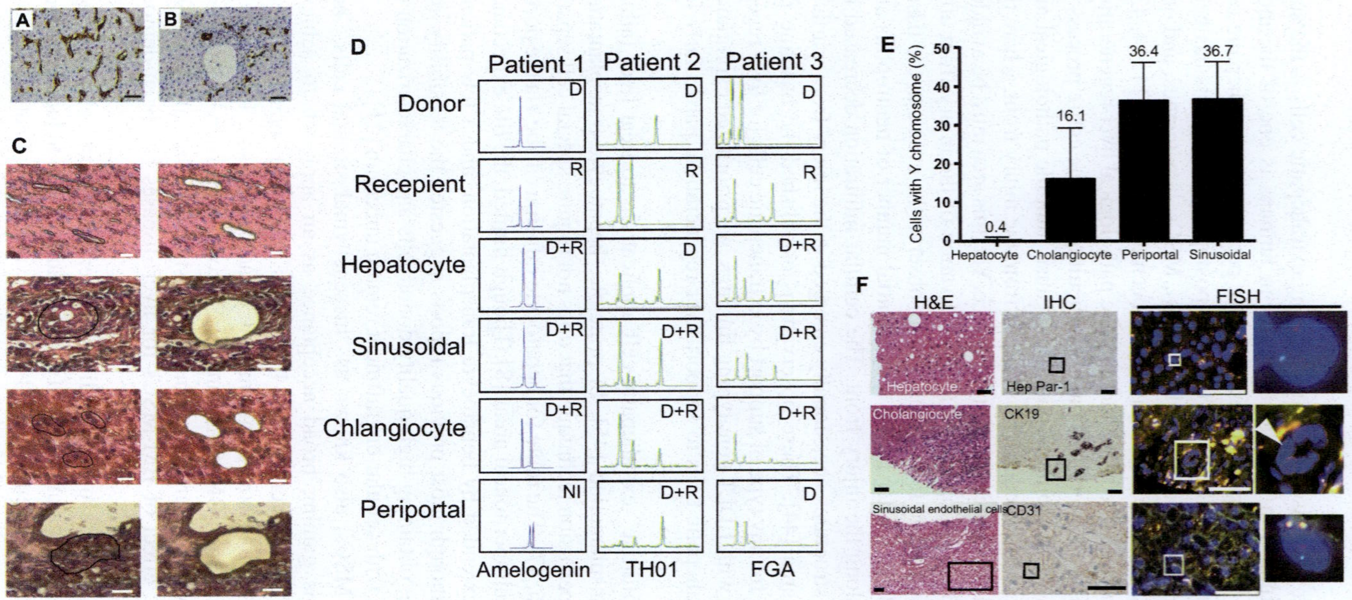

Fig. 12.3 Genotyping of LDLT samples. (**A**) CD68 and (**B**) CK7 immunostaining provide the orientation for laser microdissection of the liver sinusoid and bile duct, respectively. (**C**) Samples before (left column) and after (right column) laser microdissection. Each laser-microdissected specimen was subjected to DNA extraction and yielded sufficient amounts of DNA for PCR analysis. (**D**) Three STR markers (Amelogenin, TH01, and FGA) are represented for four different histological fractions. Donor and recipient DNA were obtained from peripheral blood mononuclear cells. (**E**) Frequency of cells carrying a Y chromosome in cases of female donors and male recipient specimens in respective cell types (n = 3). (**F**) FISH images of one sex-mismatch case. From the top row, hepatocytes, cholangiocytes, and sinusoidal endothelial cells. Cell types were identified by immunohistochemistry. Green and red fluorescence represents Y and X chromosomes, respectively. *D* donor, *R* recipient, and *NI* not informative. H&E, hematoxylin and eosin; IHC, immunohistochemistry; and FISH, fluorescent in situ hybridization. (The data are reproduced from Katagiri et al. [3])

12.4 Extrahepatic Resources for Liver Regeneration

One of the demands for liver disease treatment using extrahepatic cells is the support for hepatic local injury. In fact, hepatectomy in humans is one of the most invasive and advanced surgical procedures currently performed. Extrahepatic cell support of liver regeneration at the tissue level for postoperative patients has been reported on human hematopoietic stem cells (HSCs), MSCs, hepatocytes, and others; however, no practical example standards have been established [9, 11, 14, 43]. One major difficulty has been due to the fact that the pathological liver regeneration process has not been fully clarified. It is indeed difficult to see the entire process of liver regeneration after hepatectomy in humans because multiple pathological and imaging examinations are required, such as biopsy from the injury front, where the most active histological liver regeneration takes place. Yet, this area must be avoided due to the risk of bleeding. Using a time axis observation of the injury front after PPHx in a mouse, we were able to chronologically observe the events at the tissue level [3]. In the first 24–48 h after PPHx, a substantial number of neutrophils and monocytes infiltrated along the injury front. The cellular infiltration decreased by 72 h, but bile duct-like structures occasionally emerged in the injury front, particularly in the periportal area [44]. These observations established the baseline for steps that are adoptable for any type of support by extrahepatic cells for liver regeneration. Taken together with LDLT genotyping findings, we prioritized exploration of the potential of extrahepatic cells to support liver regeneration, particularly for such a liver injury.

One of the most "intuitive" extrahepatic cell applications to treat liver failure may be hepatocyte transplantation [45]. In principle, hepatocyte transplantation involves a collagenase perfusion technique that does not require special cell separation by fractionation [46, 47]. The route for hepatocyte infusion can be via the portal vein or intrasplenic or intraperitoneal areas [8]. Though limited, the life expectancy of some patients was extended by hepatocyte transplantation, and a few patients fully recovered [8]. Although most previous reports suggested the feasibility and safety of this procedure, obtaining high-quality donor cells and characterization of mechanisms associated with cellular engraftment remain uncertain [11].

The application of MSCs, also known as "mesenchymal stromal cells," has been established and is gaining broad acceptance as an approach to achieve cell-mediated recovery from liver failure [48]. MSCs have no associated ethical issues due to their derivation from bone marrow or adipose tissue. In a clinical setting, HSC transplantation of BM transplants likely containing a small number of BM-MSCs has already been performed and demonstrated to be safe [48, 49]. Indeed, the therapeutic effect of BM-MSCs for liver diseases has been demonstrated in humans [50, 51].

As one of the extrahepatic resources for liver regeneration, Muse cells have few ethical concerns and show low tumorigenicity in comparison with induced pluripotent stem (iPS) and embryonic stem (ES) cells [16]. The natural application of Muse cell administration is to treat liver failure by supporting tissue-level development at liver damage sites. A previous study already demonstrated that Muse cells can contribute to liver regeneration in response to carbon tetrachloride-induced acute liver failure, and our more recent study showed that Muse cells are meaningful contributors to liver damage induced by a local injury in the context of liver regeneration at the tissue level [3, 16].

12.5 Differentiation of Muse Cells that Are Integrated into the Liver

Muse cells were first reported in 2010 as a distinct population of pluripotent stem cells that account for a small percentage of MSCs [16]. The functional identification of Muse cells was reported as a stress-tolerant fraction of MSCs that survived in the presence of trypsin or in the presence of limited nutrients [52]. Although Muse cells share properties with MSCs and express mesenchymal markers such as CD105, CD90, and CD29, Muse cells can be specifically characterized by the expression of the glycolipid, stage-specific embryonic antigen-3 (SSEA-3), which is a well-known marker of undifferentiated ESs and other pluripotent cells in humans. The SSEA-3$^+$ fraction comprises ~1% of BM-MSCs in human BM aspirates and represents ~0.03% of BM-mononucleated cells [16]. Recent studies have revealed that Muse cells could be found in peripheral blood and in a wide range of connective tissues [53–55]. In addition, Muse cells have been reported to be a primary source of iPS cells [56, 57].

The PPHx model, in contrast to postoperative human specimens, provides opportunities to continually trace chronological changes in the liver, particularly in areas close to physical damage sites. From results of our LDLT genotyping studies, we speculated that extrahepatic cells such as Muse cells play certain roles in liver regeneration [3]. Moreover, accumulated findings implied that extrahepatic cells could contribute to recovery from various hepatic failures [45]. Thus, we combined the PPHx model and Muse cell administration so that the process of Muse cell integration could be observed in the context of liver regeneration.

We showed that intravenous infusion of GFP-labeled human Muse cells into the SICD mouse xenograft model demonstrated the integration of Muse cells in the damaged area adjacent to the transection line wherein GFP-labeled Muse cells appeared in the periportal regions adjacent to the actual injury at 1 week post-hepatectomy (Fig. 12.4). However, at that point, those cells did not appear to be immediately differentiated into any tissue components and were difficult to distinguish from cells that had simply proliferated around the injury as part of the inflammation process. At 2 weeks after infusion, some of the Muse cells began to form bile duct-like structures. Muse cells were also found in the sinusoid area. Integration of the Muse cells traced up to 4 weeks showed that the population of Muse cells that had integrated into the liver composed of cholangiocytes (17.7%), hepatocytes (74.3%), Kupffer cells (6.0%), and sinusoidal endothelial cells (2.0%). Importantly,

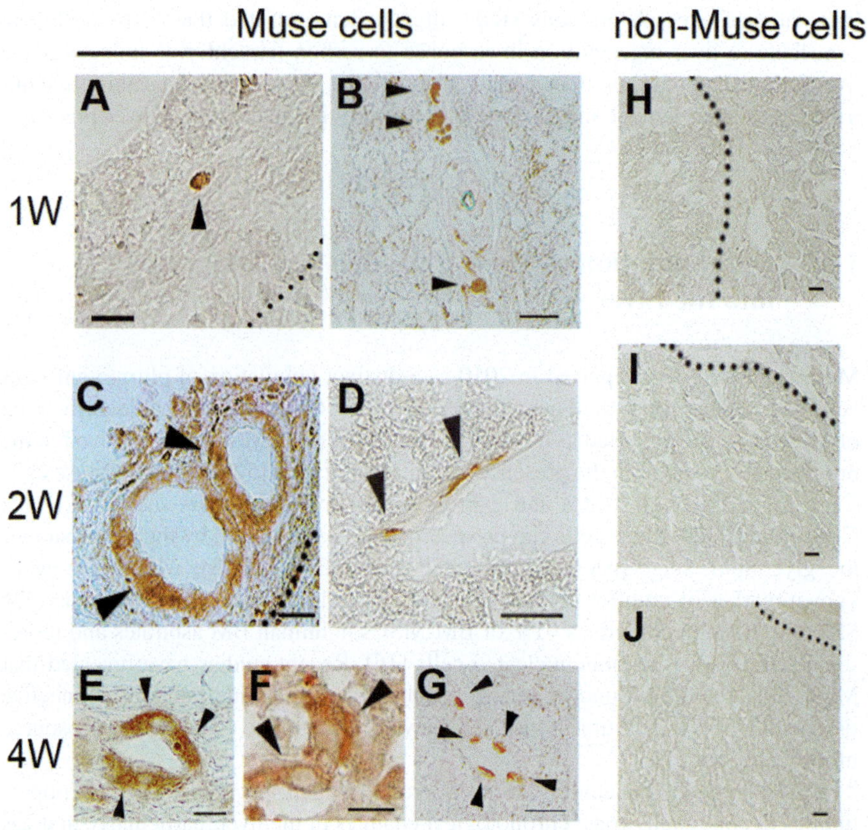

Fig. 12.4 (**A–G**) Immunohistochemical images of the liver from a GFP-labeled Muse cell-transplanted mouse. (**A**) GFP-labeled cells at the transection border of the liver. (**B**) An intact periportal area close to the transection border, where GFP-positive cells are occasionally seen. (**C**) GFP-positive bile ducts along the transection border. (**D**) GFP-positive cells in sinusoids. (**E**) GFP-positive cells form a duct-like structure. (**F**) GFP-positive hepatocytes. (**G**) GFP-positive sinusoidal cells. (**H–J**) No GFP-positive cells were seen in livers from the non-Muse cell-transplanted group. Dashed lines indicate the transection border. (The data are reproduced from Katagiri et al. [3])

these differentiation and functional markers were completely absent in all GFP^+ non-Muse cells, namely, cells other than Muse cells in BM-MSCs as control.

We also assessed the molecular and morphological features of these integrated cells. In the early phases (i.e., 1 week) after hepatectomy and Muse cell administration, the small number of Muse cells expressed human liver progenitor markers, such as CK19, delta-like protein (DLK), OV-6 (an oval cell marker), and AFP, in the periportal area adjacent to the transection line, suggesting that Muse cells may be integrated into each functional structure of the liver through the appropriate progenitor forms (Fig. 12.5). A chronological tracing of the integrated Muse cells revealed that, at 1 week, the Muse cells still expressed the liver progenitor markers.

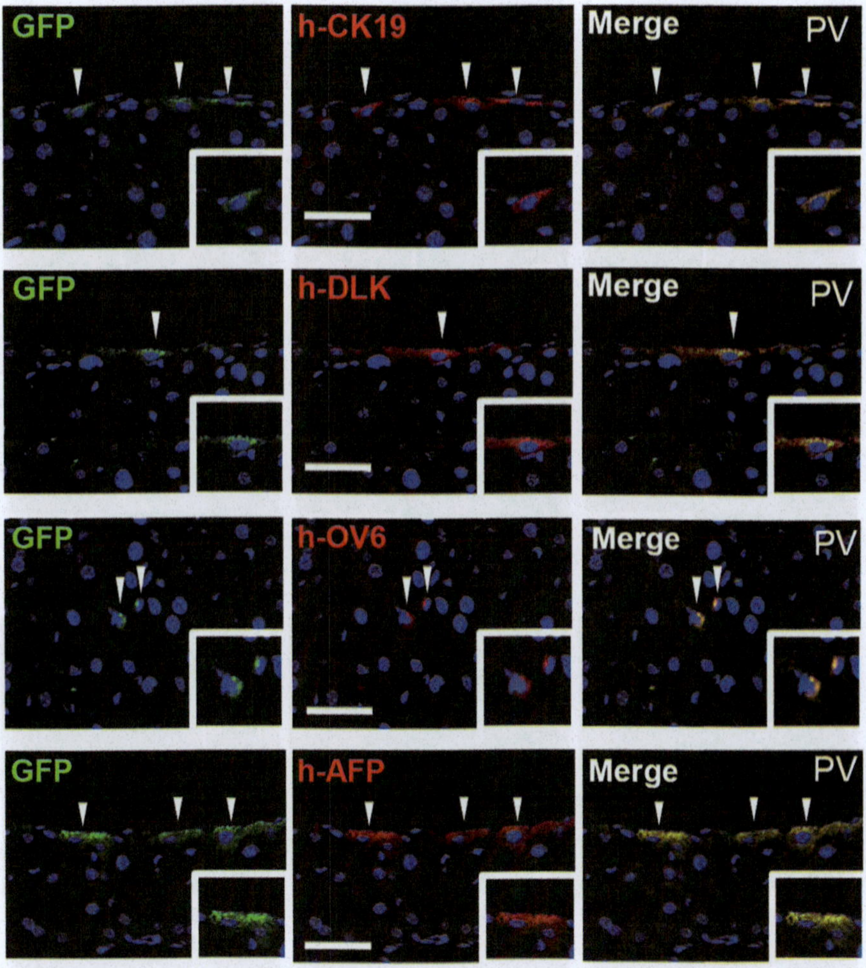

Fig. 12.5 Expression of human CK19, DLK, OV6, and AFP 1 week after GFP-labeled Muse cell administration. Arrowheads indicate staining-positive cells. Insets show high-power magnification of the region. *PV* portal vein. Scale bar, 50 μm. (The data are reproduced from Katagiri et al. [3])

At 2 weeks, the Muse cells integrated into appropriate tissue structures expressed additional markers of differentiation, such as human-specific HepPar-1, albumin, alpha-1-antitrypsin, CK7, and Lyve-1, but the majority no longer expressed liver progenitor markers (Fig. 12.6 and Table 12.1). At 4 weeks, the specificity of protein expression by the integrated Muse cells was clearly supported by species-specific RT-PCR, discriminating human Muse cells from host mouse liver cells, indicating that integrated cells expressed these markers were derived from human Muse cells [3].

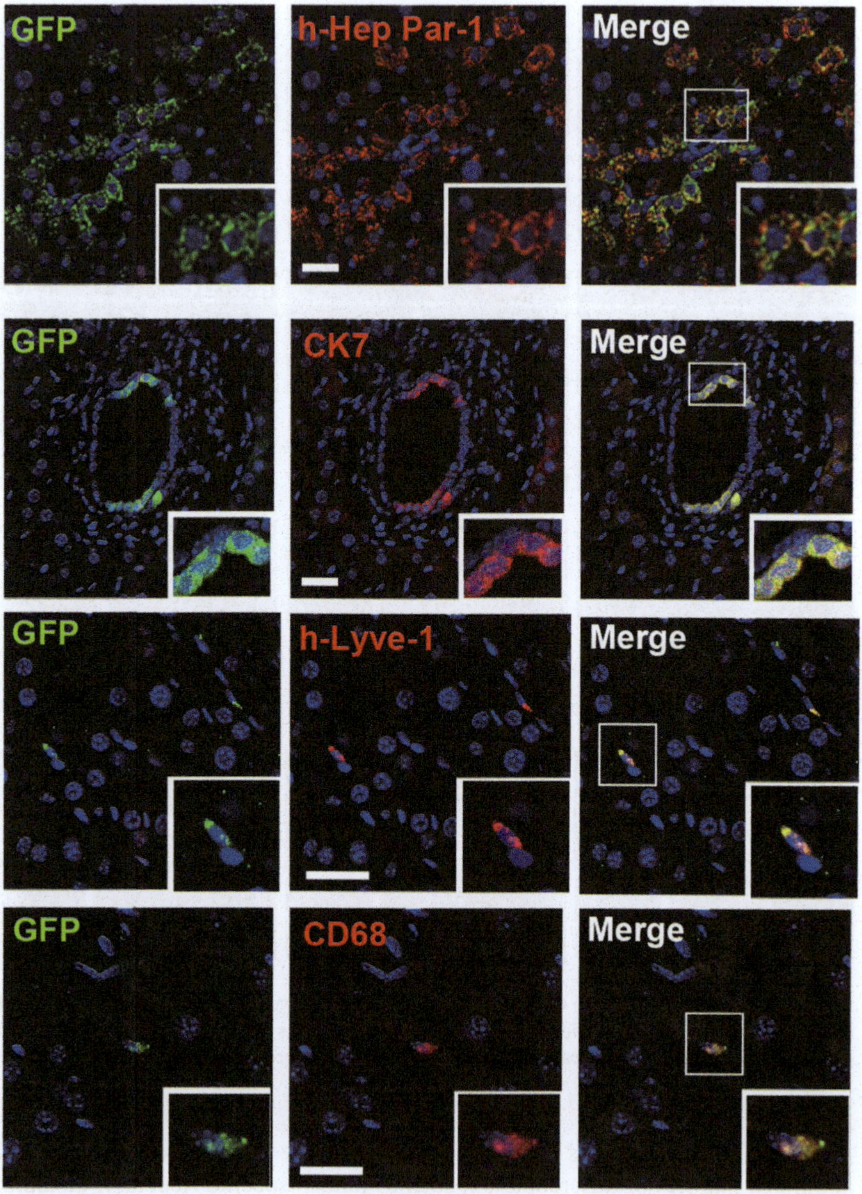

Fig. 12.6 Human HepPar-1, CK7, Lyve-1, and CD68 expression 8 weeks after GFP-labeled Muse cell administration. Insets show high-power magnification of the region. Scale bar, 20 μm (The data are reproduced from Katagiri et al. [3])

Table 12.1 Hepatoblast/hepatocyte marker expression profile at 2 days, 1 week, 2 weeks, and 4 weeks post-human Muse cell administration

	Marker	2 days	1 week	2 weeks	4 weeks
Liver progenitor	CK19	+	+	−	−
	DLK	+	+	−	−
	OV-6	+	+	−	−
	AFP	+	+	−	−
Hepatocyte	Hep Par-1	−	−	+	+
	Albumin	−	−	+	+
	α-1-antitrypsin	−	−	+	+
Cholangiocyte	CK7	−	−	+	+
SEC	Lyve-1	−	−	+	+
Kupffer cell	CD68	−	−	+	+

SEC sinusoidal endothelial cells. (Reproduced from Katagiri et al. [3])

12.6 The Role of Cell Fusion in Extrahepatic Cell Integration into the Liver

Despite these accumulated findings for Muse cell integration, these phenomena could simply be due to cell fusion that occurs during liver regeneration [58]. Extrahepatic stem cell plasticity in the context of liver regeneration has been extensively studied using fumarylacetoacetate hydrolase (*Fah*) knockout mice and discussed in a series of publications by Grompe et al. and others [59–65]. The *Fah*-deficient mice developed severe liver malfunctions following withdrawal of 2-(2-nitro-4-trifluoromethylbenzoyl)-1,3-cyclohexanedione (NTBC), which is used to treat tyrosinemia in humans via inhibition of p-OH phenylpyruvate dioxygenase [66, 67]. A *Fah* knockout mouse transplanted with sex- and genotype-mismatched bone marrow cells exhibited "repopulation" of bone marrow-derived hepatocytes in the liver through the bloodstream, which was explained by stem cell plasticity through cell fusion [68]. With these knockout and bone marrow transplant systems, the repopulation of BM cells to the damaged liver may have at least partially occurred at low frequency by fusion instead of BM-derived stem cell transdifferentiation [69, 70]. Although BM-derived cells undergo spontaneous cell fusion at very low frequencies (2–11 clones per 10^6 BM cells), fused cells could nonetheless be a dominant population if they were to acquire growth or survival advantages by compensating organ dysfunction [71]. However, an experiment involving transplantation of human hepatocytes into *Fah*-deficient mice revealed that the majority of repopulated hepatocytes were purely of human origin that had not undergone fusion [66]. With respect to the therapeutic potential for BM-derived stem cells, studies have shown that the plasticity of these cells allows conversion to liver cells without fusion [72, 73]. More recently, fusion-derived polyploidy hepatocytes are suggested not to be predisposed to convert phenotypes but instead give rise to the genetic variations that form the broad background of hepatocyte ploidy variations in the liver [74–76]. Although these studies provided substantial information on how extrahepatic cellular resources integrate into damaged liver tissue under extreme situations, these findings may not be immediately relevant for patients with potentially curable liver diseases. Thus, investigation of clinically applicable models using retrospective human materials is needed in order for clinical applications to be realistic.

To investigate whether fusion plays a role in Muse cell integration into the injured liver, we performed a FISH experiment in the Muse-transplanted hepatectomy model [3]. Using serial sections of liver samples adjacent to the transection line, we could approximate protein expression and genotyping at the cellular level. This combinational technique allowed us to examine whether GFP-labeled Muse cells that express liver markers also carry mouse chromosome markers that are indicative of fusion. Interestingly, at day 2 post-PPHx, CK19-expressing Muse cells (i.e., cholangiocytes) possessed only human chromosomes, whereas other GFP-negative CK19-expressing cells (presumably endogenous mouse cholangiocytes participating in liver regeneration) possessed only mouse chromosomes (Fig. 12.7A). This finding seems to suggest that fusion does not occur, at least immediately, in response to liver injury. Subsequent follow-up FISH analysis at 4 weeks post-PPHx revealed that a small percentage (1.9%) of hepatocytes derived from Muse cells showed mouse chromosomes, suggesting that fusion is an infrequent event (Fig. 12.7B). Our quantitative analysis of hepatocyte-differentiated Muse cells counted 74.3% of all integrated Muse cells, again suggesting that the fraction of cells that underwent fusion during liver regeneration was extremely small.

We also used FISH to investigate the possibility of fusion in humans in long-term liver regeneration using sex-mismatched (female-to-male) LDLT patient samples [21]. Liver grafts from these cases exhibited chimerism at the tissue level, indicating that extrahepatic cells differentiated into liver components. There were no cells that showed evidence of cell fusion between endogenous liver component cells and extrahepatic cells. Polyploid cells were present, but they all carried only X chromosomes (i.e., derived from only the donor liver). Interestingly, cholangiocytes carrying Y chromosomes were more frequent than other cell types, suggesting that extrahepatic cells were integrated with preference during bile duct formation. This finding, together with that for $AFP^+/CK19^+$ double-positive cholangiocytes, could imply that extrahepatic cells contribute to liver regeneration via progenitor cells. Thus, we conclude that fusion is unlikely to be a predominant mechanism in liver regeneration, and extrahepatic cells, particularly cholangiocytes, substantially contribute to the formation of liver structures as multipotent LPCs.

12.7 Potential of Muse Cells for Treating Liver Disease

Cellular transplantation of extrahepatic origin (including hepatocyte transplantation) has provided evidence that cellular support is somewhat effective for treating liver disease [45]. Development of such approaches would have higher priority if

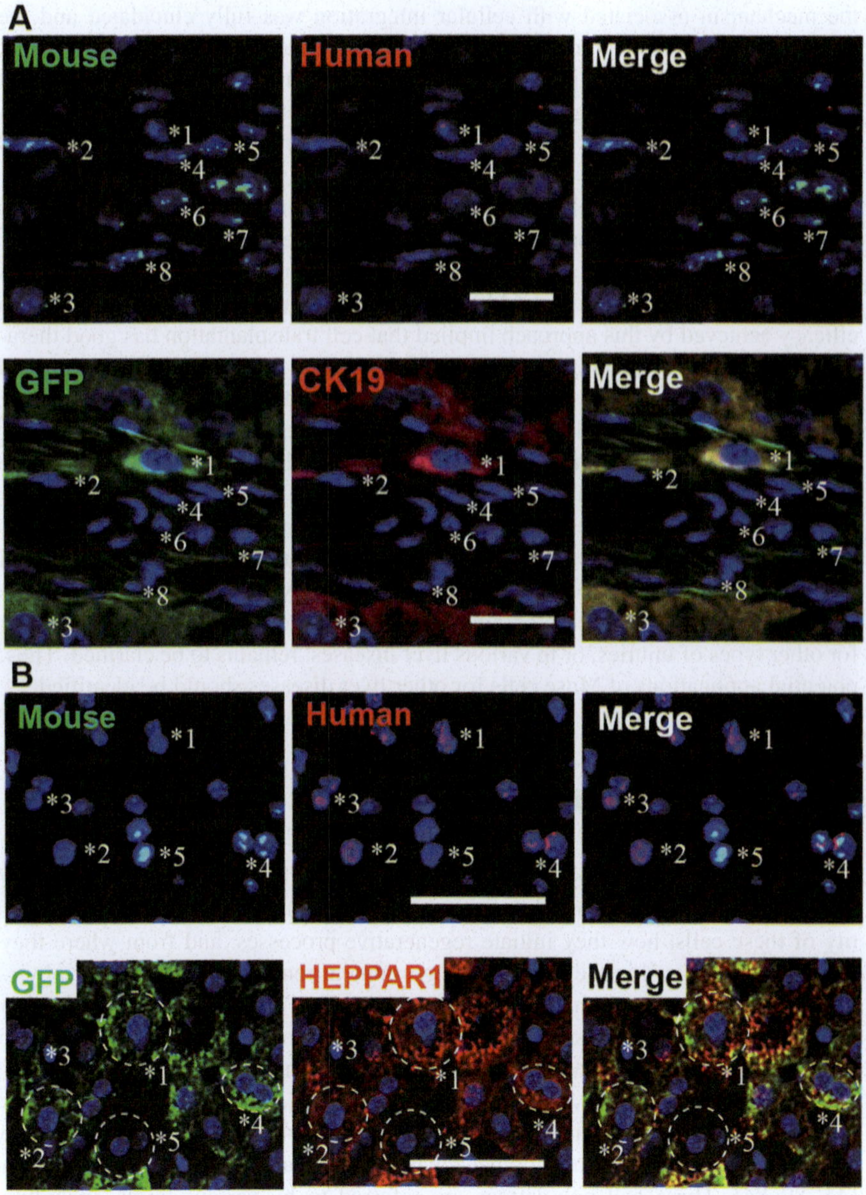

Fig. 12.7 *Muse cell marker expression following transplantation*. (**A**) Two days posttransplantation with human GFP-positive Muse cells, *1 cell is positive for GFP/CK19 and possesses only human chromosomes, where *2-*8 cells are negative for both GFP and CK19 and possess only mouse chromosomes. (**B**) At 4 weeks after transplantation of human GFP-positive Muse cells, among cells positive for both GFP and HepPar-1, *1 and *2 cells possess only human chromosomes, whereas *4 cell has both human and mouse chromosomes, *3 cell that has only human chromosomes is not reflected in the GFP/HepPar-1 section, and the *5 cell is negative for both GFP and HepPar-1 and has only mouse chromosomes. Mouse and human chromosomes are indicated by green and red signals, respectively. Scale bar, 20 μm. (The data are reproduced from Katagiri et al. [3])

the mechanism associated with cellular integration was fully elucidated and the method efficacy was guaranteed. Our study showed that the closer the injury front, Muse cells could preferentially accumulated to the damaged site. The finding that more than 70% of the integrated Muse cells in the injury front differentiated spontaneously into hepatocytes would support that this approach could have high therapeutic efficacy.

Similar to other stem cell resources, techniques to enrich Muse cell populations from BM-MSCs or, more practically, from BM cells are needed. Indeed, Muse cells represent only 2–3% of the BM-MSC population, meaning that a large number of BM cells are required to obtain Muse cells for treatment. Nonetheless, the high efficacy achieved by this approach implied that cell transplantation has good therapeutic potential. In practice, administration of Muse cells should be performed through the portal vein in patients who undergo hepatectomy. The challenge for therapeutic use of Muse cells is illustrated by the suggestion that "hepatocyte transplantation" for engraftment of extrahepatic cells to a damaged liver would require 2.5% of the entire liver weight [9].

Another important issue is what kind of liver diseases are indicated for cell therapy with Muse cells. The liver failure involved in our PPHx model was a local injury. In this model, Muse cells proliferated selectively at the liver transection line in the early time period after integration, but whether such proliferation would occur for other types of injuries, or in various liver diseases, remains to be clarified. Thus, potential applications of Muse cells for other liver diseases should be identified.

12.8 The Origin of Liver Stem Cells

The LDLT genotyping suggested that extrahepatic cells capable of differentiating into multiple cells could contribute in part to liver regeneration. However, the identity of these cells, how they initiate regenerative processes, and from where they ultimately originated in the liver microenvironment remains unclear. Under physiological conditions, new hepatocytes were shown to arise by simple replication from a small portion of existing hepatocytes during homeostatic renewal of the liver [77–79] (Fig. 12.8A). In contrast, in situations involving severe injury, such as fulminant hepatic failure [80], chronic viral hepatitis [81], alcoholic hepatitis [82], or murine PPHx induced with a harmonic scalpel [44], cells with intermediate hepatocyte-cholangiocyte phenotypes emerge and expand in the liver parenchyma [44, 80–84]. These cell populations are referred to by various terms, including "ductular hepatocytes," "atypical ductal cells," "intermediate hepatobiliary cells," or perhaps the most frequently called "LPCs" [85] (Fig. 12.8B). In rodent models, such cells were historically known as "oval cells," a term first coined by Farber et al. to describe non-parenchymal cells in the periportal region that could be observed after 2-acetylaminofluorene treatment followed by two-thirds partial hepatectomy (2-AAF/PH) in rats [86, 87]. After 2-AAF/PH, oval cells in rat models are induced, whereas in mouse models the oval cell equivalent emerged more effectively

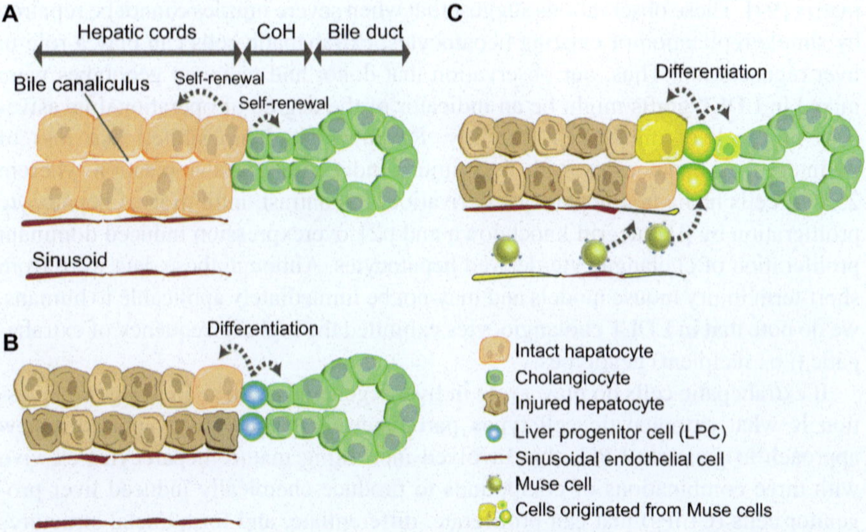

Fig. 12.8 *Stem cells for liver regeneration*. (**A**) Under physiological conditions, both new hepatocytes and cholangiocytes arise by simple replication from a small portion of existing hepatocytes and cholangiocyte, respectively, during the homeostatic renewal of the liver. (**B**) Upon severe liver injury, "liver progenitor cells (LPCs)," which have an intermediate hepatocyte-cholangiocyte phenotype, emerge and differentiate into new hepatocytes and cholangiocytes to compensate for the damaged tissues. (**C**) Muse cells can be mobilized upon liver injury from peripheral blood, bone marrow, and connective tissues. The mobilized Muse cells differentiate into liver component cells, including hepatocytes, cholangiocytes, and sinusoidal cells. Note that Muse cells differentiate into hepatocytes or cholangiocytes via "LPC"-like cells, whereas Muse cells can directly differentiate into sinusoidal endothelial cells. *CoH* Canals of Hering

following treatment with 3,5-diethoxycarbonyl-1,4-dihydrocollidine (known as DDC) or feeding of a choline-deficient, ethionine-supplemented (CDE) diet [88, 89]. The emergence of oval cells in the portal field suggests that these cells originated in the terminal branches of the intrahepatic biliary system, the Canals of Hering [90], and then expanded into the parenchyma to form duct-like structures known as "ductular reactions" [44, 91]. Indeed, an extended biliary duct remodeling has been shown for liver regeneration in response to various liver injuries because it provides a niche for LPCs [92]. In periportal area, Thy1$^+$ cells have been shown to constitute the niche for LPCs by producing FGF7 that can regulate the spread of LPCs [93].

Using inducible Cre recombinase, Furuyama et al. demonstrated that hepatocyte differentiation could be activated almost exclusively by "Sox9-positive precursors" in the duct during liver regeneration induced by carbon tetrachloride or bile duct ligation [62]. These Sox9-positive cells afforded near-complete turnover of hepatocyte mass within 6 months. Whereas these Sox9-positive cells can be considered as LPCs, their terminal differentiation into functional hepatocytes is dependent on the degree of liver damage and composition of the neighboring extracellular matrix [94]. These observations suggest that when severe injuries cannot be repaired by simple replication of existing hepatocytes, extrahepatic cells can play a role in liver regeneration. Thus, our observation that donor and recipient genotypes were mixed in LDLT grafts might be an indicator of the degree of operational invasiveness of the liver injury [3]. Recently, Raven et al. demonstrated that loss of β1-integrin in hepatocytes with liver injury induced a ductular reaction wherein 25% of cells had a non-hepatocyte derivation. In contrast, inhibition of hepatocyte proliferation by β1-integrin knockdown and p21 overexpression induced dominant proliferation of cholangiocyte-derived hepatocytes. Although these data were from short-term injury mouse models and may not be immediately applicable to humans, we do note that in LDLT cholangiocytes exhibited the highest frequency of extrahepatic (i.e., recipient) genotypes.

If extrahepatic cells do play a role in liver regeneration, then an intriguing question is what extrahepatic cell types participate. Katsuda et al. reported a new approach to generate LPCs that involved incubating mature hepatocytes ex vivo with three combinations of compounds to produce chemically induced liver progenitor cells (CLiPs) that can proliferate, differentiate, and form ductal structures [95]. The CLiPs study revealed that diploid hepatocytes that represent a minor fraction of mature hepatocytes become dominant during CLiP selection, suggesting that diploid cells are the major source for CLiPs. Since aneuploidy is a common characteristic of mature hepatocytes, the fact that diploid cells could play a major role in potential hepatocyte replacement led us to consider whether extrahepatic stem cells are a source for LPCs. Diploid stem cells, particularly HSCs, MSCs, or other stem cells, may be delivered via the blood stream from other normal tissues. In this situation, some dedicated cell fractions in the blood should increase in response to physical stress. Supporting this hypothesis, increased numbers of Muse cells were observed within 24 h after ischemic stroke of the brain and acute myocardial infarction (AMI) [54, 55]. Although ischemic stroke or AMI may not directly damage the liver, these studies indicate that in response to physical stress, the numbers of diploid Muse cells increase, as does the likelihood that these cells will localize in a damaged area where they could aid differentiation of diploid cells into needed cell types. Our genotyping of LDLT showing diploid cell-dominant potential for LPCs and circulating Muse cells in stressed conditions strongly supports the possibility that extrahepatic Muse cells are an important resource for liver regeneration (Fig. 12.8C).

12.9 Conclusion

Following transplantation of Muse cells into mice with damaged livers, only Muse cells contributed to liver regeneration. Although several issues, including enrichment method, administration pathway, and recommended indications, must be addressed before clinical application, Muse cells may be a practical candidate for cell therapy for a wide range of liver diseases.

Acknowledgment This work was partially supported by a Grant-in-Aid for Scientific Research 16 K09370 (Y.T.) and 17 K10676 (H.K.); and Keiryokai Research Foundation No. 132 (Y.S.).

References

1. Itoh T, Miyajima A (2014) Liver regeneration by stem/progenitor cells. Hepatology 59(4):1617–1626
2. Miyaoka Y, Ebato K, Kato H, Arakawa S, Shimizu S, Miyajima A (2012) Hypertrophy and unconventional cell division of hepatocytes underlie liver regeneration. Curr Biol 22(13):1166–1175
3. Katagiri H, Kushida Y, Nojima M, Kuroda Y, Wakao S, Ishida K et al (2016) A distinct subpopulation of bone marrow mesenchymal stem cells, muse cells, directly commit to the replacement of liver components. Am J Transplant 16(2):468–483
4. Liang TJ, Ghany MG (2013) Current and future therapies for hepatitis C virus infection. N Engl J Med 368(20):1907–1917
5. Rinella ME (2015) Nonalcoholic fatty liver disease: a systematic review. JAMA 313(22):2263–2273
6. Breitenstein S, Dimitroulis D, Petrowsky H, Puhan MA, Mullhaupt B, Clavien PA (2009) Systematic review and meta-analysis of interferon after curative treatment of hepatocellular carcinoma in patients with viral hepatitis. Br J Surg 96(9):975–981
7. Mitry RR, Hughes RD, Dhawan A (2011) Hepatocyte transplantation. J Clin Exp Hepatol 1(2):109–114
8. Dhawan A, Puppi J, Hughes RD, Mitry RR (2010) Human hepatocyte transplantation: current experience and future challenges. Nat Rev Gastroenterol Hepatol 7(5):288–298
9. Fox IJ, Chowdhury JR, Kaufman SS, Goertzen TC, Chowdhury NR, Warkentin PI et al (1998) Treatment of the Crigler-Najjar syndrome type I with hepatocyte transplantation. N Engl J Med 338(20):1422–1426
10. Muraca M, Gerunda G, Neri D, Vilei MT, Granato A, Feltracco P et al (2002) Hepatocyte transplantation as a treatment for glycogen storage disease type 1a. Lancet 359(9303):317–318
11. Forbes SJ, Gupta S, Dhawan A (2015) Cell therapy for liver disease: from liver transplantation to cell factory. J Hepatol 62(1 Suppl):S157–S169
12. Pontikoglou C, Deschaseaux F, Sensebe L, Papadaki HA (2011) Bone marrow mesenchymal stem cells: biological properties and their role in hematopoiesis and hematopoietic stem cell transplantation. Stem Cell Rev 7(3):569–589
13. Prockop DJ, Brenner M, Fibbe WE, Horwitz E, Le Blanc K, Phinney DG et al (2010) Defining the risks of mesenchymal stromal cell therapy. Cytotherapy 12(5):576–578
14. Tautenhahn HM, Bruckner S, Baumann S, Winkler S, Otto W, von Bergen M et al (2016) Attenuation of postoperative acute liver failure by mesenchymal stem cell treatment due to metabolic implications. Ann Surg 263(3):546–556
15. Chamberlain G, Fox J, Ashton B, Middleton J (2007) Concise review: mesenchymal stem cells: their phenotype, differentiation capacity, immunological features, and potential for homing. Stem Cells 25(11):2739–2749
16. Kuroda Y, Kitada M, Wakao S, Nishikawa K, Tanimura Y, Makinoshima H et al (2010) Unique multipotent cells in adult human mesenchymal cell populations. Proc Natl Acad Sci USA 107(19):8639–8643
17. Chiu KW, Nakano T, Chen KD, Hsu LW, Lai CY, Chiu HC et al (2013) Homogeneous phenomenon of the graft when using different genotype characteristic of recipients/donors in living donor liver transplantation. World J Hepatol 5(11):642–648
18. Chiu KW, Nakano T, Chen KD, Lai CY, Hsu LW, Chiu HC et al (2013) Pyrosequencing to identify homogeneous phenomenon when using recipients/donors with different CYP3A5*3 genotypes in living donor liver transplantation. PLoS One 8(8):e71314

19. Hove WR, van Hoek B, Bajema IM, Ringers J, van Krieken JH, Lagaaij EL (2003) Extensive chimerism in liver transplants: vascular endothelium, bile duct epithelium, and hepatocytes. Liver Transpl 9(6):552–556
20. Kleeberger W, Rothamel T, Glockner S, Flemming P, Lehmann U, Kreipe H (2002) High frequency of epithelial chimerism in liver transplants demonstrated by microdissection and STR-analysis. Hepatology 35(1):110–116
21. Ng IO, Chan KL, Shek WH, Lee JM, Fong DY, Lo CM et al (2003) High frequency of chimerism in transplanted livers. Hepatology 38(4):989–998
22. Makuuchi M (2013) Surgical treatment for HCC–special reference to anatomical resection. Int J Surg 11(Suppl 1):S47–S49
23. Hasegawa Y, Nitta H, Sasaki A, Takahara T, Itabashi H, Katagiri H et al (2015) Long-term outcomes of laparoscopic versus open liver resection for liver metastases from colorectal cancer: a comparative analysis of 168 consecutive cases at a single center. Surgery 157(6):1065–1072
24. Page AJ, Weiss MJ, Pawlik TM (2014) Surgical management of noncolorectal cancer liver metastases. Cancer 120(20):3111–3121
25. Roll GR, Parekh JR, Parker WF, Siegler M, Pomfret EA, Ascher NL et al (2013) Left hepatectomy versus right hepatectomy for living donor liver transplantation: shifting the risk from the donor to the recipient. Liver Transpl 19(5):472–481
26. Takahara T, Wakabayashi G, Hasegawa Y, Nitta H (2015) Minimally invasive donor hepatectomy: evolution from hybrid to pure laparoscopic techniques. Ann Surg 261(1):e3–e4
27. Akamatsu N, Sugawara Y, Nagata R, Kaneko J, Aoki T, Sakamoto Y et al (2014) Adult right living-donor liver transplantation with special reference to reconstruction of the middle hepatic vein. Am J Transplant 14(12):2777–2787
28. Duclos J, Bhangui P, Salloum C, Andreani P, Saliba F, Ichai P et al (2016) Ad Integrum functional and volumetric recovery in right lobe living donors: is it really complete 1 year after donor hepatectomy? Am J Transplant 16(1):143–156
29. Kogure K, Ishizaki M, Nemoto M, Kuwano H, Makuuchi M (1999) A comparative study of the anatomy of rat and human livers. J Hepato-Biliary-Pancreat Surg 6(2):171–175
30. Malarkey DE, Johnson K, Ryan L, Boorman G, Maronpot RR (2005) New insights into functional aspects of liver morphology. Toxicol Pathol 33(1):27–34
31. Hori T, Ohashi N, Chen F, Baine AM, Gardner LB, Jermanus S et al (2011) Simple and sure methodology for massive hepatectomy in the mouse. Ann Gastroenterol 24(4):307–318
32. Mitchell C, Willenbring H (2008) A reproducible and well-tolerated method for 2/3 partial hepatectomy in mice. Nat Protoc 3(7):1167–1170
33. Song GW, Lee SG (2014) Living donor liver transplantation. Curr Opin Organ Transplant 19(3):217–222
34. Brown RS Jr (2008) Pros and cons of living donor liver transplant. Gastroenterol Hepatol (N Y) 4(9):622–624
35. Katagiri HNS, Nitta H, Wakabayashi G (2014) Potential involvement of extrahepatic cells in liver regeneration. J Iwate Med Assoc 66(2):67–74
36. Haga J, Shimazu M, Wakabayashi G, Tanabe M, Kawachi S, Fuchimoto Y et al (2008) Liver regeneration in donors and adult recipients after living donor liver transplantation. Liver Transpl 14(12):1718–1724
37. Ibrahim S, Chen CL, Wang CC, Wang SH, Lin CC, Liu YW et al (2005) Liver regeneration and splenic enlargement in donors after living-donor liver transplantation. World J Surg 29(12):1658–1666
38. Kato Y, Shimazu M, Wakabayashi G, Tanabe M, Morikawa Y, Hoshino K et al (2001) Significance of portal venous flow in graft regeneration after living related liver transplantation. Transplant Proc 33(1–2):1484–1485
39. Kido M, Ku Y, Fukumoto T, Tominaga M, Iwasaki T, Ogata S et al (2003) Significant role of middle hepatic vein in remnant liver regeneration of right-lobe living donors. Transplantation 75(9):1598–1600
40. Michalopoulos GK, DeFrances MC (1997) Liver regeneration. Science 276(5309):60–66

41. Tanimizu N, Mitaka T (2014) Re-evaluation of liver stem/progenitor cells. Organogenesis 10(2):208–215
42. Isse K, Lesniak A, Grama K, Maier J, Specht S, Castillo-Rama M et al (2013) Preexisting epithelial diversity in normal human livers: a tissue-tethered cytometric analysis in portal/periportal epithelial cells. Hepatology 57(4):1632–1643
43. Schmelzle M, Duhme C, Junger W, Salhanick SD, Chen Y, Wu Y et al (2013) CD39 modulates hematopoietic stem cell recruitment and promotes liver regeneration in mice and humans after partial hepatectomy. Ann Surg 257(4):693–701
44. Suzuki Y, Katagiri H, Wang T, Kakisaka K, Kume K, Nishizuka SS et al (2016) Ductular reactions in the liver regeneration process with local inflammation after physical partial hepatectomy. Lab Investig J Tech Methods Pathol 96(11):1211–1222
45. Huebert RC, Rakela J (2014) Cellular therapy for liver disease. Mayo Clin Proc 89(3):414–424
46. Mitry RR, Hughes RD, Aw MM, Terry C, Mieli-Vergani G, Girlanda R et al (2003) Human hepatocyte isolation and relationship of cell viability to early graft function. Cell Transplant 12(1):69–74
47. Strom SC, Davila J, Grompe M (2010) Chimeric mice with humanized liver: tools for the study of drug metabolism, excretion, and toxicity. Methods Mol Biol 640:491–509
48. Dezawa M (2006) Insights into autotransplantation: the unexpected discovery of specific induction systems in bone marrow stromal cells. Cell Mol Life Sci 63(23):2764–2772
49. Ferraro F, Lymperi S, Mendez-Ferrer S, Saez B, Spencer JA, Yeap BY et al (2011) Diabetes impairs hematopoietic stem cell mobilization by altering niche function. Sci Transl Med 3(104):104ra1
50. Lin BL, Chen JF, Qiu WH, Wang KW, Xie DY, Chen XY et al (2017) Allogeneic bone marrow-derived mesenchymal stromal cells for hepatitis B virus-related acute-on-chronic liver failure: a randomized controlled trial. Hepatology 66(1):209–219
51. Suk KT, Yoon JH, Kim MY, Kim CW, Kim JK, Park H et al (2016) Transplantation with autologous bone marrow-derived mesenchymal stem cells for alcoholic cirrhosis: phase 2 trial. Hepatology 64(6):2185–2197
52. Wakao S, Akashi H, Kushida Y, Dezawa M (2014) Muse cells, newly found non-tumorigenic pluripotent stem cells, reside in human mesenchymal tissues. Pathol Int 64(1):1–9
53. Dezawa M (2016) Muse cells provide the pluripotency of mesenchymal stem cells: direct contribution of muse cells to tissue regeneration. Cell Transplant 25(5):849–861
54. Hori E, Hayakawa Y, Hayashi T, Hori S, Okamoto S, Shibata T et al (2016) Mobilization of pluripotent multilineage-differentiating stress-enduring cells in ischemic stroke. J Stroke Cerebrovasc Dis 25(6):1473–1481
55. Tanaka T, Nishigaki K, Minatoguchi S, Nawa T, Yamada Y, Kanamori H et al (2018) Mobilized muse cells after acute myocardial infarction predict cardiac function and remodeling in the chronic phase. Circ J 82(2):561–571
56. Wakao S, Kitada M, Dezawa M (2013) The elite and stochastic model for iPS cell generation: multilineage-differentiating stress enduring (Muse) cells are readily reprogrammable into iPS cells. Cytometry A 83(1):18–26
57. Wakao S, Kitada M, Kuroda Y, Shigemoto T, Matsuse D, Akashi H et al (2011) Multilineage-differentiating stress-enduring (Muse) cells are a primary source of induced pluripotent stem cells in human fibroblasts. Proc Natl Acad Sci USA 108(24):9875–9880
58. Dahlke MH, Popp FC, Larsen S, Schlitt HJ, Rasko JE (2004) Stem cell therapy of the liver—fusion or fiction? Liver Transpl 10(4):471–479
59. Grompe M, al-Dhalimy M, Finegold M, Ou CN, Burlingame T, Kennaway NG et al (1993) Loss of fumarylacetoacetate hydrolase is responsible for the neonatal hepatic dysfunction phenotype of lethal albino mice. Genes Dev 7(12A):2298–2307
60. Grompe M, Lindstedt S, al-Dhalimy M, Kennaway NG, Papaconstantinou J, Torres-Ramos CA et al (1995) Pharmacological correction of neonatal lethal hepatic dysfunction in a murine model of hereditary tyrosinaemia type I. Nat Genet 10(4):453–460
61. Grompe M, Strom S (2013) Mice with human livers. Gastroenterology 145(6):1209–1214

62. He Z, Zhang H, Zhang X, Xie D, Chen Y, Wangensteen KJ et al (2010) Liver xeno-repopulation with human hepatocytes in Fah-/-Rag2-/- mice after pharmacological immunosuppression. Am J Pathol 177(3):1311–1319
63. Huch M, Dorrell C, Boj SF, van Es JH, Li VS, van de Wetering M et al (2013) In vitro expansion of single Lgr5+ liver stem cells induced by Wnt-driven regeneration. Nature 494(7436):247–250
64. Li L, Zeng Z, Qi Z, Wang X, Gao X, Wei H et al (2015) Natural killer cells-produced IFN-gamma improves bone marrow-derived hepatocytes regeneration in murine liver failure model. Sci Rep 5:13687
65. Qi Z, Wang X, Wei H, Sun R, Tian Z (2015) Infiltrating neutrophils aggravate metabolic liver failure in fah-deficient mice. Liver Int 35(3):774–785
66. Azuma H, Paulk N, Ranade A, Dorrell C, Al-Dhalimy M, Ellis E et al (2007) Robust expansion of human hepatocytes in Fah-/-/Rag2-/-/Il2rg-/- mice. Nat Biotechnol 25(8):903–910
67. Grompe M, Overturf K, al-Dhalimy M, Finegold M (1998) Therapeutic trials in the murine model of hereditary tyrosinaemia type I: a progress report. J Inherit Metab Dis 21(5):518–531
68. Wang X, Willenbring H, Akkari Y, Torimaru Y, Foster M, Al-Dhalimy M et al (2003) Cell fusion is the principal source of bone-marrow-derived hepatocytes. Nature 422(6934):897–901
69. Vassilopoulos G, Wang PR, Russell DW (2003) Transplanted bone marrow regenerates liver by cell fusion. Nature 422(6934):901–904
70. Wang X, Montini E, Al-Dhalimy M, Lagasse E, Finegold M, Grompe M (2002) Kinetics of liver repopulation after bone marrow transplantation. Am J Pathol 161(2):565–574
71. Terada N, Hamazaki T, Oka M, Hoki M, Mastalerz DM, Nakano Y et al (2002) Bone marrow cells adopt the phenotype of other cells by spontaneous cell fusion. Nature 416(6880):542–545
72. Harris RG, Herzog EL, Bruscia EM, Grove JE, Van Arnam JS, Krause DS (2004) Lack of a fusion requirement for development of bone marrow-derived epithelia. Science 305(5680):90–93
73. Jang YY, Collector MI, Baylin SB, Diehl AM, Sharkis SJ (2004) Hematopoietic stem cells convert into liver cells within days without fusion. Nat Cell Biol 6(6):532–539
74. Duncan AW, Dorrell C, Grompe M (2009) Stem cells and liver regeneration. Gastroenterology 137(2):466–481
75. Duncan AW, Hickey RD, Paulk NK, Culberson AJ, Olson SB, Finegold MJ et al (2009) Ploidy reductions in murine fusion-derived hepatocytes. PLoS Genet 5(2):e1000385
76. Duncan AW, Taylor MH, Hickey RD, Hanlon Newell AE, Lenzi ML, Olson SB et al (2010) The ploidy conveyor of mature hepatocytes as a source of genetic variation. Nature 467(7316):707–710
77. Malato Y, Naqvi S, Schurmann N, Ng R, Wang B, Zape J et al (2011) Fate tracing of mature hepatocytes in mouse liver homeostasis and regeneration. J Clin Invest 121(12):4850–4860
78. Wang B, Zhao L, Fish M, Logan CY, Nusse R (2015) Self-renewing diploid Axin2(+) cells fuel homeostatic renewal of the liver. Nature 524(7564):180–185
79. Yanger K, Knigin D, Zong Y, Maggs L, Gu G, Akiyama H et al (2014) Adult hepatocytes are generated by self-duplication rather than stem cell differentiation. Cell Stem Cell 15(3):340–349
80. Demetris AJ, Seaberg EC, Wennerberg A, Ionellie J, Michalopoulos G (1996) Ductular reaction after submassive necrosis in humans. Special emphasis on analysis of ductular hepatocytes. Am J Pathol 149(2):439–448
81. Libbrecht L, Desmet V, Van Damme B, Roskams T (2000) Deep intralobular extension of human hepatic 'progenitor cells' correlates with parenchymal inflammation in chronic viral hepatitis: can 'progenitor cells' migrate? J Pathol 192(3):373–378
82. Sancho-Bru P, Altamirano J, Rodrigo-Torres D, Coll M, Millan C, Jose Lozano J et al (2012) Liver progenitor cell markers correlate with liver damage and predict short-term mortality in patients with alcoholic hepatitis. Hepatology 55(6):1931–1941
83. Roskams TA, Theise ND, Balabaud C, Bhagat G, Bhathal PS, Bioulac-Sage P et al (2004) Nomenclature of the finer branches of the biliary tree: canals, ductules, and ductular reactions in human livers. Hepatology 39(6):1739–1745

84. Turanyi E, Dezso K, Csomor J, Schaff Z, Paku S, Nagy P (2010) Immunohistochemical classification of ductular reactions in human liver. Histopathology 57(4):607–614
85. Miyajima A, Tanaka M, Itoh T (2014) Stem/progenitor cells in liver development, homeostasis, regeneration, and reprogramming. Cell Stem Cell 14(5):561–574
86. Farber E (1956) Similarities in the sequence of early histological changes induced in the liver of the rat by ethionine, 2-acetylamino-fluorene, and 3'-methyl-4-dimethylaminoazobenzene. Cancer Res 16(2):142–148
87. Evarts RP, Nagy P, Marsden E, Thorgeirsson SS (1987) A precursor-product relationship exists between oval cells and hepatocytes in rat liver. Carcinogenesis 8(11):1737–1740
88. Preisegger KH, Factor VM, Fuchsbichler A, Stumptner C, Denk H, Thorgeirsson SS (1999) Atypical ductular proliferation and its inhibition by transforming growth factor beta1 in the 3,5-diethoxycarbonyl-1,4-dihydrocollidine mouse model for chronic alcoholic liver disease. Lab Investig J Tech Methods Pathol 79(2):103–109
89. Akhurst B, Croager EJ, Farley-Roche CA, Ong JK, Dumble ML, Knight B et al (2001) A modified choline-deficient, ethionine-supplemented diet protocol effectively induces oval cells in mouse liver. Hepatology 34(3):519–522
90. Theise ND, Saxena R, Portmann BC, Thung SN, Yee H, Chiriboga L et al (1999) The canals of Hering and hepatic stem cells in humans. Hepatology 30(6):1425–1433
91. Roskams TA, Libbrecht L, Desmet VJ (2003) Progenitor cells in diseased human liver. Semin Liver Dis 23(4):385–396
92. Kaneko K, Kamimoto K, Miyajima A, Itoh T (2015) Adaptive remodeling of the biliary architecture underlies liver homeostasis. Hepatology 61(6):2056–2066
93. Takase HM, Itoh T, Ino S, Wang T, Koji T, Akira S et al (2013) FGF7 is a functional niche signal required for stimulation of adult liver progenitor cells that support liver regeneration. Genes Dev 27(2):169–181
94. Espanol-Suner R, Carpentier R, Van Hul N, Legry V, Achouri Y, Cordi S et al (2012) Liver progenitor cells yield functional hepatocytes in response to chronic liver injury in mice. Gastroenterology 143(6):1564–75.e7
95. Katsuda T, Kawamata M, Hagiwara K, Takahashi RU, Yamamoto Y, Camargo FD et al (2017) Conversion of terminally committed hepatocytes to culturable bipotent progenitor cells with regenerative capacity. Cell Stem Cell 20(1):41–55

Chapter 13
Current Cell-Based Therapies in the Chronic Liver Diseases

Taketo Nishina, Kyoko Tomita Hoshikawa, and Yoshiyuki Ueno

Abstract Liver diseases account for one of the leading causes of deaths in global health care. Furthermore, chronic liver failure such as liver cirrhosis is, namely, responsible for these fatal conditions. However, only liver transplantation is an established treatment for this end-stage condition, although the availability of this salvage treatment option is quite limited. Thus, the novel therapy such as artificial liver devices or cellular administration has been regarded as feasible. Especially cellular therapies have been proposed in decades. The technical advancement and progress of understanding of cellular differentiation have contributed to the development of basis of cellular therapy. This attractive therapeutic option has been advanced from original embryonic stem cells to more effective cellular fractions such as Muse cells. Indeed several cellular therapies including bone marrow-derived stem cells or peripheral blood-derived stem cells were initiated; the recent most organized clinical trials could not demonstrate its efficacy. Thus, truly innovative cellular therapy is needed to meet the scientific demands, and Muse cell administration is the remaining approach to this. In this article, we will discuss the current development and status of cellular therapy toward chronic liver failure.

Keywords Liver cirrhosis · Albumin · Hepatocyte · Regeneration · Liver failure · Transaminase · Bilirubin · Fibrosis

T. Nishina
Department of Gastroenterology, Yamagata University Faculty of Medicine, Yamagata, Japan

K. T. Hoshikawa · Y. Ueno (✉)
Department of Gastroenterology, Yamagata University Faculty of Medicine, Yamagata, Japan

CREST, Yamagata, Japan
e-mail: y-ueno@med.id.yamagata-u.ac.jp

M. Dezawa (ed.), *Muse Cells*, Advances in Experimental Medicine and Biology 1103, https://doi.org/10.1007/978-4-431-56847-6_13

13.1 Introduction

It is well known that chronic liver diseases including viral infection (hepatitis B virus, hepatitis C virus), autoimmune hepatitis (AIH), primary biliary cholangitis (PBC), nonalcoholic steatohepatitis (NASH), alcoholic hepatitis (AH), and some others may lead to the development of liver cirrhosis and ultimately end-stage liver failure [1]. Although the causes of liver cirrhosis differ according to area or county, no life-saving treatment for this condition other than liver transplantation currently exists. A recent etiological survey has confirmed that HBV infection, HCV infection, alcoholic liver disease, and NASH are the leading causes of liver failure [1]. However, the development of effective antiviral treatments has changed the clinical situation. It is estimated that the incidence of HCV-related cirrhosis will decrease dramatically within the next few decades [2]. Also, the introduction of effective nucleos(t)ide analogues for HBV infection has successfully suppressed the replication of HBV, resulting in clinically significant and durable suppression of hepatic fibrosis and inflammation in the long term. Thus, the current major causes of liver cirrhosis are remaining liver diseases such as NASH, for which no fundamental treatments have been established. Moreover, there are still many patients with established liver failure who continuously suffer from complications such as ascites, hepatic encephalopathy, spontaneous bacterial peritonitis, and ultimately hepatocellular carcinoma (HCC) [1]. Currently, liver transplantation, either cadaveric or from living donors, is recognized as the only option for end-stage liver disease. However, its use is limited by a shortage of donors, a high incidence of surgical complications, and high medical costs. In this situation, the development of medical therapies other than liver transplantation would be desirable [3]. The present strategy of medical therapy for liver cirrhosis is (1) resolution of hepatic fibrosis, (2) recovery of hepatic function (both synthetic and metabolic), and (3) reducing the incidence of complications [4]. Recently, various anti-fibrotic drugs have been investigated in clinical trials, including the apoptosis signal-regulating kinase 1 (ASK1) inhibitor, selonsertib [5]. To reverse the decline in the synthetic function of the liver, several nutritional therapies such as branched-chain amino acid (BCAA) supplementation have been applied, although their effects have been proven only for patients with comparative reversal of decompensated cirrhosis. Furthermore, artificial liver support including extracorporeal xenogeneic hepatocyte-based approaches has demonstrated limited effects in patients with chronic liver failure. As a consequence, these forms of artificial liver support are merely regarded as temporary bridging therapies to liver transplantation. Moreover, no rational approach has been established for prevention of liver cirrhosis in patients with end-stage liver diseases. Against this background, the development and introduction of novel therapies for end-stage liver diseases would seem to be desirable. Among them, cell-based therapy has been regarded as very promising. The purpose of cell therapy is for grafted cells to migrate to damaged organs and participate in tissue recovery. For this purpose, cell therapy would seem to be a more effective approach than the use of artificial extracorporeal devices for hepatic disease. The specific characteristics of the liver, such as its ample blood supply, a marked capacity for regeneration, and comparatively easy access from the body surface, are all amenable to the development of novel forms of cellular therapy for intractable end-stage liver disease. For example, as access routes for cellular infusion, transplanted cells can be injected peripherally, intra-arterially (via the hepatic artery), or via the portal vein. However, when considering the possible complications of cell therapy, administration of cells via a peripheral vein may decrease the risk of such complications. In this review, we discuss the current status of cell-based therapy for end-stage liver diseases.

13.2 Stem Cells

Stem cells are known to have various specific abilities such as self-renewal and differentiation. There are several types of stem cells in mammals, and embryonic stem (ES) cells are the most prototypic. However, due to their limited accessibility and ethical issues, the clinical application of ES cells has a number of specific hurdles. Another type of stem cell is the mesenchymal stem cell (MSC). The application of MSCs as a source of cell therapy has been investigated worldwide for numerous conditions. MSCs have the advantage of easy accessibility; they can be obtained even from medical waste tissues such as adipose tissue, umbilical tissue, and dental pulp. Another type of stem cell is hematopoietic stem cells, which are reported to differentiate into hepatocyte-like cells under certain conditions [6]. A number of studies using animal models and some human investigational trials have described their application for hepatic regenerative therapy [7–28]. Unfortunately, however, most of those studies were hampered by significant bias [29]. Table 13.1 summarizes the major clinical trials of cell therapy for liver cirrhosis. The majority of cellular sources have been autologous bone marrow or allogenic umbilical cord. Clinical trials of this form of cell therapy have obtained data based on laboratory tests (albumin, total bilirubin, aspartate aminotransferase (AST), alanine aminotransferase (ALT), etc.), residual hepatic function (Child-Pugh score), MELD (model for end-stage liver disease) score, clinical symptoms (hepatic encephalopathy, anemia, edema, ascites fluid), and occasionally histological evaluation. In these clinical trials, umbilical cord MSCs were administered via a peripheral vein, and the number of cells infused was usually around 5.0×10^6 cells per kg body weight, being given three times at 4-week intervals. Some of the studies reported an improvement in the serum levels of albumin and total bilirubin, a decrease of the MELD score, or an improvement of clinical symptoms such as ascites at the end of the observation period [25, 29]. Clinical trials using bone marrow-derived MSCs as the cell source have made use of autologous bone marrow and administration via various routes such as the hepatic artery, portal vein, peripheral veins, or intrahepatic vessels. The number of transplanted cells in those studies ranged between 3.4×10^8 and 0.75×10^6/patient [29]. Although some studies reported an improvement of surrogate markers, the results were not consistent [20, 29, 30]. A few explorative clinical studies of MSC administration resulted in partial improvement of hepatic reserve in patients with alcoholic cirrhosis [22, 31]. However, a recent

Table 13.1 Summary of clinical trial regarding cell therapy for liver cirrhosis

No.	Authors	Years	Cell source	Autologous	N	Etiology	Route	Number of cells infused	Period (week)	Design	Trial	Findings
1	Terai [22]	2006	Bone marrow	Autologous	8	HBV, HCV, others	Peripheral	2.21–8.05 × 10^9/pts	24	Cohort	NA	Improvement in CP score
2	Mohamadnejad [14]	2007	Bone marrow	Autologous	4	HBV, PBC, others	Hepatic artery	5.25 × 10^6/pts	24	Cohort	Ph. 1	No beneficial effect
3	Kharaziha [12]	2009	Bone marrow	Autologous	8	HBV, HCV, others	Peripheral	30–50 × 10^6/pts	24	Cohort	Ph. I	Improvement in MELD score
4	Amer ME [7]	2011	Bone marrow	Autologous	40	HCV	Peripheral	20 × 10^6/pts	24	RCT	NA	Improvement in CP and MELD score
5	Peng L [15]	2011	Bone marrow	Autologous	158	HBV	Hepatic artery	3.4 × 10^8/pts	192	Case control	NA	Improvement in MELD score
6	Shi M [26]	2012	Umbilical cord	Allogenic	43	HCV	Peripheral	5.0 × 10^6/kg, 3 times	48	Case control	NA	Improvement in MELD score
7	Zhang Z [25]	2012	Umbilical cord	Allogenic	45	HBV	Peripheral	5.0 × 10^6/kg, 3 times	48	Case control	NA	Improvement in MELD score
8	Amin MA [8]	2013	Bone marrow	Autologous	20	HCV	Intrasplenic	10 × 10^6/pts	24	Cohort	NA	Decrease in T-Bil, AST, ALT
9	El-Ansary [11]	2013	Bone marrow	Autologous	25	HCV	Peripheral	1.0 × 10^6/kg	24	Case control	Ph. II	Improvement in MELD score
10	Spahr L [20]	2013	Bone marrow	Autologous	58	Alcohol	Hepatic artery	0.47 × 10 /kg	12	RCT	NA	No beneficial effect
11	Wang L [23]	2013	Umbilical cord	Allogenic	7	PBC	Peripheral	5.0 × 10^6/kg, 3 times	48	Cohort	NA	Improvement in MELD score
12	Jang YO [27]	2014	Bone marrow	Autologous	11	Alcohol	Hepatic artery	5.0 × 10^7/pts	12	Cohort	Ph. II	Histological improvement in CP score

13	Salama H [17]	2014	Bone marrow	Autologous	40	HCV	Peripheral	1.0×10^6/kg	24	RCT	NA	Decrease in T-Bil. Increase in Alb
14	Xu L [24]	2014	Bone marrow	Autologous	39	HBV	Hepatic artery	0.75×10^6/pts	24	RCT	NA	Decrease of T-Bil. Increase of Alb
15	Andreone P [9]	2015	Bone marrow	Autologous	17	HCV, alcohol,	Hepatic artery	0.05–1.0×10^6/kg	48	Cohort	Ph. I	Improvement in MELD score
16	Suk KT [21]	2016	Bone marrow	Autologous	68	Alcohol	Hepatic artery	5.0×10^7/pts	48	RCT	Ph. II	Improvement in CP and MELD score
17	Lanthier N [28]	2017	Bone marrow	Autologous	58	Alcohol	Hepatic artery	0.47×10^8/kg	4	RCT	NA	No beneficial effect
18	Newsome PN [32]	2018	Bone marrow	Autologous	81	Alcohol, HCV,	Peripheral	0.2×10^6 / kg	22.5	RCT	Ph. II	No beneficial effect

well-conducted randomized trial concluded that there was no beneficial effect of MSC administration combined with administration of granulocyte colony-stimulating factor [32]. This disappointing result further emphasizes the need for novel cell-based therapies for chronic liver failure [33]. At least, we need to summarize the reasons for this trial failure in scientific views. This includes the fundamental questions such as the candidacy of MSC as the cellular source toward the organ like the liver, which is one of the largest organs consisting mammalian body.

13.3 Future Novel Cellular Therapies Including Muse Cell Administration Toward Chronic Liver Failures (Fig. 13.1)

Current forms of cellular therapy require harvesting of MSCs from bone marrow and a certain period of time to prepare a sufficient number of pure cells, which limits the clinical application of this approach, especially in emergency cases such as acute liver failure. ES cells were initially reported as the potential cell source for administration. However, tumorigenicity and ethical issues for using fertilized egg are major barriers for feasibility of ES cells, and there are numerous issues to be overcome before reaching clinical trials with this cell source. Inducible pluripotent stem (iPS) cells have been engineered to overcome these difficulties and can theoretically differentiate into various types of cells, tissues, and organs [18, 34–36]. Although this approach is reported safe so far, there are still significant concerns about the artificial introduction of exogenous genes such as retroviral vectors [37]. Since the long-term efficacy and safety of iPS cell administration have not been proved, we need to be very careful about its clinical application as a standard form of care.

Multilineage-differentiating stress-enduring (Muse) cells are a form of mesenchymal stem cell with several of the novel characteristics of non-tumorigenic pluripotent characters [38, 39]. The previous reports demonstrated the capabilities of pluripotent differential abilities of Muse cells into liver-constituting cells such as hepatocytes [6, 40–42]. As Muse cells are able to recognize the sites of tissue damage/injury, thus contributing to tissue repair and promoting the improvement of organ function, their application to cellular therapy has naturally attracted attention. Besides differentiation capability, Muse cells also have other technical advantages over traditional MSCs. Muse cells were reported to home specifically into damaged tissue after intravenous injection and keep engrafted as tissue-specific cells for a longer period over several months, while majority of MSCs other than Muse cells, namely, non-Muse MSCs, basically do not home into damaged tissue nor they engraft as differentiated cells in the tissue [43]. The most recent study revealed that the sphingosine-1-phosphate is the major migratory factor of Muse cells, which will in turn be utilized for more efficient future isolation methods [44]. Interestingly, this humoral factor has been reported to be important by independent researchers [13]. Moreover, there is no need for Muse cells to introduce exogenous genes for acquiring

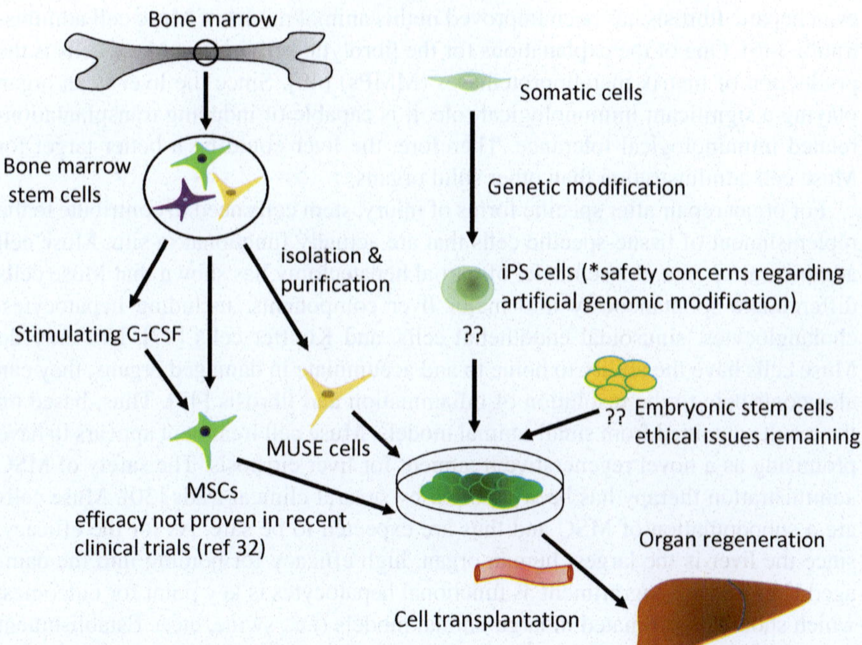

Fig. 13.1 Current concept of cell-based therapy for chronic liver failure
The major possible cellular source was either mesenchymal stem cells (MSCs), Muse cells, or iPS cells. The former two cells were essentially isolated from bone marrow, whereas the latter could be transformed by genetic modification. Muse cells could be subspecialized population of MSCs (see text). Although these concepts have been proposed, none of these have proven their efficacy in phase II clinical trials in chronic liver failure

pluripotency, which is essential for iPS cells, because Muse cells are already pluripotent. Besides safety profiles, Muse cells have superiority over MSC, ES cells, or iPS cells; without any prior gene introductions, Muse cells can selectively home into damaged tissue and efficiently replenish tissue-specific cells by intravenous injection. Exploiting this property, Muse cell administration has been reportedly effective in animal models with cerebral infarction, nephropathy, myocardial infarction, and liver resection [45–49]. The recent report by Iseki et al. used the Muse cell administration in mice model of liver cirrhosis [46]. In this report, the authors demonstrated that intravenously injected Muse cells have been recruited selectively to the liver and not to other organs. Moreover, Muse cells spontaneously differentiated in the damaged liver tissue into hepatocyte marker-positive cells without fusing with host hepatocytes [46]. These differentiated cells express major hepatocyte markers such as HepPar-1, albumin, and ant1-trypsin. They also expressed cytochrome (CPY) 1A2, an enzyme for detoxication, and glucose-6-phosphatase, an enzyme for glyocolysis, as representative markers of hepatocytes. As a consequence, the elevation of serum albumin levels and the decrease of total bilirubin levels were delivered by intravenous administration of Muse cells [46]. Surprisingly, even hepatic fibrosis has been improved in this animal model of Muse cell administration [46]. One of the explanations for the fibrolytic activities of Muse cells is the production of matrix metalloproteinases (MMPs) [40]. Since the liver is an organ playing a significant immunological role, it is capable of inducing transplantation-related immunological tolerance. Therefore, the liver could be a better target for Muse cell administration than other solid organs.

For organ repair after specific forms of injury, stem cells need to contribute to the replenishment of tissue-specific cells that are actually functional in situ. Muse cell administration in a mouse model of partial hepatectomy has shown that Muse cells differentiate spontaneously into major liver components, including hepatocytes, cholangiocytes, sinusoidal endothelial cells, and Kupffer cells [47]. Not only do Muse cells have the ability to home to and accumulate in damaged organs, they can also contribute to the resolution of inflammation and fibrosis [47]. Thus, based on the results obtained from small-animal models, Muse cell treatment appears to have promising as a novel regenerative treatment for liver cirrhosis. The safety of MSC administration therapy has been reported by several clinical trials [30]. Muse cells are a subpopulation of MSC and thus are expected to be safe. As for the efficacy, since the liver is the largest human organ, high efficacy for homing into the damaged liver and for engraftment as functional hepatocytes is key point for outcomes, which should be estimated in large-animal models (i.e., swine, etc.). Establishment of suitable models involving large animals with chronic liver failure and fibrosis will help to clarify a life-size efficacy and safety of Muse cell administration, leading in turn to human clinical trials. In this viewpoint, the most recent press release announcing the launch of clinical trials of Muse cell administration to evaluate its efficacy and safety in acute myocardial infarction has given a definite conviction of the application of this fascinating cell administration toward liver diseases in the near future [50].

Acknowledgment This work was supported in part by the Grant-in-Aid for Scientific Research (B) (Grant #JP16H05283) from JSPS and CREST (grant #17gm0610001h0006) and Research Program on Hepatitis from AMED.

References

1. Fukui H, Saito H, Ueno Y, Uto H, Obara K, Sakaida I et al (2016) Evidence-based clinical practice guidelines for liver cirrhosis 2015. J Gastroenterol 51(7):629–650
2. Naggie S, Muir AJ (2017) Oral combination therapies for hepatitis C virus infection: successes, challenges, and unmet needs. Annu Rev Med 68:345–358
3. Forbes SJ, Newsome PN (2012) New horizons for stem cell therapy in liver disease. J Hepatol 56(2):496–499
4. Terai S, Tsuchiya A (2017) Status of and candidates for cell therapy in liver cirrhosis: overcoming the "point of no return" in advanced liver cirrhosis. J Gastroenterol 52(2):129–140
5. Loomba R, Lawitz E, Mantry PS, Jayakumar S, Caldwell SH, Arnold H et al (2018) The ASK1 inhibitor selonsertib in patients with nonalcoholic steatohepatitis: a randomized, phase 2 trial. Hepatology 67(2):549–559

6. Alison MR, Poulsom R, Jeffery R, Dhillon AP, Quaglia A, Jacob J et al (2000) Hepatocytes from non-hepatic adult stem cells. Nature 406(6793):257
7. Amer ME, El-Sayed SZ, El-Kheir WA, Gabr H, Gomaa AA, El-Noomani N et al (2011) Clinical and laboratory evaluation of patients with end-stage liver cell failure injected with bone marrow-derived hepatocyte-like cells. Eur J Gastroenterol Hepatol 23(10):936–941
8. Amin MA, Sabry D, Rashed LA, Aref WM, el-Ghobary MA, Farhan MS et al (2013) Short-term evaluation of autologous transplantation of bone marrow-derived mesenchymal stem cells in patients with cirrhosis: Egyptian study. Clin Transpl 27(4):607–612
9. Andreone P, Catani L, Margini C, Brodosi L, Lorenzini S, Sollazzo D et al (2015) Reinfusion of highly purified CD133+ bone marrow-derived stem/progenitor cells in patients with end-stage liver disease: a phase I clinical trial. Dig Liver Dis 47(12):1059–1066
10. Dubuquoy L, Louvet A, Lassailly G, Truant S, Boleslawski E, Artru F et al (2015) Progenitor cell expansion and impaired hepatocyte regeneration in explanted livers from alcoholic hepatitis. Gut 64(12):1949–1960
11. El-Ansary M, Abdel-Aziz I, Mogawer S, Abdel-Hamid S, Hammam O, Teaema S et al (2012) Phase II trial: undifferentiated versus differentiated autologous mesenchymal stem cells transplantation in Egyptian patients with HCV induced liver cirrhosis. Stem Cell Rev 8(3):972–981
12. Kharaziha P, Hellstrom PM, Noorinayer B, Farzaneh F, Aghajani K, Jafari F et al (2009) Improvement of liver function in liver cirrhosis patients after autologous mesenchymal stem cell injection: a phase I-II clinical trial. Eur J Gastroenterol Hepatol 21(10):1199–1205
13. King A, Houlihan DD, Kavanagh D, Haldar D, Luu N, Owen A et al (2017) Sphingosine-1-phosphate prevents egress of hematopoietic stem cells from liver to reduce fibrosis. Gastroenterology 153(1):233–248 e16
14. Mohamadnejad M, Namiri M, Bagheri M, Hashemi SM, Ghanaati H, Zare Mehrjardi N et al (2007) Phase 1 human trial of autologous bone marrow-hematopoietic stem cell transplantation in patients with decompensated cirrhosis. World J Gastroenterol 13(24):3359–3363
15. Peng L, Xie DY, Lin BL, Liu J, Zhu HP, Xie C et al (2011) Autologous bone marrow mesenchymal stem cell transplantation in liver failure patients caused by hepatitis B: short-term and long-term outcomes. Hepatology 54(3):820–828
16. Sakaida I, Terai S, Yamamoto N, Aoyama K, Ishikawa T, Nishina H et al (2004) Transplantation of bone marrow cells reduces CCl4-induced liver fibrosis in mice. Hepatology 40(6):1304–1311
17. Salama H, Zekri AR, Medhat E, Al Alim SA, Ahmed OS, Bahnassy AA et al (2014) Peripheral vein infusion of autologous mesenchymal stem cells in Egyptian HCV-positive patients with end-stage liver disease. Stem Cell Res Ther 5(3):70
18. Si-Tayeb K, Noto FK, Nagaoka M, Li J, Battle MA, Duris C et al (2010) Highly efficient generation of human hepatocyte-like cells from induced pluripotent stem cells. Hepatology 51(1):297–305
19. Sokal EM, Lombard CA, Roelants V, Najimi M, Varma S, Sargiacomo C et al (2017) Biodistribution of liver-derived mesenchymal stem cells after peripheral injection in a hemophilia a patient. Transplantation 101(8):1845–1851
20. Spahr L, Chalandon Y, Terraz S, Kindler V, Rubbia-Brandt L, Frossard JL et al (2013) Autologous bone marrow mononuclear cell transplantation in patients with decompensated alcoholic liver disease: a randomized controlled trial. PLoS One 8(1):e53719
21. Suk KT, Yoon JH, Kim MY, Kim CW, Kim JK, Park H et al (2016) Transplantation with autologous bone marrow-derived mesenchymal stem cells for alcoholic cirrhosis: phase 2 trial. Hepatology 64(6):2185–2197
22. Terai S, Ishikawa T, Omori K, Aoyama K, Marumoto Y, Urata Y et al (2006) Improved liver function in patients with liver cirrhosis after autologous bone marrow cell infusion therapy. Stem Cells 24(10):2292–2298
23. Wang L, Li J, Liu H, Li Y, Fu J, Sun Y et al (2013) Pilot study of umbilical cord-derived mesenchymal stem cell transfusion in patients with primary biliary cirrhosis. J Gastroenterol Hepatol 28(Suppl 1):85–92

24. Xu L, Gong Y, Wang B, Shi K, Hou Y, Wang L et al (2014) Randomized trial of autologous bone marrow mesenchymal stem cells transplantation for hepatitis B virus cirrhosis: regulation of Treg/Th17 cells. J Gastroenterol Hepatol 29(8):1620–1628
25. Zhang Z, Lin H, Shi M, Xu R, Fu J, Lv J et al (2012) Human umbilical cord mesenchymal stem cells improve liver function and ascites in decompensated liver cirrhosis patients. J Gastroenterol Hepatol 27(Suppl 2):112–120
26. Shi M, Zhang Z, Xu R, Lin H, Fu J, Zou Z et al (2012) Human mesenchymal stem cell transfusion is safe and improves liver function in acute-on-chronic liver failure patients. Stem Cells Transl Med 1(10):725–731
27. Jang YO, Kim YJ, Baik SK, Kim MY, Eom YW, Cho MY et al (2014) Histological improvement following administration of autologous bone marrow-derived mesenchymal stem cells for alcoholic cirrhosis: a pilot study. Liver Int 34(1):33–41
28. Lanthier N, Lin-Marq N, Rubbia-Brandt L, Clement S, Goossens N, Spahr L (2017) Autologous bone marrow-derived cell transplantation in decompensated alcoholic liver disease: what is the impact on liver histology and gene expression patterns? Stem Cell Res Ther 8(1):88
29. Moore JK, Stutchfield BM, Forbes SJ (2014) Systematic review: the effects of autologous stem cell therapy for patients with liver disease. Aliment Pharmacol Ther 39(7):673–685
30. Trounson A, McDonald C (2015) Stem cell therapies in clinical trials: progress and challenges. Cell Stem Cell 17(1):11–22
31. Saito T, Okumoto K, Haga H, Nishise Y, Ishii R, Sato C et al (2011) Potential therapeutic application of intravenous autologous bone marrow infusion in patients with alcoholic liver cirrhosis. Stem Cells Dev 20(9):1503–1510
32. Newsome PN, Fox R, King AL, Barton D, Than NN, Moore J et al (2018) Granulocyte colony-stimulating factor and autologous CD133-positive stem-cell therapy in liver cirrhosis (REALISTIC): an open-label, randomised, controlled phase 2 trial. Lancet Gastroenterol Hepatol 3(1):25–36
33. Lanthier N (2018) Haemopoietic stem cell therapy in cirrhosis: the end of the story? Lancet Gastroenterol Hepatol. 3(1):3–5
34. Takayama K, Morisaki Y, Kuno S, Nagamoto Y, Harada K, Furukawa N et al (2014) Prediction of interindividual differences in hepatic functions and drug sensitivity by using human iPS-derived hepatocytes. Proc Natl Acad Sci USA 111(47):16772–16777
35. Rashid ST, Corbineau S, Hannan N, Marciniak SJ, Miranda E, Alexander G et al (2010) Modeling inherited metabolic disorders of the liver using human induced pluripotent stem cells. J Clin Invest 120(9):3127–3136
36. Iwamuro M, Komaki T, Kubota Y, Seita M, Kawamoto H, Yuasa T et al (2010) Hepatic differentiation of mouse iPS cells in vitro. Cell Transplant 19(6):841–847
37. Normile D (2017) iPS cell therapy reported safe. Science 355(6330):1109–1110
38. Kuroda Y, Kitada M, Wakao S, Nishikawa K, Tanimura Y, Makinoshima H et al (2010) Unique multipotent cells in adult human mesenchymal cell populations. Proc Natl Acad Sci USA 107(19):8639–8643
39. Kuroda Y, Wakao S, Kitada M, Murakami T, Nojima M, Dezawa M (2013) Isolation, culture and evaluation of multilineage-differentiating stress-enduring (Muse) cells. Nat Protoc 8(7):1391–1415
40. Iseki M, Kushida Y, Wakao S, Akimoto T, Mizuma M, Motoi F et al (2017) Muse cells, nontumorigenic pluripotent-like stem cells, have liver regeneration capacity through specific homing and cell replacement in a mouse model of liver fibrosis. Cell Transplant 26(5):821–840
41. Ogura F, Wakao S, Kuroda Y, Tsuchiyama K, Bagheri M, Heneidi S et al (2014) Human adipose tissue possesses a unique population of pluripotent stem cells with nontumorigenic and low telomerase activities: potential implications in regenerative medicine. Stem Cells Dev 23(7):717–728
42. Wakao S, Kitada M, Kuroda Y, Shigemoto T, Matsuse D, Akashi H et al (2011) Multilineage-differentiating stress-enduring (Muse) cells are a primary source of induced pluripotent stem cells in human fibroblasts. Proc Natl Acad Sci USA 108(24):9875–9880

43. Dezawa M (2016) Muse cells provide the pluripotency of mesenchymal stem cells: direct contribution of Muse cells to tissue regeneration. Cell Transplant 25(5):849–861
44. Tanaka T, Nishigaki K, Minatoguchi S, Nawa T, Yamada Y, Kanamori H et al (2018) Mobilized Muse cells after acute myocardial infarction predict cardiac function and remodeling in the chronic phase. Circ J 82(2):561–571
45. Uchida N, Kushida Y, Kitada M, Wakao S, Kumagai N, Kuroda Y et al (2017) Beneficial effects of systemically administered human Muse cells in adriamycin nephropathy. J Am Soc Nephrol 28(10):2946–2960
46. Uchida H, Morita T, Niizuma K, Kushida Y, Kuroda Y, Wakao S et al (2016) Transplantation of unique subpopulation of fibroblasts, Muse cells, ameliorates experimental stroke possibly via robust neuronal differentiation. Stem Cells 34(1):160–173
47. Katagiri H, Kushida Y, Nojima M, Kuroda Y, Wakao S, Ishida K et al (2016) A distinct subpopulation of bone marrow mesenchymal stem cells, Muse cells, directly commit to the replacement of liver components. Am J Transplant 16(2):468–483
48. Yamauchi T, Kuroda Y, Morita T, Shichinohe H, Houkin K, Dezawa M et al (2015) Therapeutic effects of human multilineage-differentiating stress enduring (MUSE) cell transplantation into infarct brain of mice. PLoS One 10(3):e0116009
49. Kinoshita K, Kuno S, Ishimine H, Aoi N, Mineda K, Kato H et al (2015) Therapeutic potential of adipose-derived SSEA-3-positive Muse cells for treating diabetic skin ulcers. Stem Cells Transl Med 4(2):146–155
50. Life Science Institute, Inc., Press release http://www.lsii.co.jp/pdf/20180115-1.pdf (Japanese)

Chapter 14
Artificial Pigmented Human Skin Created by Muse Cells

Takeshi Yamauchi, Kenshi Yamasaki, Kenichiro Tsuchiyama, and Setsuya Aiba

Abstract The skin composes physiological and chemical barrier and renews skin component cells throughout the human life. Melanocytes locate in the basal layer of the epidermis and produce melanin to protect the skin from ultraviolet. Melanin plays key roles in determining human skin and hair color. Melanocyte dysfunction observed in albinism and vitiligo not only causes cosmetic problems but also increases risk of skin cancer. As rejuvenate therapy, embryonic stem (ES) cells and induced pluripotent stem (iPS) cells have been reported to generate melanocytes. Other than ES and iPS cells, human skin tissues maintain pluripotent stem cells, named multilineage-differentiating stress-enduring (Muse) cells. We employ Muse cells isolated from human fibroblasts and adipose tissue to differentiate into melanocytes (Muse-MC). Muse-MC express melanocyte-related molecules, such as tyrosinase and DCT, and show tyrosinase activity. We also succeeded to differentiate Muse cells into fibroblasts and keratinocytes and created three-dimensional (3D) reconstituted skin with Muse cell-derived melanocytes, fibroblasts, and keratinocytes. The 3D reconstituted skin of Muse cell-derived cells coordinately showed epidermis layers and Muse-MC localized in the basal layer of the epidermis. Thus Muse cells in the human skin can be a source of rejuvenation medicine for the skin reconstruction.

Keywords Melanocyte · Fibroblast · Keratinocyte · Melanin · 3D skin · a-MSH

T. Yamauchi · K. Yamasaki (✉) · K. Tsuchiyama · S. Aiba
Department of Dermatology, Tohoku University Graduate School of Medicine, Sendai, Japan
e-mail: t-yamauchi@derma.med.tohoku.ac.jp; kyamasaki@med.tohoku.ac.jp

M. Dezawa (ed.), *Muse Cells*, Advances in Experimental Medicine and Biology 1103, https://doi.org/10.1007/978-4-431-56847-6_14

14.1 Skin as a Target of Stem Cell Therapy

14.1.1 Reconstituted Skin as Physiological Barrier and for Alternative Test Materials

The skin's most important function is to form an effective barrier between the inside and outside of the organism [1]. The skin is composed of the epidermis, dermis, and subcutaneous tissue. Keratinocytes are the predominant cell type in the epidermis (> 90%). And melanocytes, Langerhans cells, and Merkel cells are also present in the epidermis. Melanocytes locate in the bottom layer of the epidermis and produce melanin to protect the skin from ultraviolet (UV). Merkel cells have mechanoreceptors and are known as tactile epithelial cells. Langerhans cells are dendritic cells. The work of these cells is uptake and processing of antigens. The epidermis is supported by the dermis and subcutaneous tissue. The dermis is a layer of the skin between the epidermis and subcutaneous tissue and is composed of extracellular matrix along with fibroblasts, macrophages and endothelial cells forming blood vessels. The major component of subcutaneous layer is adipose tissue. If the layers of the skin are lost, human body is easily infected by microbes including bacteria, viruses, and fungus from outside. When the large area of the skin is lost, a human would be dead because of the loosing body fluid and hypothermia. Thus, burned skin or large skin defect by injury needs to be covered in timely manner. The 3D reconstituted skins composed with keratinocytes and fibroblasts are used in medical treatments in the case of these situations. The reconstituted skin not only covers the surface of the body but also promotes to heal the wound area by releasing various cytokines and growth factors [2].

In basic and translational researches, the 3D reconstituted skins are used for trans-epidermal water loss analysis, ceramide analysis, drug toxicity tests, and melanin production inhibition test. The 3D reconstituted skins are also used to test materials and drugs as alternative methods to reduce animal experiments. The advantages of the 3D reconstituted skins are that we can produce skins of equal quality, size, thickness, and cell density and that we can modify the quality of 3D skins adjusted for experimental designs.

14.1.2 Melanocytes Produce Melanin to Survive on the Earth Where Ultraviolet Covers

Human melanocytes exist in the basal layer of the epidermis and produce melanin in melanocyte-specific organelle melanosomes. Melanocytes deliver melanin in melanosome to neighboring keratinocytes, and melanin forms so-called supranuclear melanin cap to protect nucleus from DNA damage by UV rays [3, 4]. In addition, melanin plays key roles in determining human skin and hair color. Melanocyte dysfunction results in a variety of pigment disorders, such as albinism and vitiligo, which cause not only cosmetic problems but also increase the risk of skin cancers due to incomplete protection from UV rays [5]. Albinism is a group of inherited disorders of melanin biosynthesis characterized by a generalized reduction in pigmentation of the hair, skin, and eyes. At least four genes are responsible for the four major types of the albinism (*TYR, OCA2, TYRP1, and MATP*) [6]. Although the worldwide prevalence of albinism considerably varies, approximately 1 at 17,000 birth has albinism, and it is estimated that about 1 in 70 people carries a mutation of genes relating to albinism, while vitiligo is an acquired cutaneous disorder of pigmentation, with an incidence of 0.5% to 2% worldwide. Immunological disorders including aberrant autoimmune reaction and melanocyte dysfunctions are likely to be involved in vitiligo though the causes of vitiligo are not fully elucidated. As vitiligo can spread to the whole body and have a major effect on quality of life, preferably treatment should begin as early as possible when the disease is active [7].

14.1.3 Rejuvenate Therapy Relating to Melanocyte Cell Biology

Current treatment for vitiligo includes topical treatments of corticosteroid and vitamin D to control abnormal immune reaction, UV irradiation to enhance pigmentation, and autologous skin grafts to transfer normal melanocytes from non-lesional skin [8, 9]. Autologous human skin melanocyte transplantation such as smash grafting and mini-grafting is the most practical cell therapy for vitiligo to date [10]. Because human melanocytes sporadically exist in the basal layer of the epidermis, melanocytes are difficult to isolate from the mixture of epidermal keratinocytes and other cells in the epidermis and hard to amplify in a large scale in vitro for clinical application [11, 12]. To overcome these limitation of the rejuvenate therapy relating to melanocyte disorders, several groups attempted to generate melanocytes from pluripotent stem cells including embryonic stem (ES) cells, induced pluripotent stem (iPS) cells, and mesenchymal cells in fibroblasts [13–17]. ES cells can differentiate any kind of cells theoretically but have the ethical issues because ES cells are obtained from human zygotes. The iPS cells are great source to produce pluripotent stem cells by gene manipulation but have the risk of tumorigenesis as well as ES cells, which requires careful monitoring of cell quality and procedures for clinical use. Additional caution is required to determine whether melanocytes derived from these stem cells develop into malignancy, since melanocyte induction requires phorbol ester, which is known as carcinogen. Thus, efficient and safer source of pluripotent stem cells are desirable for rejuvenate therapy relating to melanocyte disorders.

This chapter discusses on potential of skin rejuvenate therapy by multilineage-differentiating stress enduring (Muse) cells, which are found as a novel type of non-tumorigenic endogenous pluripotent stem cells that are collectable from the human skin, adipose tissue, and bone marrow [18–24]. We describe here methods to generate functional melanocytes, keratinocytes, and fibroblast from Muse cells in vitro and techniques to reconstruct 3-dimensional skin with Muse-derived melanocytes, keratinocytes, and fibroblast.

14.2 Skin as a Source of Stem Cell Therapy

14.2.1 Muse Cells Reside in Connective and Mesenchymal Tissues in Normal Skin Among Mesenchymal Stem Cells

Muse cells, which were first reported in 2010 by Dezawa et al., are non-tumorigenic endogenous pluripotent stem cells and reside in connective and mesenchymal tissue, such as the dermis and adipose tissue, and the bone marrow [18]. Muse cells exist sporadically in these tissues and do not associate with any known structural niches [25] (Fig. 14.1a, b). Muse cells are collectable from cultured mesenchymal stem cells (MSCs), which are adult stem cells with a lower risk of tumorigenesis and used for the treatment of diseases such as acute graft versus host disease and lipodystrophy. Muse cells express pluripotency markers, such as transcription factors Oct3/4, Sox2, and Nanog, as well as a pluripotent surface marker, stage-specific embryonic antigen-3 (SSEA-3). SSEA-3 is glycosphingolipid, which is highly expressed in early stage of embryo body, and is assumed to be involved in defining stage of stem cell [26, 27]. Muse cells can be collected as SSEA-3-positive cells from various sources, such as commercially available fibroblasts, adipose-derived stem cells (ASCs), and bone marrow cells [25] (Fig. 14.1c, d). Muse cells are able to differentiate into three germ layer-derived cells from a single cell (Fig. 14.1e, f and g). Since Muse cells naturally reside in human normal tissues, they are thought to be non-tumorigenic. It is reported that gene expression patterns of cell cycle-related molecules in Muse cells differ from those in ES or iPS cells. In general, cell cycle-related molecules express higher in ES or iPS cells than Muse cells, and Muse cells express these genes at levels similar to normal somatic cells [19]. Moreover, Muse cells have low telomerase activity compared to iPS cells and Hela cells [19, 20]. When human bone marrow-derived Muse cells were injected into mice testes, Muse cells did not develop teratoma up to 6 months, whereas ES cells and iPS cells developed teratoma in several weeks [18]. These suggest that Muse cells are pluripotent stem cells with little potential to develop tumors.

14.2.2 Sources and Origin of Muse Cells in Skin

So far, Muse cells have been obtained from bone marrow, adipose tissue, dermal fibroblasts, dental pulp, and umbilical cord [18–24]. Among them, dermal fibroblasts and adipose tissue are useful sources to collect Muse cells because the tissues are easily obtained from human body surface. Fibroblasts can be obtained from the human dermis by skin biopsy, and adipose stem cells (ASCs) can be isolated from adipose tissue excised or removed by surgical procedures including liposuction. Human skin dermal fibroblasts as well as commercially available human dermal fibroblasts contain several percent of SSEA-3-positive Muse cells [19, 28] (Fig. 14.1a, c). Similarly, several percent SSEA-3-positive Muse cells are identified

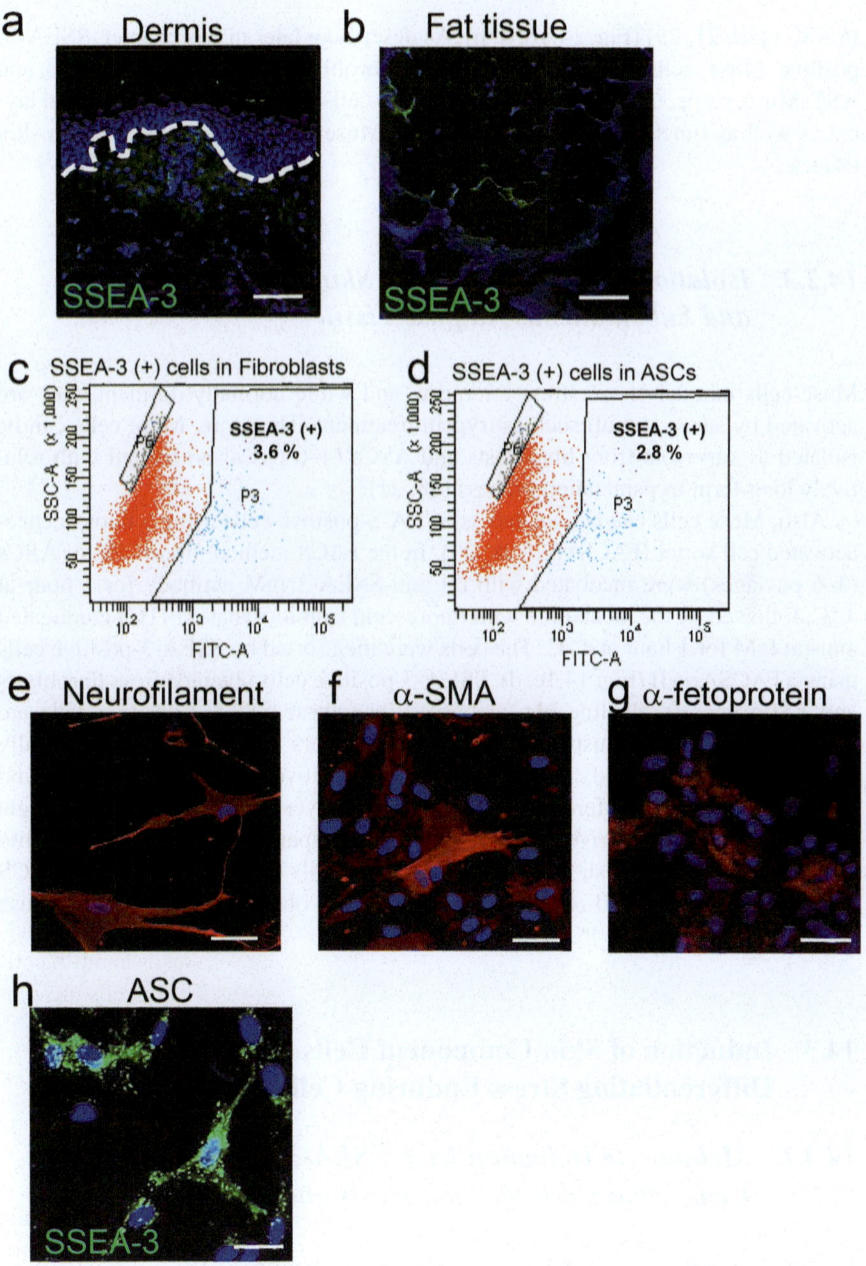

Fig. 14.1 Muse cells exist in human connective tissue. (**a**, **b**) SSEA-3-positive Muse cells exist in the human dermis and subcutaneous tissue in vivo. (**c**, **d**) The examples of cell sorting of SSEA-3-positive cells from dermal fibroblasts (c) and human ASCs (d). (**e–g**) Muse cells can differentiate into cells positive for neurofilament as ectodermal lineage, alpha smooth muscle (α-SMA) as mesodermal lineage, and alpha fetoprotein (α-fetoprotein) as endodermal lineage. (**h**) ASCs contain SSEA-3-positive Muse cells in the population. Scale bars = 50 μm

in ASCs [20, 21, 29] (Fig. 14.1b, d, h). As describing later in this chapter, SSEA-3-positive Muse cells derived from dermal fibroblasts and ASCs (Fb-Muse and ASC-Muse, respectively) can differentiate into cells associated with three germ layers as well as functional melanocytes. Thus, Muse cells exist sporadically in skin tissues.

14.2.3 Isolation of Muse Cells from Skin Fibroblasts and Subcutaneous Adipose Tissue

Muse cells uniquely have stress tolerance, and while normally dormant, they are activated by stress stimuli such as trypsin treatment. Therefore, Muse cells can be isolated as survivors from fibroblasts and ASCs (4–6 passages) treated with relatively long-term trypsinization process [18, 21].

Also, Muse cells can be isolated as SSEA-3-positive cells using a fluorescence-activated cell sorter (FACS) [20, 28, 29]. In the FACS method, fibroblasts or ASCs (4–6 passages) were incubated with rat anti-SSEA-3 IgM antibody for 1 hour at 4 °C, followed by the incubation with fluorescein isothiocyanate (FITC)-conjugated anti-rat IgM for 1 hour at 4 °C. The cells were then sorted by SSEA-3-positive cells using a FACSAria II (Fig. 14.1c, d). SSEA-3-positive cells obtained from fibroblasts and ASCs express alkaline phosphatase. When these SSEA-3-positive cells are cultured in single cell suspension, they form clusters which are morphologically similar to embryonic body of ES cell. SSEA-3-positive cells from fibroblasts and ASCs are also able to differentiate into three germ layers derived cells from a single cell when single cell-derived clusters formed in suspension were transferred onto gelatin-coated culture dish to allow expansion of cells (Fig. 14.1e, f and g). FACS methods are efficient and reproducible procedure to obtain SSEA-3-positive Muse cells.

14.3 Induction of Skin Component Cells from Multilineage-Differentiating Stress Enduring Cells

14.3.1 Melanocyte Induction from SSEA-3-Positive Muse Cells Obtained from Fibroblasts and ASCs

We have carefully examined several combinations of factors relating to melanocyte biology and selected the essential factors to induce melanocyte from Muse cells (Muse-MC). We defined the melanocyte-inducing medium (MIM) that contains ten essential factors: Wnt3a, stem cell factor (SCF), endothelin-3 (ET-3), 12-O-tetradecanoyl-phorbol 13-acetate (TPA), basic fibroblast growth factor (b-FGF), cholera toxin, L-ascorbic acid, linoleic acid, dexamethasone, and insulin-transferrin-selenium (ITS) [17, 28]. Wnt3a, SCF, and ET-3 are required to differentiate stem cells into melanocytes [30]. TPA, cholera toxin, and b-FGF enhance proliferation of melanocytes [31, 32]. Wnt3a, SCF, ET-3, TPA, b-FGF, and cholera toxin are cAMP inducers, and these factors are considered to induce melanocyte-related factors in Muse cells through MITF-M activation. MITF-M regulates the differentiation and development of melanocytes and is also responsible for pigment cell-specific transcription of the melanogenesis enzyme genes [33]. The tyrosinase activity of melanocytes is activated by L-ascorbic acid [34]. Linoleic acid, dexamethasone, and ITS are components to maintain matured melanocytes and are included in most of the documented melanocyte culture media [14]. Linoleic acid and ITS are also used as supplemental factors in low serum medium. Yamane et al. reported that dexamethasone promoted the generation of melanocytic cells from mouse ES cells [35]. Cooperation of these ten factors produces a necessary and sufficient condition to differentiate Muse cells into melanocytes so far.

Melanocyte-inducing system was set as follows. Muse cells were seeded at a density of 15,000 cells per well of a 6-well plate and cultured for 1 day in α-MEM without serum. These cells were then cultured in MIM (50% high-glucose DMEM, 30% low-glucose DMEM, and 20% MCDB201 medium containing 50 ng/mL Wnt3a, 50 ng/mL SCF, 100 nM ET-3, 50 nM TPA, 4 ng/mL b-FGF, 20 pM cholera toxin, 100 mM L-ascorbic acid, 1 mg/mL linoleic acid, 50 mM dexamethasone, and ITS) for 6 weeks. The cells were subcultured when cells grew to 80% confluence.

During the 6-week culture of Muse cells in MIM, the cells also increased in number to $9.58 \pm 0.17 \times 10^6$ cells as shown in the representative growth curve [29]. Muse cells in MIM gradually extended dendrites and became morphologically similar to normal human epidermal melanocytes (NHEMs) (Fig. 14.2a). After 6 weeks of culture in MIM, the cells were positive for the L-DOPA reaction assay, suggesting that Muse cells acquired tyrosinase activity (Fig. 14.2b). The expression of melanocyte-related genes was examined in these cells after 6-week culture. Muse cells originally express MITF and KIT. They started to express tyrosinase-related protein 1 (TYRP1) after 3-week, PMEL (gp100) and dopachrome tautomerase (DCT) after 5-week, and finally tyrosinase (TYR) at 6-week culture (Fig. 14.2b). The expression levels of these genes were comparable to those of cultured NHEMs. Protein expressions of these melanocyte-related molecules were confirmed by observation of immunofluorescence staining. PMEL/gp100, a marker of the melanosome, also increased in cells cultured in MIM. The percentage of PMEL/gp100+ cells in Muse-MC after 6-week culture was 43.1 ± 17.1%, while Muse cells were negative for PMEL/gp100+ cells, and the percentage of PMEL/gp100+ cells in cultured NHEMs was 96.4 ± 5.0% [29]. The representative immunofluorescence staining of PMEL/gp100 in NHEMs and Muse-MC are shown in Fig. 14.2d, e, respectively.

Functional melanocytes express melanocortin 1 receptor (MC1R) and increase eumelanin synthesis by stimulation of MC1R ligand α-MSH [29]. *MC1R* expression lacked in original Muse cells but was detected after 4 weeks in MIM, continued to increase by 5 weeks, and reached a plateau by 6 weeks as similar levels as NHEMs. When we stimulated Muse-MC at 6 weeks with α-MSH and cultured them for further 3 weeks with α-MSH, the visible darkness of the cell pellets was obviously increased, which was quantitated by image processing and intracellular melanin measured by

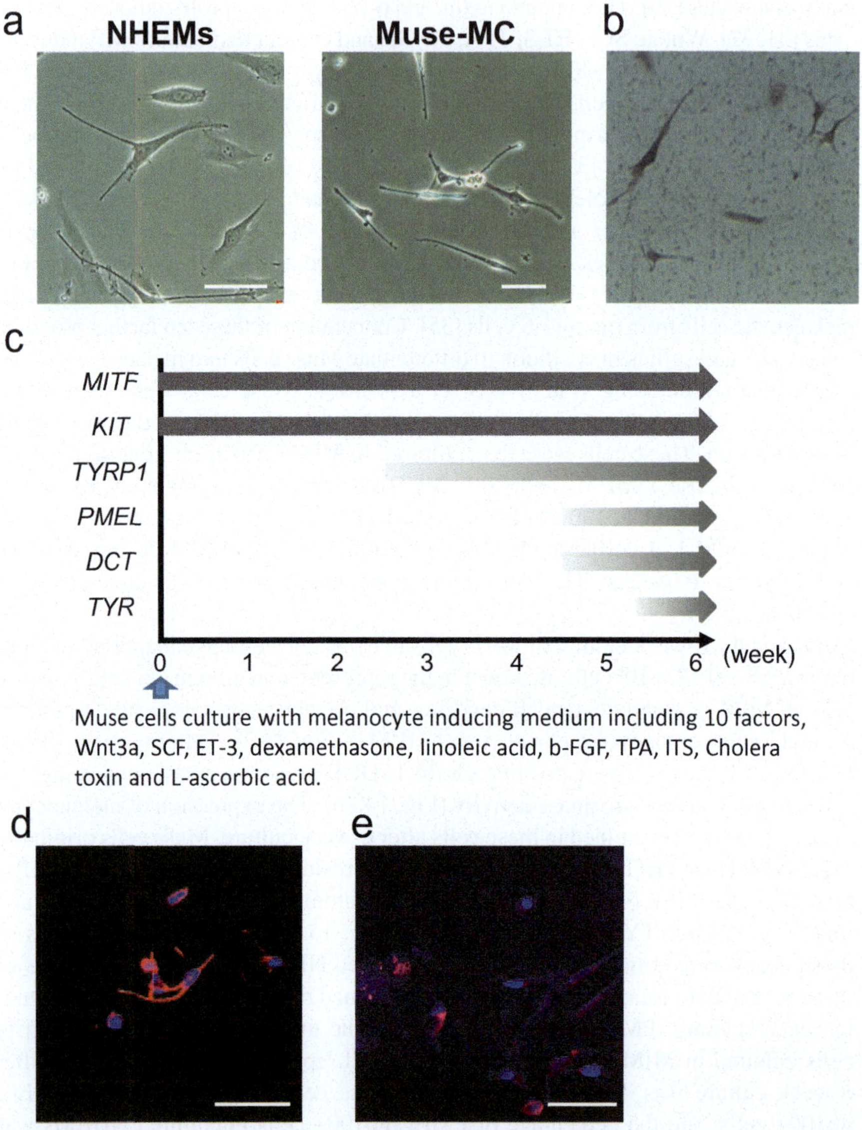

Fig. 14.2 Differentiation and migration of Muse cell-derived melanocytes. (**a**) Phase contrast microscopic images of normal human epidermal melanocytes (NHEMs) and Muse cell-derived melanocytes (Muse-MC). (**b**) L-DOPA assay reaction of Muse-MC. (**b**) The scheme of time-dependent expressions of melanocyte-related genes. The arrow indicates the starting point of stimulation to differentiate Muse cells into melanocytes by ten factors. Muse cells had been cultured with melanocyte-inducing medium for 6 weeks. (**d**, **e**) Representative gp100 immunofluorescence images of NHEMs (d) and ASC-Muse cells after 6 weeks of culture (e). Scale bars = 50 μm

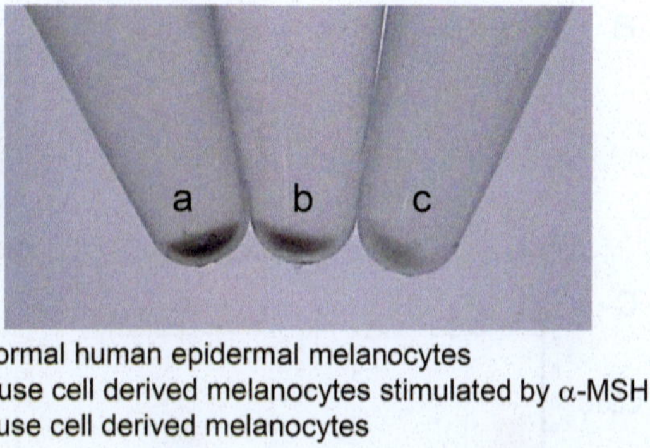

Fig. 14.3 The Muse-MC increase melanin content in cells by α-MSH stimulation. (a, b, c) The cell pellets of (**a**) cultured NHEMs, (**b**) Muse-MC stimulated with α-melanocyte-stimulating hormone (α-MSH), and (**b**) Muse-MC without α-MSH stimulation are shown

OD 450 nm (Fig. 14.3) [29]. These indicated that Muse cells differentiated into functional melanocytes (Muse-MC) by culturing in MIM containing ten factors.

14.3.2 Generation of Muse Cell-Derived Fibroblasts

In order to differentiate Muse cells into fibroblasts, Muse cells were cultured with DMEM including transforming growth factor β2 on floating culture for 1 week. TGF-β2 differentiates stem cells into mesoderm lineage. After forming embryonic bodies, these clusters were attached to a dish and cultured in DMEM including ascorbic acid and fetal bovine serum for 1 week. Ascorbic acid activates proliferation of fibroblasts, which in turn differentiates Muse cells into fibroblasts (Muse-Fb) [36]. Morphology of Muse-Fb was similar to normal human fibroblasts (Fig. 14.4a). Most of Muse-Fb are expressed on collagen 1 and collagen 3 (Fig. 14.4b). CD10 and 73, which are expressed on mesenchymal stem cell and fibroblasts, are also expressed in Muse-Fb (Fig. 14.4c). These data show that Muse cell can differentiate into fibroblasts [36].

14.3.3 Generation of Muse Cell-Derived Keratinocyte

In order to differentiate Muse cells into keratinocytes, Muse cells were cultured on Matrigel-coating dishes (Matrigel purchased from Becton Dickinson) with defined keratinocyte serum-free medium (DKSFM) (Invitrogen) including bone morphogenetic protein 4 (BMP4) and all-trans retinoic acid (ATRA). BMP4 promotes the ventralization of ectoderm and inhibits differentiation of ectoderm into neuron. ATRA induces pluripotent stem cells into ectodermal lineage. After 4-week culture, Muse cells turned their morphology similar to normal epidermal keratinocytes (Muse-KC) (Fig. 14.5a). Muse-KC expressed keratinocyte-related molecules desmoglein 3 and keratin 14 (K14) (Fig. 14.5b, c). In this method, a half of Muse-KC

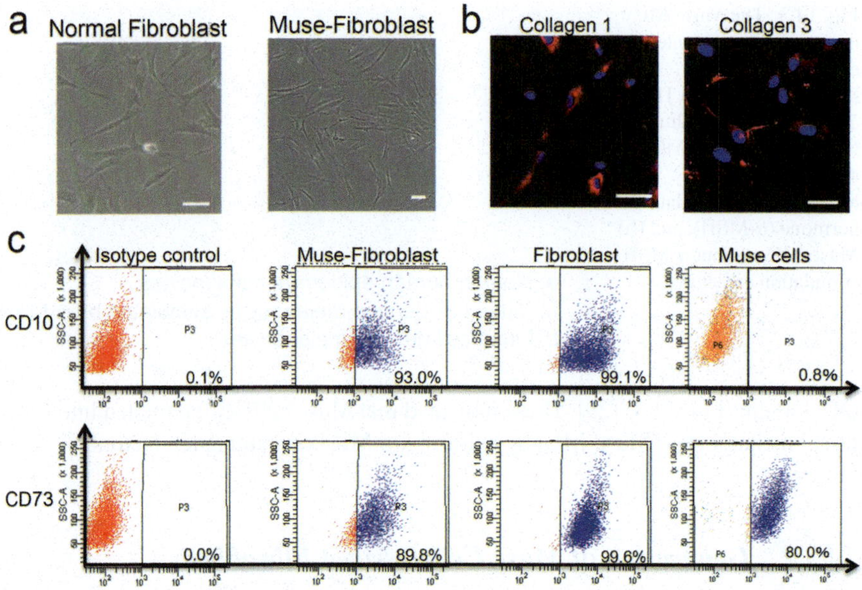

Fig. 14.4 Muse cells differentiation into fibroblasts. (**a**) The cell morphologies of normal fibroblasts and Muse-fibroblasts. (**b**) Immunofluorescence staining of collagen 1 and 3 in Muse-fibroblasts. (**c**) FACS analysis of CD10 and CD73 expression on Muse-fibroblasts, normal fibroblasts, and Muse cells. Scale bars = 50 μm

in the monolayer condition expressed both K14 and K18 (46.9 ± 0.6 and 43.1 ± 0.5%, respectively.) [36].

Itoh et al. cultured iPS cells on Matrigel with 10 ng/mL BMP4 (ca. 0.4 nM) and 1 μM ATRA containing DKSFM [37, 38]. Guenou et al. reported that ES cells differentiated into keratinocytes using BMP4 (0.5 nM) but without ATRA [39]. Muse cells required BMP4 similarly to iPS cells and ES cells to differentiate into keratinocytes. Muse cells did not change their morphology toward keratinocytes with lower BMP4 concentration up to 0.2 nM even though 1 μM ATRA was added (data not shown). Because BMP4 gene expression in Muse cells are lower than ES cell and iPS cell [40], Muse cells may require the higher BMP4 as well as ATRA to differentiate into keratinocytes efficiently. This data and previous reports confirmed that BMP4 and ATRA are essential molecules for pluripotent stem cells including Muse, iPS, and ES cells to differentiate into keratinocytes [36].

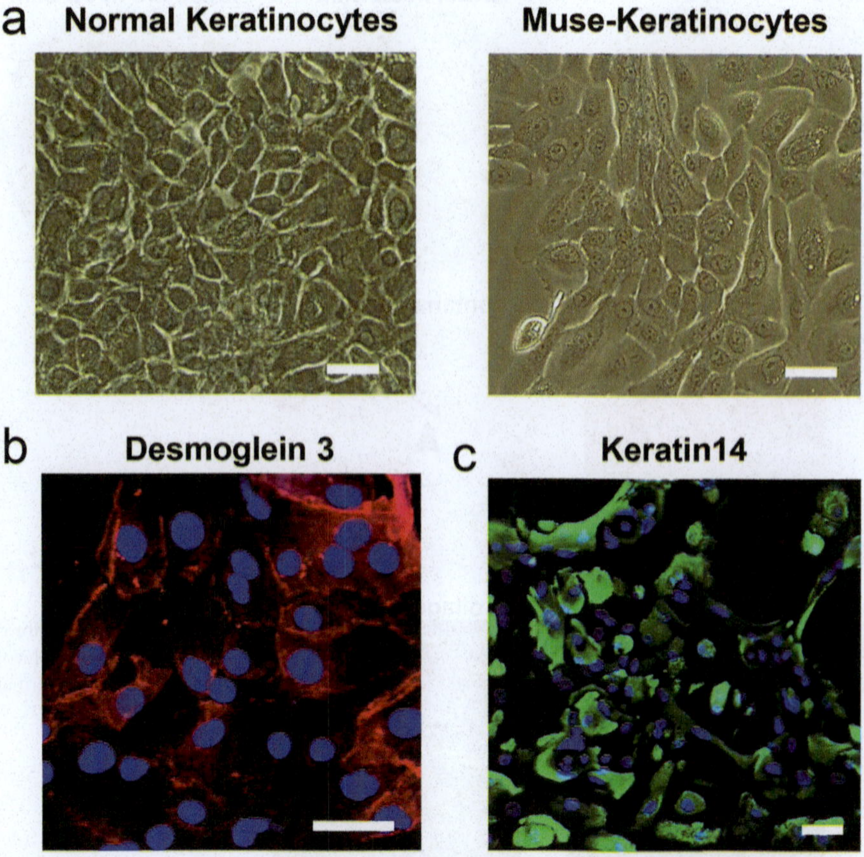

Fig. 14.5 Muse cell differentiation into keratinocyte. (**a**) The cell morphologies of normal keratinocytes and Muse-keratinocytes. (**b**, **c**) Immunofluorescence staining of desmoglein 3 (b) and keratin14 (c) in Muse-keratinocytes. Scale bars = 50 μm

14.3.4 Generation and Functional Evaluation of Human Pigmented 3D Skin Reconstituted with Muse Cell-Derived Skin Component Cells In Vitro

Human 3D reconstituted skins are used for experiments mimicking skin in vivo, such as for disease models and for testing drug toxicities. In the clinical treatment, 3D reconstituted skins are used to treat burn lesions and large skin defect after excision of large hairy nevus. The 3D reconstituted skins including melanocytes will be good treatment option for vitiligo to replace the hypopigmented skin lesions. After we succeeded to differentiate Muse cells to melanocytes (Muse-MC), keratinocytes (Muse-KC), and fibroblasts (Muse-Fb), we attempted to generate 3D skins reconstituted with these Muse-derived cells. A gel layer was created mixing type 1 collagen and Muse-Fb to mimic the dermis and cultured for 4 days. After making the dermal equivalent,

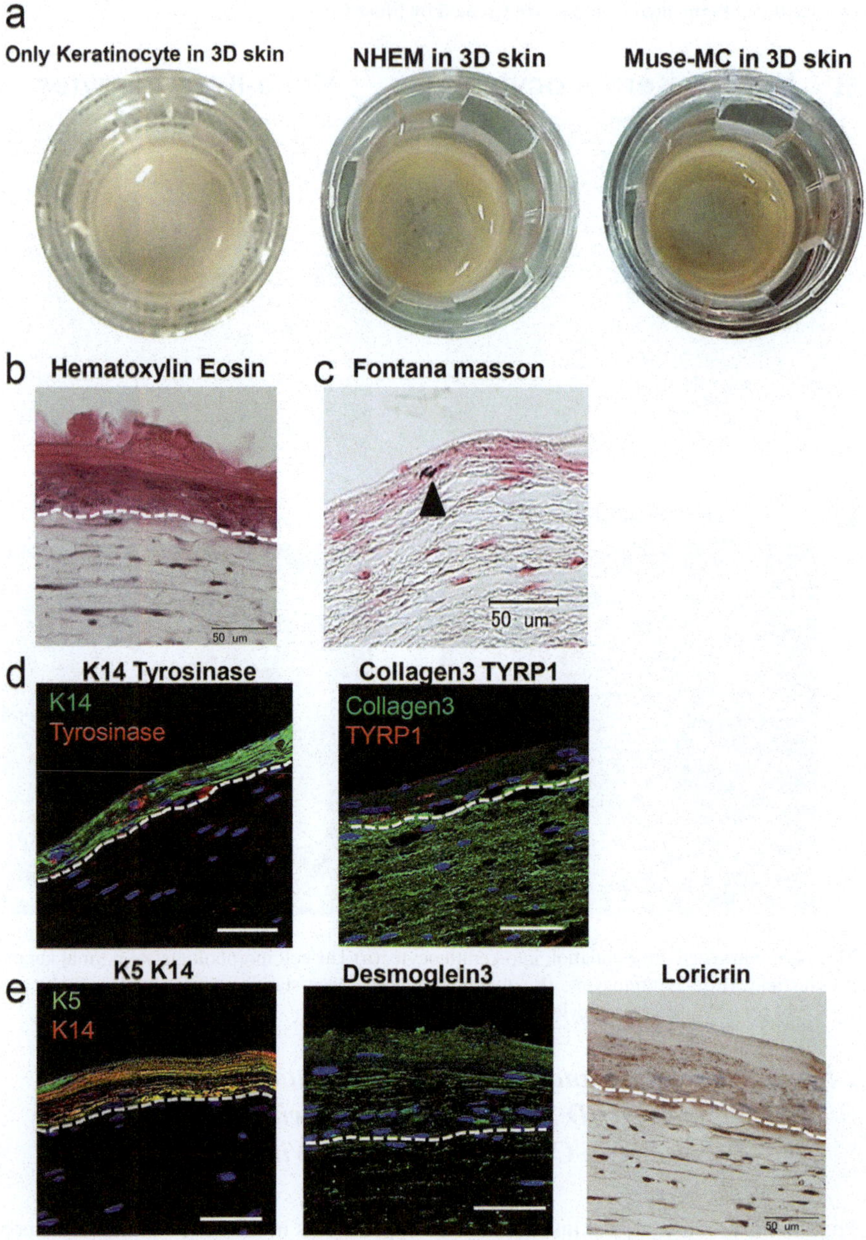

Fig. 14.6 Generation of reconstituted skin using Muse cell-derived skin components. (**a**) The gross view of 3D skins composed with Muse-KC and Muse-Fb (*left*), with Muse-KC, Muse-Fb, and normal human epidermal melanocytes (NHEM) (*middle*), and with Muse-KC, Muse-Fb, and Muse-MC (*right*). (**b**) The hematoxylin-eosin staining image of Muse cell-derived 3D reconstituted skins. (**c**) The Fontana Masson Staining image of Muse cell-derived 3D reconstituted skins. The arrow indicates the melanin deposition. (**d**) Immunofluorescence staining shows melanocyte markers tyrosinase and TYRP1-possitive Muse-MC locate in the basal layer of the epidermis (K14) and above the dermis (collagen 3) in Muse cell-derived 3D reconstituted skin. (**e**) Immunofluorescence staining or immunohistological analysis of keratin 5 (K5), keratin 14 (K14), desmoglein 3, and loricrin in Muse cell-derived reconstituted skin. Nuclei were counterstained with DAPI. Scale bars = 50 μm

Muse-KC and Muse-MC are seeded at the ratio of 5:1 onto the collagen gel layers. The sheets were cultured in keratinocyte medium with a gradually increased Ca2+ concentration for 7 days and cultured at the air-liquid interface for 5 days [36]. The 3D skin reconstituted with Muse-derived cells showed visible pigmentation (Fig. 14.6a). It was observed that 3D reconstituted skin containing Muse-MC have pigmentation similar to 3D reconstituted skin containing NHEM.

The skin is to form a permeability barrier to protect organs inside of the body. In other words, 3D reconstituted skins are needed to form a firm epidermis like human skin. 3D reconstituted skins generated with immortalized keratinocytes HaCat form loose and irregularly differentiated epidermis compared to those generated with primary keratinocyte [41]. The tissue architecture and cellular localization were determined by hematoxylin and eosin stain and immunofluorescence. In microscopic analysis, we observed that Muse-MC resides on the basal layer of the epidermis and transfers melanin to neighboring keratinocytes, suggesting that Muse-MC maintained melanocyte properties in the 3D reconstituted skin (Fig. 14.6b). Fontana Masson Staining confirmed that Muse-MC produced melanin in the epidermis (Fig. 14.6c). Additionally, Muse-MC expressed melanocyte-related proteins, tyrosinase and TYRP1, and were coordinately expressed in the basal layer of the K14 positive epidermis, but not in the dermis shown by collagen 3-positive area (Fig. 14.6d). The 3D reconstituted skins have a stratified and differentiated epidermis in hematoxylin eosin staining, as well as coordinate expression of keratinocyte-related proteins, keratin 5, keratin 14, loricrin, and desmoglein 3 (Fig. 14.6e). The 3D reconstituted skins were examined using immunofluorescence staining for melanocytes, fibroblasts, and keratinocytes.

14.4 Other Stem Cells as Source for Melanocytes Induction

14.4.1 DPSCs and Fibroblast Might Be Differentiated into Melanocytes

ES cells and iPS cells can differentiate into melanocytes and keratinocytes. It has also been reported that iPS cells could be differentiated into fibroblasts. A few reports also show that other mesenchymal stem cells can differentiate into melanocytes. Melanocytes can be generated from dental pulp stem cell (DSPC) and fibroblasts [42, 43]. Stevens et al. induced Mart-1-positive cells from DPSCs [42], and Paino et al. reported that DPSCs differentiated into melanocytes without any stimulation after culturing more than 150 days [43]. Although it is not clear whether Muse cells are contained in the DPSCs, Muse cells can differentiate into melanocytes within 6 weeks cultivation with ten factors and expressed the melanocyte-related markers, such as TYR, TYRP1, and S100. Yang et al. reported direct conversion of human fibroblasts to functional melanocytes by defined factors [44]. Since they were not isolated Muse cells from human fibroblasts, their reports might show that Muse cells existing in human fibroblasts are differentiated into melanocytes.

14.4.2 Skin-Derived Precursors (SKPs) Are Another Type of Skin Stem Cells Different from Muse Cells

Toma et al. reported that skin-derived precursors (SKPs), which exist in human foreskin tissue, are multipotent stem cells derived from neural crest cells [45, 46]. The niche of SKPs is in human dermal papilla. SKPs express Sox2 and Nanog and have self-renewal ability. SKPs express specific markers, such as Snail and Slug, but Muse cells do not [19]. Because of the different localizations in the skin and different specific marker molecules, Muse cells are different from SKPs and are independent population of pluripotent stem cells in the skin. Moreover, Muse cells can differentiate into ectoderm-, mesoderm-, and endoderm-derived cells. SKPs can differentiate into both ectoderm- and mesoderm-derived cells but not differentiate into endoderm-derived cells [45, 46]. These data indicate that SKPs are not pluripotent stem cells. Therefore, Muse cells are distinct from SKPs.

14.5 Future Perspective

Our study demonstrated that a substantial number of melanocytes at relatively high purity (approximately 40%) can be obtained from adipose-MSCs from human subcutaneous adipose tissue; this system may be eminently suitable for studying melanocyte differentiation in vitro and for examining the effects of chemical or molecular interventions or perturbations of this process.

Muse cells are non-tumorigenic endogenous pluripotent stem cells that show high differentiation ability across oligolineage boundaries from mesodermal to endodermal or ectodermal. Notably, the current technique does not require exogenous gene introduction to convert dermal-Muse cells into functional melanocytes, which lower hurdles for clinical application. In addition, Muse-MC supply a sufficient number of high-quality cultured melanocytes for clinical use.

Moreover, Muse cells also could be differentiated into keratinocytes and fibroblasts. Now the annual chronic wound care products market is expanding rapidly due to increasing aging population and a sharp rise in the incidence of diabetes and obesity worldwide [47]. There is an urgent need to regenerate these diseases. Kinoshita et al. reported that treatment of Muse cells accelerated diabetic skin ulcer healing [48]. The cell growth of Muse cell is not very high, and they do not form teratoma [18]. Additionally, their differentiation does not need to manipulate exogenous genes. Hu et al. reported that Muse cells demonstrate the potential to circumvent certain limitations of ES cells and iPS cells for skin regeneration [49]. These studies demonstrated the potential of somatic Muse cells as pluripotent stem cells, which would be a promising source for regenerative medicine in the skin.

References

1. Proksch E, Brandner JM, Jensen JM (2008) The skin: an indispensable barrier. Exp Dermatol 17(12):1063–1072 PubMed PMID: 19043850
2. Shirakata Y, Tokumaru S, Yamasaki K, Sayama K, Hashimoto K (2003) So-called biological dressing effects of cultured epidermal sheets are mediated by the production of EGF family. TGF-beta and VEGF J Dermatol Sci 32(3):209–215 PubMed PMID: 14507446
3. Kondo T, Hearing VJ (2011) Update on the regulation of mammalian melanocyte function and skin pigmentation. Expert Rev Dermatol 6(1):97–108 PubMed PMID: 21572549. Pubmed Central PMCID: PMC3093193
4. Slominski A, Tobin DJ, Shibahara S, Wortsman J (2004) Melanin pigmentation in mammalian skin and its hormonal regulation. Physiol Rev 84(4):1155–1228 PubMed PMID: 15383650
5. Mabula JB, Chalya PL, McHembe MD, Jaka H, Giiti G, Rambau P et al (2012) Skin cancers among Albinos at a University teaching hospital in Northwestern Tanzania: a retrospective review of 64 cases. BMC Dermatol 12:5 PubMed PMID: 22681652. Pubmed Central PMCID: PMC3483204
6. Gronskov K, Ek J, Brondum-Nielsen K (2007) Oculocutaneous albinism. Orphanet J Rare Dis 2:43 PubMed PMID: 17980020. Pubmed Central PMCID: PMC2211462
7. Ghafourian A, Ghafourian S, Sadeghifard N, Mohebi R, Shokoohini Y, Nezamoleslami S et al (2014) Vitiligo: symptoms, pathogenesis and treatment. Int J Immunopathol Pharmacol 27(4):485–489 PubMed PMID: 25572727
8. Alikhan A, Felsten LM, Daly M, Petronic-Rosic V (2011) Vitiligo: a comprehensive overview part I. Introduction, epidemiology, quality of life, diagnosis, differential diagnosis, associations, histopathology, etiology, and work-up. J Am Acad Dermatol 65(3):473–491 PubMed PMID: 21839315
9. Felsten LM, Alikhan A, Petronic-Rosic V (2011) Vitiligo: a comprehensive overview part II: treatment options and approach to treatment. J Am Acad Dermatol 65(3):493–514 PubMed PMID: 21839316
10. Tsuchiyama K, Watabe A, Sadayasu A, Onodera N, Kimura Y, Aiba S (2016) Successful treatment of segmental vitiligo in children with the combination of 1-mm minigrafts and phototherapy. Dermatology 232(2):237–241 PubMed PMID: 26836583
11. van Geel N, Ongenae K, Naeyaert JM (2001) Surgical techniques for vitiligo: a review. Dermatology 202(2):162–166 PubMed PMID: 11306848
12. Fioramonti P, Onesti MG, Marchese C, Carella S, Ceccarelli S, Scuderi N (2012) Autologous cultured melanocytes in vitiligo treatment comparison of two techniques to prepare the recipient site: erbium-doped yttrium aluminum garnet laser versus dermabrasion. Dermatol Surg 38(5):809–812 PubMed PMID: 22335629
13. Nissan X, Larribere L, Saidani M, Hurbain I, Delevoye C, Feteira J et al (2011) Functional melanocytes derived from human pluripotent stem cells engraft into pluristratified epidermis. Proc Natl Acad Sci U S A 108(36):14861–14866 PubMed PMID: 21856949. Pubmed Central PMCID: PMC3169131
14. Fang D, Leishear K, Nguyen TK, Finko R, Cai K, Fukunaga M et al (2006) Defining the conditions for the generation of melanocytes from human embryonic stem cells. Stem Cells 24(7):1668–1677 PubMed PMID: 16574754
15. Yang R, Jiang M, Kumar SM, Xu T, Wang F, Xiang L et al (2011) Generation of melanocytes from induced pluripotent stem cells. J Invest Dermatol 131(12):2458–2466 PubMed PMID: 21833016. Pubmed Central PMCID: PMC3213325
16. Motohashi T, Aoki H, Chiba K, Yoshimura N, Kunisada T (2007) Multipotent cell fate of neural crest-like cells derived from embryonic stem cells. Stem Cells 25(2):402–410 PubMed PMID: 17038669
17. Ohta S, Imaizumi Y, Okada Y, Akamatsu W, Kuwahara R, Ohyama M et al (2011) Generation of human melanocytes from induced pluripotent stem cells. PLoS One 6(1):e16182 PubMed PMID: 21249204. Pubmed Central PMCID: PMC3020956

18. Kuroda Y, Kitada M, Wakao S, Nishikawa K, Tanimura Y, Makinoshima H et al (2010) Unique multipotent cells in adult human mesenchymal cell populations. Proc Natl Acad Sci U S A 107(19):8639–8643 PubMed PMID: 20421459. Pubmed Central PMCID: PMC2889306
19. Wakao S, Kitada M, Kuroda Y, Shigemoto T, Matsuse D, Akashi H et al (2011) Multilineage-differentiating stress-enduring (muse) cells are a primary source of induced pluripotent stem cells in human fibroblasts. Proc Natl Acad Sci U S A 108(24):9875–9880 PubMed PMID: 21628574. Pubmed Central PMCID: PMC3116385
20. Ogura F, Wakao S, Kuroda Y, Tsuchiyama K, Bagheri M, Heneidi S et al (2014) Human adipose tissue possesses a unique population of pluripotent stem cells with nontumorigenic and low telomerase activities: potential implications in regenerative medicine. Stem Cells Dev 23(7):717–728 PubMed PMID: 24256547
21. Heneidi S, Simerman AA, Keller E, Singh P, Li X, Dumesic DA et al (2013) Awakened by cellular stress: isolation and characterization of a novel population of pluripotent stem cells derived from human adipose tissue. PLoS One 8(6):e64752 PubMed PMID: 23755141. Pubmed Central PMCID: PMC3673968
22. Kuroda Y, Wakao S, Kitada M, Murakami T, Nojima M, Dezawa M (2013) Isolation, culture and evaluation of multilineage-differentiating stress-enduring (Muse) cells. Nat Protoc 8(7):1391–1415 PubMed PMID: 23787896
23. Amiri F, Halabian R, Salimian M, Shokrgozar MA, Soleimani M, Jahanian-Najafabadi A et al (2014) Induction of multipotency in umbilical cord-derived mesenchymal stem cells cultivated under suspension conditions. Cell Stress Chaperones 19(5):657–666 PubMed PMID: 24464492. Pubmed Central PMCID: PMC4147073
24. Liu Q, Zhang RZ, Li D, Cheng S, Yang YH, Tian T et al (2016) Muse cells, a new type of pluripotent stem cell derived from human fibroblasts. Cell Reprogram 18(2):67–77 PubMed PMID: 27055628
25. Dezawa M (2016) Muse cells provide the pluripotency of mesenchymal stem cells: direct contribution of Muse cells to tissue regeneration. Cell Transplant 25(5):849–861 PubMed PMID: 26884346
26. Fox NW, Damjanov I, Knowles BB, Solter D (1984) Stage-specific embryonic antigen 3 as a marker of visceral extraembryonic endoderm. Dev Biol 103(1):263–266 PubMed PMID: 6143701
27. Thomson JA, Itskovitz-Eldor J, Shapiro SS, Waknitz MA, Swiergiel JJ, Marshall VS et al (1998) Embryonic stem cell lines derived from human blastocysts. Science 282(5391):1145–1147 PubMed PMID: 9804556
28. Tsuchiyama K, Wakao S, Kuroda Y, Ogura F, Nojima M, Sawaya N et al (2013) Functional melanocytes are readily reprogrammable from multilineage-differentiating stress-enduring (muse) cells, distinct stem cells in human fibroblasts. J Invest Dermatol 133(10):2425–2435 PubMed PMID: 23563197
29. Yamauchi T, Yamasaki K, Tsuchiyama K, Koike S, Aiba S (2017) A quantitative analysis of multilineage-differentiating stress-enduring (Muse) cells in human adipose tissue and efficacy of melanocytes induction. J Dermatol Sci 86(3):198–205 PubMed PMID: 28292562
30. White RM, Zon LI (2008) Melanocytes in development, regeneration, and cancer. Cell Stem Cell 3(3):242–252 PubMed PMID: 18786412
31. Hirobe T, Shinpo T, Higuchi K, Sano T (2010) Life cycle of human melanocytes is regulated by endothelin-1 and stem cell factor in synergy with cyclic AMP and basic fibroblast growth factor. J Dermatol Sci 57(2):123–131 PubMed PMID: 20045284
32. Ito K, Morita T, Sieber-Blum M (1993) In vitro clonal analysis of mouse neural crest development. Dev Biol 157(2):517–525 PubMed PMID: 7684712
33. Shibahara S, Takeda K, Yasumoto K, Udono T, Watanabe K, Saito H et al (2001) Microphthalmia-associated transcription factor (MITF): multiplicity in structure, function, and regulation. J Investig Dermatol Symp Proc 6(1):99–104 PubMed PMID: 11764295

34. Lee SA, Son YO, Kook SH, Choi KC, Lee JC (2011) Ascorbic acid increases the activity and synthesis of tyrosinase in B16F10 cells through activation of p38 mitogen-activated protein kinase. Arch Dermatol Res 303(9):669–678 PubMed PMID: 21667118
35. Yamane T, Hayashi S, Mizoguchi M, Yamazaki H, Kunisada T (1999) Derivation of melanocytes from embryonic stem cells in culture. Dev Dyn 216(4–5):450–458 PubMed PMID: 10633864
36. Yamauchi T, Yamasaki K, Tsuchiyama K, Koike S, Aiba S (2017) The potential of Muse cells for regenerative medicine of skin: procedures to reconstitute skin with Muse cell-derived keratinocytes, fibroblasts, and melanocytes. J Invest Dermatol 137(12):2639–2642 PubMed PMID: 28736234
37. Itoh M, Kiuru M, Cairo MS, Christiano AM (2011) Generation of keratinocytes from normal and recessive dystrophic epidermolysis bullosa-induced pluripotent stem cells. Proc Natl Acad Sci U S A 108(21):8797–8802 PubMed PMID: 21555586. Pubmed Central PMCID: 3102348
38. Itoh M, Umegaki-Arao N, Guo Z, Liu L, Higgins CA, Christiano AM (2013) Generation of 3D skin equivalents fully reconstituted from human induced pluripotent stem cells (iPSCs). PLoS One 8(10):e77673 PubMed PMID: 24147053. Pubmed Central PMCID: 3795682
39. Guenou H, Nissan X, Larcher F, Feteira J, Lemaitre G, Saidani M et al (2009) Human embryonic stem-cell derivatives for full reconstruction of the pluristratified epidermis: a preclinical study. Lancet 374(9703):1745–1753 PubMed PMID: 19932355
40. Wakao S, Kitada M, Kuroda Y, Dezawa M (2012) Isolation of adult human pluripotent stem cells from mesenchymal cell populations and their application to liver damages. Methods Mol Biol 826:89–102 PubMed PMID: 22167642
41. Gotz C, Pfeiffer R, Tigges J, Blatz V, Jackh C, Freytag EM et al (2012) Xenobiotic metabolism capacities of human skin in comparison with a 3D epidermis model and keratinocyte-based cell culture as in vitro alternatives for chemical testing: activating enzymes (Phase I). Exp Dermatol 21(5):358–363 PubMed PMID: 22509833
42. Stevens A, Zuliani T, Olejnik C, LeRoy H, Obriot H, Kerr-Conte J et al (2008) Human dental pulp stem cells differentiate into neural crest-derived melanocytes and have label-retaining and sphere-forming abilities. Stem Cells Dev 17(6):1175–1184 PubMed PMID: 18393638
43. Paino F, Ricci G, De Rosa A, D'Aquino R, Laino L, Pirozzi G et al (2010) Ecto-mesenchymal stem cells from dental pulp are committed to differentiate into active melanocytes. Eur Cell Mater 20:295–305 PubMed PMID: 20931491
44. Yang R, Zheng Y, Li L, Liu S, Burrows M, Wei Z et al (2014) Direct conversion of mouse and human fibroblasts to functional melanocytes by defined factors. Nat Commun 5:5807 PubMed PMID: 25510211. Pubmed Central PMCID: 4335710
45. Toma JG, Akhavan M, Fernandes KJ, Barnabe-Heider F, Sadikot A, Kaplan DR et al (2001) Isolation of multipotent adult stem cells from the dermis of mammalian skin. Nat Cell Biol 3(9):778–784 PubMed PMID: 11533656
46. Toma JG, McKenzie IA, Bagli D, Miller FD (2005) Isolation and characterization of multipotent skin-derived precursors from human skin. Stem Cells 23(6):727–737 PubMed PMID: 15917469
47. Sen CK, Gordillo GM, Roy S, Kirsner R, Lambert L, Hunt TK et al (2009) Human skin wounds: a major and snowballing threat to public health and the economy. Wound repair and regeneration : official publication of the wound healing society [and] the European tissue repair. Society 17(6):763–771 PubMed PMID: 19903300. Pubmed Central PMCID: 2810192
48. Kinoshita K, Kuno S, Ishimine H, Aoi N, Mineda K, Kato H et al (2015) Therapeutic potential of adipose-derived SSEA-3-positive Muse cells for treating diabetic skin ulcers. Stem Cells Transl Med 4(2):146–155 PubMed PMID: 25561682. Pubmed Central PMCID: 4303359
49. Hu MS, Longaker MT (2017) A MUSE for skin regeneration. J Invest Dermatol 137(12):2471–2472 PubMed PMID: 29169463

Chapter 15
Muse Cells and Aortic Aneurysm

Katsuhiro Hosoyama and Yoshikatsu Saiki

Abstract The aorta is a well-organized, multilayered structure comprising several cell types, namely, endothelial cells (ECs), vascular smooth muscle cells (VSMCs), and fibroblasts, as well as an extracellular matrix (ECM), which includes elastic and collagen fibers. Aortic aneurysms (AAs) are defined as progressive enlargements of the aorta that carries an incremental risk of rupture as the diameter increases over time. The destruction of the aortic wall tissue is triggered by atherosclerosis, inflammation, and oxidative stress, leading to the activation of matrix metalloproteinases (MMPs), and inflammatory cytokines and chemokines, resulting in the loss of the structural back bone of VSMCs, ECM, and ECs. To date, cell-based therapy has been applied to animal models using several types of cells, such as VSMCs, ECs, and mesenchymal stem cells (MSCs). Although these cells indeed deliver beneficial outcomes for AAs, particularly by paracrine and immunomodulatory effects, the attenuation of aneurysmal dilation with a robust tissue repair is insufficient. Meanwhile, multilineage-differentiating stress-enduring (Muse) cells are known to be endogenous non-tumorigenic pluripotent-like stem cells that are included as several percent of MSCs. Since Muse cells are pluripotent-like, they have the ability to differentiate into cells representative of all three germ layers from a single cell and to self-renew. Moreover, Muse cells are able to home to the site of damage following simple intravenous injection and repair the tissue by replenishing new functional cells through spontaneous differentiation into tissue-compatible cells. Given these unique properties, Muse cells are expected to provide an efficient therapeutic efficacy for AA by simple intravenous injection. In this chapter, we summarize several studies on Muse cell therapy for AA including our recent data, in comparison with other kinds of cell therapies.

Keywords Aortic aneurysm · Muse cell · Vascular progenitor cell · Endothelial cell differentiation · Smooth muscle cell differentiation · Vasa vasorum

---

K. Hosoyama · Y. Saiki (✉)
Division of Cardiovascular Surgery, Tohoku University Graduate School of Medicine, Sendai, Japan
e-mail: yoshisaiki@med.tohoku.ac.jp

M. Dezawa (ed.), *Muse Cells*, Advances in Experimental Medicine and Biology 1103, https://doi.org/10.1007/978-4-431-56847-6_15

15.1 Aortic Aneurysm

15.1.1 Background

An aortic aneurysm (AA) is a silent but life-threatening disease. The incidence is said to be 1.3–8.9% in men and 1.0–2.2% in women in the abdominal aorta [1–3]. AA larger than 50 mm in diameter in the abdominal aorta and 60 mm in diameter in the thoracic aorta increases the risk of rupture [4, 5]. Prosthetic graft replacement is applied to prevent foreseeable rupture for these AAs, but the invasiveness of open surgery is often associated with mortality and morbidity [6]. A less invasive endovascular aortic repair technique recently emerged as a durable procedure but cannot be applied to all AA patients because of anatomic limitations and pre-existing pathologic conditions, such as a shaggy aorta and aortic calcification [7]. In addition, even the aneurysm smaller than the size of surgical indication is known to develop 10% dilatation in average in a year and can be dilated enough to have some risk of rupture in the future [8]. Thus, an alternative less invasive strategy is sought for the treatment of these inoperable AAs and early-stage AAs.

15.1.2 Etiology

Although some AAs are the direct consequences of specific causes such as trauma, infection (brucellosis, salmonellosis, tuberculosis), inflammatory diseases (Behcet disease, Takayasu arteritis), and connective tissue disorders (Marfan syndrome, Loeys-Dietz syndrome, Ehlers-Danlos syndrome) [9–11], most aneurysms are categorized to non-specific AAs [12]. Since the non-specific disorder is conventionally associated with severe atherosclerotic damage of the aortic wall, AAs had been traditionally regarded as a consequence of atherosclerosis [13]. However, some of the recent reports suggest that pathogenic mechanisms of AAs differ, at least in part, from those responsible for athero-occlusive disease [14, 15]. Given that not all patients with atherosclerosis develop AAs, additional factors are presumably involved in aneurysm development. Among environmental risk factors, tobacco smoking has a strong clinical association with AA development; the prevalence of abdominal AAs in cigarette smokers is more than four times higher than that of non-smokers [16]. Besides, male gender, age, hypertension, chronic obstructive pulmonary disease, hyperlipidemia, and family history of the disorder are considered to be risk factors for AAs [17–19].

15.1.3 Pathophysiology

The aorta is a well-organized, multilayered structure comprising several cell types, namely, endothelial cells (ECs), vascular smooth muscle cells (VSMCs), and fibroblasts, as well as an extracellular matrix (ECM), which includes elastic and collagen fibers. Elastic fibers are composed of elastin and associated proteins and responsible for the viscoelastic properties. Elastic fibers associated with smooth muscle cells are most abundant in the media of the aortic wall. Collagen, in polymeric form, is also a significant component of the media and the surrounding fibrous adventitia, providing tensile strength and structural integrity of the vascular wall. Beside elastic and collagen fibers, proteoglycans are also implicated in the organization of the aortic wall [20].

Elastic and collagen fibers are the main determinants of the mechanical properties of the aorta, and alterations of these ECM components play a crucial role in the development of AAs. The fragmentation of the elastic fibers and medial attenuation seem to be an early step in AA formation [21]. Collagen fibers are responsible for the resistance of the aorta in the absence of medial elastin. Increased collagen turnover has been reported in human abdominal AAs, suggesting the existence of a repair process [22]. The alteration of ECM is caused by multistep destruction of the tissue; atherosclerosis, inflammation, and/or oxidative stress cause the infiltration of lymphocytes and monocytes/macrophages, leading to the activation of matrix metalloproteinases (MMPs), and inflammatory cytokines and chemokines [23–27]. Elastin and collagen fibers are degraded by the activated MMPs [28–33]. The tissue inhibitors of matrix metalloproteinases (TIMPs) are also increased in the AA tissue [34]; however, the balance between proteases and antiproteases is in favor of proteolysis [35, 36]. Many studies have been reported that interleukin (IL)-1β, IL-6, tumor necrosis factor (TNF)-α, and monocyte chemotactic protein (MCP)-1 were upregulated in the AA wall of human or experimental animal aortic aneurysm [37–39]. These cytokines and chemokines induce apoptosis of VSMCs and reduction of elastic media, which is regarded as a key event in the development of AA [40, 41]. VSMCs are known to synthesize elastin polypeptide [42], and its gene expression is modulated by transforming growth factor (TGF)-β1 and insulin-like growth factor (IGF)-1 [43, 44]. VSMCs participate in vascular wall remodeling through localized expression of various ECM proteins as well as proteases and their inhibitors. Additionally, VSMCs have a protective role against inflammation and proteolysis [45]. EC apoptosis is also induced in AA development, resulting in thrombin generation, platelet adhesion, and fibrin-rich thrombus formation [46–48], which is a major source of proteases in the aneurysmal wall that provoke arterial wall thinning and the absence of luminal healing [25, 26]. The loss of ECs is known to trigger thrombus formation, wall atrophy, and AA expansion [49]. This series of reactions converges to thin the aortic wall due to the loss of the structural back bone of ECM, VSMCs, and ECs, which eventually results in an irreversible aneurysmal dilation that carries an incremental risk of rupture as the diameter increases over time.

15.1.4 Treatment

As mentioned above, surgical intervention, such as prosthetic graft replacement and endovascular repair, is considered to prevent foreseeable rupture when the diameter of an AA exceeds 50 mm in the abdominal aorta and 60 mm in the thoracic aorta [5]. However, to deal with inoperable AAs or early-stage AAs, a less invasive strategy is required. Given that control of chronic inflammation is crucial for prevention of AA progression, a number of experimental investigations and clinical studies have attempted to treat AA using various drugs and factors to control the inflammation; for example, a combination of angiotensin-converting enzyme inhibitors and statins prevented abdominal AA rupture in a case-control study [50, 51], doxycycline decreased aneurysmal expansion rate in an experimental model [52] and in a randomized double-blinded clinical trial [53], nonsteroidal anti-inflammatory drugs decreased abdominal AA expansion rate in a case-control study [54], and c-Jun N-terminal kinase inhibitor regressed abdominal AA in a murine model [55]. Indeed, these studies suggested some benefits; however, the results are insufficient to support clinical recommendations. In addition, these therapies have other disadvantages, such as side effects caused by systemic administration of these agents and requirement of invasive administration route to deliver them to AA tissue locally.

Recently, cell-based therapy has been applied to animal models using several types of cells, such as VSMCs [45, 56, 57], ECs [49], and mesenchymal stem cells (MSCs), including adipose-derived stem cells [58–64] and skeletal muscle-derived stem cells [65]. These cells indeed deliver beneficial outcomes for AAs, particularly by paracrine and immunomodulatory effects for ECM preservation, while on the other hand, engraftment of these cell types into AA tissue is generally limited, and the attenuation/inhibition of aneurysmal dilation with a robust tissue repair is insufficient. Consequently, the number of transplanted cells is relatively large to achieve substantial efficacy, and sometimes the delivery methods are inevitably invasive, such as local injection [59, 62, 63] and sheet transplantation [58]. Therefore, the cells with the ability to differentiate into multiple cell types of the aortic tissue and selectively engraft into AAs using a simple approach, such as intravenous injection, would be a powerful tool for treating AAs.

15.1.5 Feasibility of Muse Cells for AA Treatment

Multilineage-differentiating stress-enduring (Muse) cells were first reported by Kuroda et al. in 2010, as endogenous non-tumorigenic pluripotent-like stem cells that reside in the connective tissue of nearly every organ as well as in the bone marrow (BM) and peripheral blood [66, 67]. Muse cells are collected as cells positive for the pluripotent surface marker stage-specific embryonic antigen (SSEA)-3, which is expressed on pluripotent cells in normal development in the fertilized egg to epiblast stem cells, as well as on human embryonic stem (ES) cells [68]. Muse cells are also included as several percent of commercially available cultured mesenchymal cells such as fibroblasts and BM-MSCs. Since Muse cells are pluripotent-like, they have the ability to differentiate into cells representative of all three germ layers from a single cell and to self-renew. Moreover, as demonstrated in liver- and skeletal muscle-damaged models, Muse cells home to the site of damage following simple intravenous injection and repair the tissue by replenishing new functional cells through spontaneous differentiation into tissue-compatible cells [69]. In addition, Muse cells secrete factors that enable them to survive under stressful conditions and to suppress immunologic reactions [70, 71]. Because of these unique properties, Muse cells are expected to provide an efficient therapeutic efficacy for AA by simple intravenous injection, due to their stress tolerance, which enables them to survive in the destructive AA tissue, their ability to selectively home to the damaged site, and their tissue reparative capability through spontaneous differentiation into multiple cell types that comprise the aorta.

15.2 In Vitro Properties of Muse Cells

15.2.1 Spontaneous Differentiation

Using a fluorescence-activated cell sorter (FACS; Aria II, Becton Dickinson, Franklin Lakes, NJ), Muse cells are segregated as SSEA-3-positive fraction (comprising ~5% of human BM-MSCs) from commercially available human BM-MSCs (Lonza Japan, Tokyo, Japan) [66]. Muse cells have an ability to form clusters similar to ES cell-derived embryoid bodies in a single-cell suspension culture and to spontaneously differentiate into cells representative of three germ layers from those clusters. This property is not observed in non-Muse cells. Our group reported that cells expanded from a single Muse cell-derived cluster on gelatin-coated dishes contained αSMA^{+} cells (~13.4% of the total expanded cells) and CD31^{+} cells (~0.7%), suggesting that Muse cells have the potential to generate VSMCs and ECs (Fig. 15.1) [72]. On the other hand, non-Muse cells (SSEA-3^{-} cells) are unable to generate clusters, and the spontaneous differentiation ability is not observed in non-Muse cells in vitro.

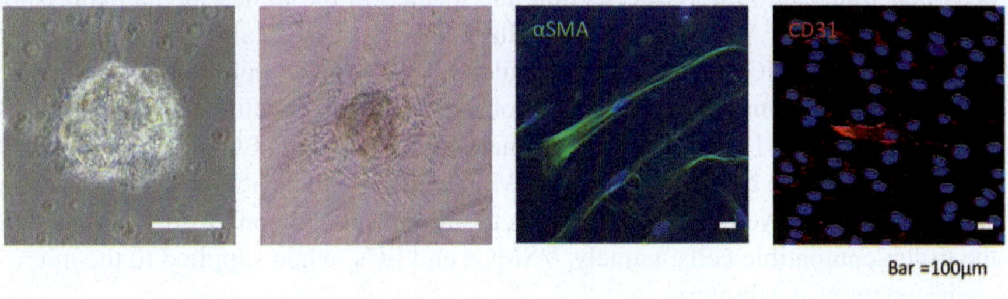

Fig. 15.1 Representative view of a cluster formed in a single-cell suspension culture of Muse cells, and adhesion culture of the cluster onto a gelatin-coated dish to allow cells to expand. The expanded cells were positive for αSMA and CD31

15.2.2 Comparison with Vascular Progenitor Cells

In our previous report [72], we demonstrated the intrinsic differentiation potential of human bone marrow-Muse cells into vascular component cells with quantitative polymerase chain reaction (qPCR) and compared the potential with that in endothelial progenitor cells (EPCs) and $CD34^+$ progenitor cells. In addition to vascular differentiation factors, the expression of stress-tolerant factors was also examined. As demonstrated, EPCs are able to differentiate into ECs under several cytokine inductions [73]. Similarly, $CD34^+$ cells, corresponding to hematopoietic stem cells as well as to vascular progenitor cells, are also able to differentiate into both ECs and VSMCs under the presence of cytokines [74]. Of the EC markers, the expression of FOXC1 was highest in Muse cells compared to EPCs and $CD34^+$ cells, while that of KLF2 and MEF2C in Muse cells was moderate. Expression of ELK1, MYH10, and CAMK2δ, markers for de-differentiated VSMCs, was highest in Muse cells compared to EPCs and $CD34^+$ cells. Factors relevant to stress tolerance, such as HSPA8, PDIA3, and MDH1, were significantly higher in Muse cells than in EPCs and $CD34^+$ progenitor cells. These findings demonstrated that, unlike EPCs and $CD34^+$ progenitor cells, the differentiation potential of Muse cells is not confined to a single cell type but spans the vascular component cells, VSMCs and ECs. Notably, the expression of VSMC dedifferentiation-related markers was generally higher in Muse cells than EPCs and $CD34^+$ progenitor cells. Since VSMCs are a pivotal element that regulates structural strength of aortic tissue, the high ability of Muse cells to differentiate into VSMCs is one of the advantages of Muse cells for treating AAs.

15.2.3 Culture with Explanted Aneurysm

Previously, our group reported that Muse cells have a property of differentiation and maturation into aortic component cells in the AA microenvironment [72]. In coculture of GFP^+-Muse cells with explanted murine AA tissue section, GFP^+-Muse cells penetrated into the tissue section and located in the tunica media expressing αSMA at 2 weeks. At 3 weeks, GFP^+-Muse cells expressed calponin, a marker for differentiated VSMCs (Fig. 15.2A). Similarly, $CD31^+/GFP^+$ cells and $CD34^+/GFP^+$ cells, markers for immature ECs, were observed at day 7, while it was not until 3 weeks that $CD141^+$/GFP+ cells, a marker for mature ECs, were observed (Fig. 15.2B). These results indicate that Muse cells would be able to survive even in such an inflammative environment as AA tissue and spontaneously differentiate into the tissue-compatible cells, namely, VSMCs and ECs, when supplied to the microenvironment of AA in vitro.

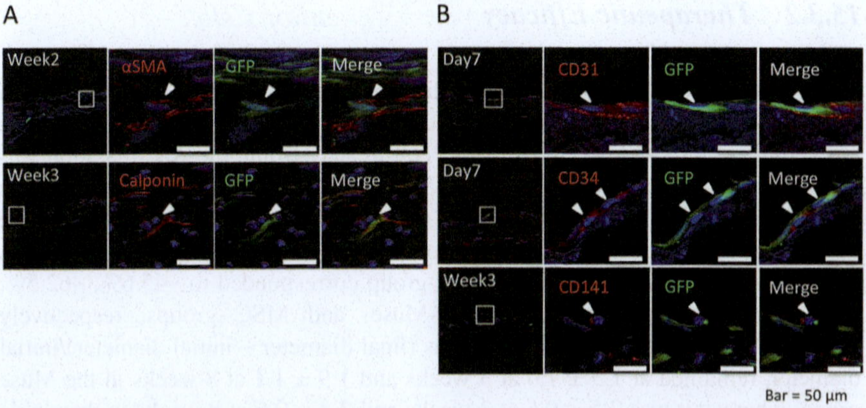

Fig. 15.2 (**A**, **B**) Representative sagittal view of frozen cross-sections stained with αSMA, calponin CD31, CD34, and CD141. Arrowheads indicate double-positive cells

15.3 Muse Cells in Animal Model Aortic Aneurysm

15.3.1 Experimental Models

The angiotensin-II-infused apolipoprotein E-knockout murine model is widely used as an experimental model of AA [75]. In our previous study, to avoid the effect of immune rejection of human bone marrow-Muse cells, we used 8-week-old male severe combined immunodeficient (SCID) mice (C.B-17/lcr-scid/scidJcl, CLEA Japan, Tokyo, Japan) [72]. According to the previous report by Bi et al. [76], the abdominal AA was induced by periaortic application of $CaCl_2$ (0.5 mol/L; Otsuka, Tokyo, Japan) and porcine pancreatic elastase (0.5 unit/μl, Wako Pure Chemical Industries, Osaka, Japan). Briefly, under general anesthesia and microscopic guidance (MZ6; Leica, Wetzlar, Germany), the infrarenal aorta in a living SCID mouse was exposed, and the lumbar arteries originated from the aorta were then ligated with 11-0 nylon. The isolated region of the aorta was then circumferentially wrapped with a piece of sterile gauze, which was soaked with $CaCl_2$ and elastase solution and incubated for 20 min. Thereafter, the gauze was removed, and the arterial segment was washed twice with PBS. The mice were allowed to recover from the procedures. As early as on postoperative day 3, we could confirm AA formation defined by the size greater than 1.5 times dilation based on ultrasound inspection and then proceeded to further experiments.

15.3.2 Therapeutic Efficacy

We subsequently administrated human bone marrow-Muse cells (20,000 cells in 0.2 ml PBS) intravenously at three time points, on day 0, day 7, and 2 weeks, via the tail vein (Fig. 15.3A) [72]. In the Muse cell-treated group, the AA dilatation was significantly attenuated at 3 weeks, compared with the non-treated, non-Muse (cells other than Muse cells in MSCs), or MSC groups with three injections as done in the Muse group. The significant attenuation of AA dilation was maintained at 8 weeks. At 8 weeks, the size of AA in the Muse group corresponded to ~45.6%, ~62.5%, and ~55.6% in the non-treated, non-Muse, and MSC groups, respectively (Fig. 15.3B). The dilation rate, defined as (final diameter – initial diameter)/initial diameter, remained at 1.5 ± 1.0 at 3 weeks and 1.9 ± 1.1 at 8 weeks in the Muse group, whereas it was 2.5 ± 0.6 at 3 weeks and 2.8 ± 0.6 at 8 weeks in the single injected-Muse group (Fig. 15.3C, D). Although the difference between single and multiple administration of Muse cells was not statistically different in either time points, the size of AA dilation tended to be smaller in multiple injections compared with single injection for the entire observation period up to 8 weeks.

15.3.3 In Vivo Differentiation Ability

In the study, we also evaluated integration of Muse cells in the AA by histological assessment [72]. In the Muse group, αSMA$^+$/GFP$^+$ cells were detected among GFP$^+$ cells at a frequency of 71.9 ± 27.0 cells/mm^2 at 3 weeks and 57.3 ± 36.3 cells/mm^2 at 8 weeks, suggesting spontaneous differentiation of Muse cells into VSMCs after integration into AA (Fig. 15.4A). In contrast, the number of αSMA$^+$/GFP$^+$ cells in the non-Muse, MSC, and even the single injected-Muse group was significantly smaller than that in the Muse group. Similarly, in the Muse group, CD31$^+$/GFP$^+$ cells were detected at a frequency of 25.8 ± 14.5 cells/mm^2 at 3 weeks and 22.5 ± 11.2 cells/mm^2 at 8 weeks, suggesting differentiation of Muse cells into ECs in AA tissue (Fig. 15.4B). In contrast, only a small number of CD31$^+$/GFP$^+$ cells were counted in the non-Muse and MSC groups. In the single injected-Muse group, the number of CD31$^+$/GFP$^+$ cells was similar to the multiple injected-Muse group at 3 weeks, but that declined significantly lower at 8 weeks. Regarding medial elastin content, quantitated by the percent area of elastin as compared with total medial tissue area via Elastica-Masson staining, the ratio of the Muse group was 13.7 ± 4.4%, significantly higher than that of the non-Muse (6.4 ± 1.5%), MSC (4.8 ± 4.2%), and non-treated groups (5.2 ± 2.1%), while there was no significant difference between the single (10.5 ± 3.0%) and multiple injected-Muse groups (Fig. 15.4C). The medial elastin area of the Muse group was 1.3, 2.9, 2.1, and 2.6 times larger than that in the single injected-Muse, non-Muse, MSC, and non-treated groups, respectively. These findings indicated that the attenuation of the AAs evolved from VSMCs and ECs that spontaneously differentiated from migrated Muse cells and possibly subsequent preservation of ECM including elastic fibers.

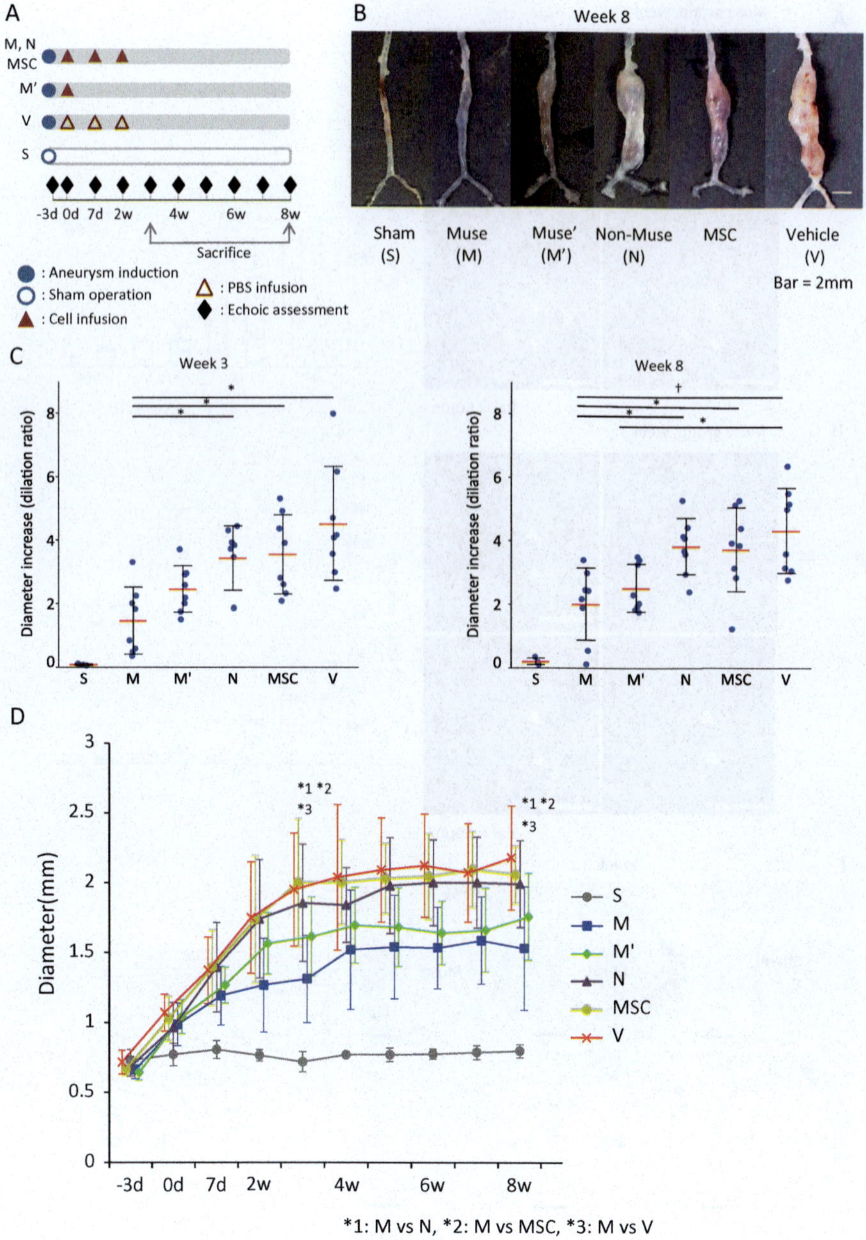

Fig. 15.3 (**A**) Schematic representation of our in vivo study protocol. (**B**) Macroscopic view of the aortic aneurysm at 8 weeks. (**C**) Dilation ratio of the aortic aneurysm. Red lines represent the mean of each group. (**D**) Diameter measurement of the aortic aneurysm by ultrasound inspection. S, sham group (n = 8); M, Muse group (multiple injection) (n = 16); M', Muse group (single injection) (n = 16); N, non-Muse group (n = 16); MSC, MSC group (n = 16); V, vehicle group (n = 16). Error bars represent the standard deviation. *$p < 0.05$, †$p < 0.01$

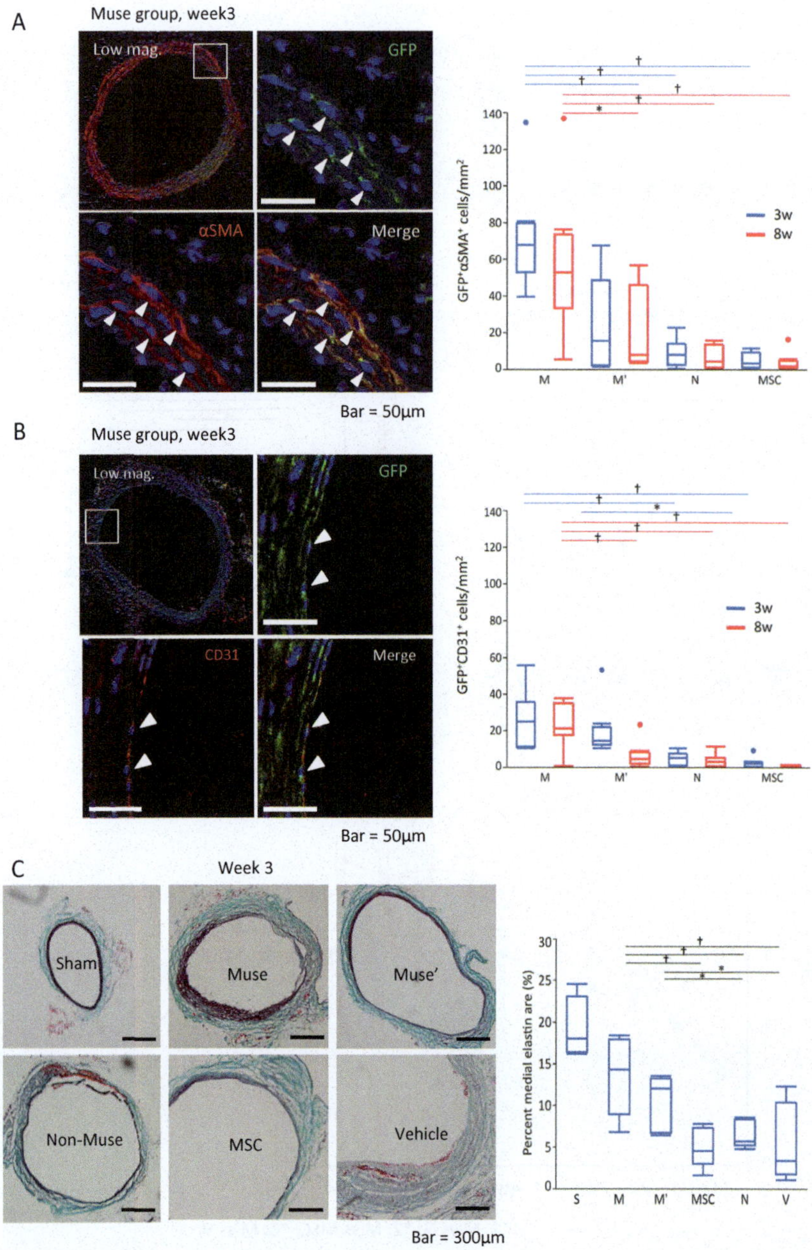

Fig. 15.4 (**A**) αSMA and GFP immunostaining of the AAs at 3 weeks, along with quantitative analysis. (**B**) CD31 and GFP immunostaining and quantitative analysis of AAs at 3 weeks. (**C**) Elastica-Masson staining of aortic aneurysms at 3 weeks. Arrowheads indicate double-positive cells. S, sham group (n = 8); M, Muse group (multiple injection) (n = 16); M', Muse group (single injection) (n = 16); N, non-Muse group (n = 16); MSC, MSC group (n = 16); V, vehicle group (n = 16). The middle horizontal line represents the median. $*p < 0.05$, $\dagger p < 0.01$

15.3.4 Anti-Inflammatory Effect

Anti-inflammatory effect of Muse cells was also suggested in our previous study [72]. In the assessment using F4/80⁺ staining, an indicator of macrophages, the Muse group had a lower infiltration of inflammatory cells compared with the non-treated group. Although the other groups of single injected-Muse, non-Muse, and MSCs also had a significantly lower infiltration compared with the non-treated group, the number of F4/80⁺ cells was the lowest in the Muse group both at 3 weeks and at 8 weeks. Similar to the conventional MSC treatment, it is suggested that anti-inflammation property might have contributed to the therapeutic mechanism of Muse cells to some extent.

15.3.5 Tissue Distribution After Systemic Administration

In other experimental disease models, it has been validated that systemically administrated Muse cells accumulate specifically to the damaged organs [77–81]. Similarly, we assessed the tissue distribution of intravenously injected human Muse cells by determining the expression level of the human-specific Alu sequence in the AA model murine organs at 8 weeks [72]. The expression of human Alu in the abdominal aorta in the Muse group was 13.2-fold higher than that in the non-Muse group. Human Alu was not detected at the other organs even in the Muse group, except for low levels of expression at the lung and spleen. In the non-Muse group, a very low level of human Alu was detected only in the abdominal aorta and not in other organs.

Several factors have been reported to control migration of MSCs, such as stromal cell-derived factor-1 (SDF-1)–C-X-C chemokine receptor type 4 (CXCR4) axis [82]. B Muse cells are a subpopulation of MSCs; Iseki et al. reported the involvement of CXCR4 in Muse cell migration [81]. In their in vitro migration assay, migration of Muse cells was partially suppressed by AMD3100, a CXCR4 antagonist, whereas migration of non-Muse cells was completely abrogated. The authors concluded that CXCR4 is suggested to be one of the key receptors to mediate Muse cell migration, although other critical factors also appear to be responsible. Meanwhile, plasma sphingosine-1-phosphate (S1P) mobilizes MSCs from the bone marrow [83, 84]. Tanaka et al. reported that plasma S1P levels were positively correlated with the number of Muse cells in the peripheral blood in AMI patients [85]. S1P may be one of the factors that triggers preferential migration of Muse cells into the damaged tissue after systemic administration.

15.3.6 Mode of Migration

Diaz-Flores et al. reported that macrophages and circulating progenitor cells migrating through vasa vasorum contribute to the development of athero-occlusive disease [86]. On the other hand, to this point, the mode of the migration of transplanted cells for AA has not been elucidated. In our previous study, we demonstrated that injected Muse cells migrate into the AA from the adventitial side, possibly via vasa vasorum [72]. After injection of 20,000 GFP^+ Muse cells at day 0, we harvested the aneurysmal tissue at day 3 or day 5 and inspected the migration dynamics by multiphoton laser microscopy (Fig. 15.5). GFP^+ Muse cells were detected only at the outermost layer of the vasculature at day 3, and, in some parts, GFP^+ Muse cells accumulated around a vasa vasorum-like vasculature structure in the tunica externa (Fig. 15.5A, arrowhead). At day 5, GFP^+ Muse cells still remained at the tunica externa but expanded toward the tunica media and luminal layers of the aortic wall. These findings suggested that intravenously injected Muse cells migrated into the AA tissue through the vasa vasorum-like vasculature structure, homed into the tunica externa first, and then migrated to the luminal side through tunica media.

15.4 Comparison to Other Cell Therapies

AA pathology is initially characterized by the destruction of elastic lamellae [87]. Accordingly, managing excessive destruction of the ECM has been centered in the nonsurgical therapy for AA. Previous reports of VSMC [45, 56, 57] and MSC [58–65] treatments demonstrated that anti-MMP and anti-inflammatory effects were major actions and the contribution of these cells through differentiation into vascular component cells was considered a minor event because the transplanted cells barely survived in the host aorta for a long period, nor did they differentiate into vascular cells. In contrast, our study suggested that the attenuation of the AAs evolved from VSMCs and ECs that spontaneously differentiated from homed Muse cells, and possibly subsequent preservation of ECM, including elastic fibers. This unique therapeutic mechanism of Muse cells may enable us to achieve curative effect by the noninvasive administration of relatively small number of cells. In addition, high expression of stress-tolerant factors in Muse cells might be related to the high proportion of cell survival and integration into AA tissue, where pro-apoptotic and pro-inflammatory factors are abundant. Minimization of cell number is important for practical concerns because it will decrease both the time and cost required to prepare cells and decrease the risks of adverse effects, such as embolic events. Furthermore, allogenic Muse cells would be more feasible for clinical application than autologous cells. Previously, Muse cells were demonstrated to have immunomodulatory effects [70, 88]. If allogenic Muse cells are effective for treating AA, they could be an "off-the-shelf" cell source for AA therapy.

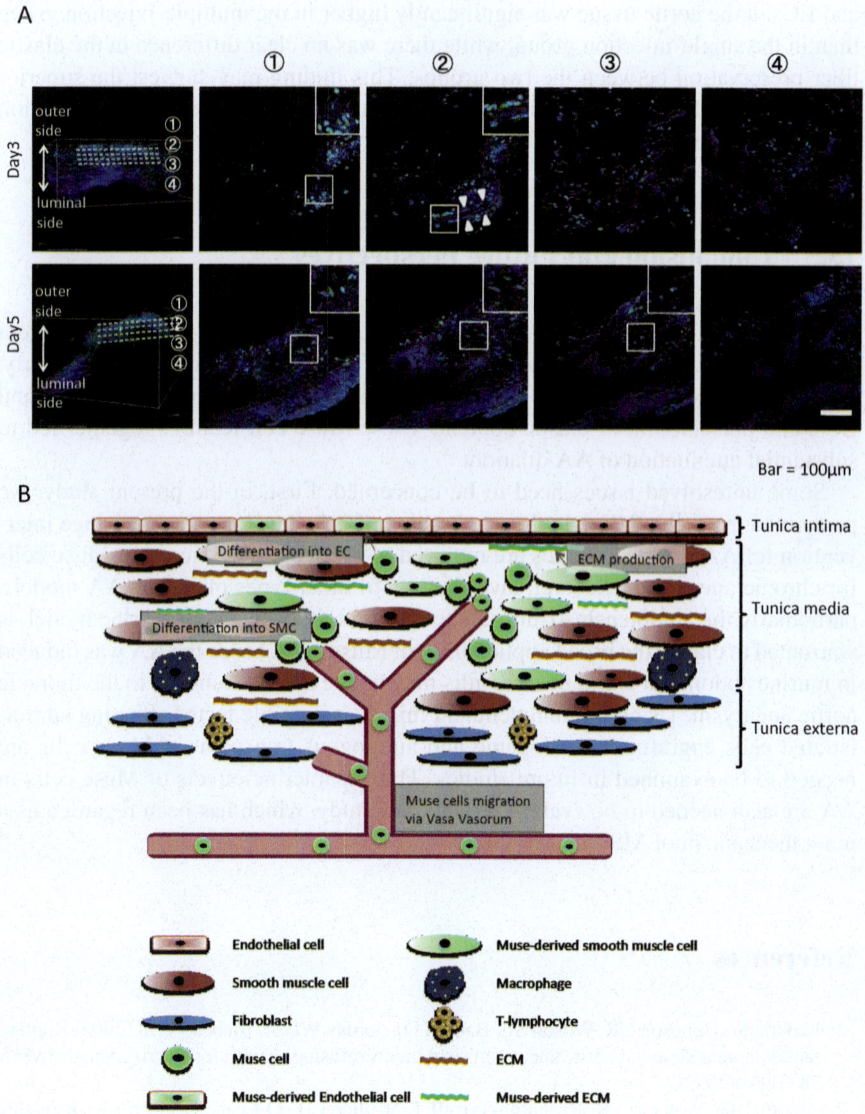

Fig. 15.5 (**A**) Representative multiphoton laser microscopy image of AAs harvested at day 3 and day 5 (n = 3 each). The *left* column is a 3D-reconstructed image of the aortic sample indicating the location of each axial view (*dotted white lines* 1–4), captured every 20 μm from the outer side to the luminal side. *Arrowheads* indicate vasa vasorum-like vasculature structure in the tunica externa. (**B**) Hypothetic model of Muse cells treating AAs

Fu et al. reported the superiority of multiple administrations of MSCs compared with a single administration of MSCs [60]. In our study, multiple injections of Muse cells attenuated aneurysmal dilation in the earlier period, but no statistically significant difference was observed compared with single-injection group at either 3 or 8 weeks. Interestingly, the proportion of Muse cells that differentiated into VSMCs and ECs in the aortic tissue was significantly higher in the multiple-injection group than in the single-injection group, while there was no clear difference in the elastic fiber preservation between the two groups. This finding may suggest the superior importance of the restoration of elastic fibers in attenuation of aneurysmal dilation than replenishment of new vascular cells.

15.5 Conclusion and Future Perspectives

The efficacy of Muse cells for treatment of AA has been suggested in a current experimental study showing the advantages of the AA-specific homing property, inhibition of excess inflammation, spontaneous differentiation into VSMCs and ECs, and preservation of elastic content. These Muse cell features together led to substantial attenuation of AA dilation.

Some unresolved issues need to be concerned. First, in the present study, we performed the cell infusion in the acute phase of AA, day 3, as an early-stage intervention for AA. Further studies are required to elucidate the efficacy of Muse cells for chronic aneurysms. Second, investigation of other types of animal AA models, particularly the angiotensin-II-infused apolipoprotein E-knockout murine model, is warranted to clarify the broad applicability of Muse cells. Since the AA was induced in murine abdominal aorta, these results may not be able to translate to the thoracic aortic aneurysm. Using immunodeficient mice is a possible factor affecting administrated cells engraftment. Allogenic and autologous transplant of Muse cells are needed to be examined in future studies. Third, paracrine effects of Muse cells in AA are also needed to be evaluated in further study, which has been regarded as a main mechanism of MSCs.

References

1. Lederle FA, Johnson GR, Wilson SE, Ballard DJ, Jordan WD Jr, Blebea J et al (2002) Rupture rate of large abdominal aortic aneurysms in patients refusing or unfit for elective repair. JAMA 287:2968–2972
2. Alcorn HG, Wolfson SK Jr, Sutton-Tyrrell K, Kuller LH, O'Leary D (1996) Risk factors for abdominal aortic aneurysms in older adults enrolled in the cardiovascular health study. Arterioscler Thromb Vasc Biol 16:963–970
3. Singh K, Bonaa KH, Jacobsen BK, Bjork L, Solberg S (2001) Prevalence of and risk factors for abdominal aortic aneurysms in a population-based study : the Tromso study. Am J Epidemiol 154:236–244
4. Hollier LH, Taylor LM, Ochsner J (1992) Recommended indications for operative treatment of abdominal aortic aneurysms. Report of a subcommittee of the Joint Council of the Society for vascular surgery and the North American chapter of the International Society for Cardiovascular Surgery. J Vasc Surg 15:1046–1056
5. Erbel R, Aboyans V, Boileau C, Bossone E, Bartolomeo RD, Eggebrecht H et al (2014) ESC guidelines on the diagnosis and treatment of aortic diseases: document covering acute and

chronic aortic diseases of the thoracic and abdominal aorta of the adult. The task force for the diagnosis and treatment of aortic diseases of the European Society of Cardiology (ESC). Eur Heart J 35:2873–2926
6. De Bruin JL, Baas AF, Buth J, Prinssen M, Verhoeven EL, Cuypers PW et al (2010) Long-term outcome of open or endovascular repair of abdominal aortic aneurysm. New Engl J Med 362:1881–1889
7. Guideline. STFfCP (2009) clinical practice guideline for patients undergoing endovascular repair of abdominal aortic aneurysms (AAA). J Vasc Nurs 27:48–63
8. Isselbacher EM (2005) Thoracic and abdominal aortic aneurysms. Circulation 111:816–828
9. Erentug V, Bozbuga N, Omeroglu SN, Ardal H, Eren E, Guclu M et al (2003) Rupture of abdominal aortic aneurysms in Behcet's disease. Ann Vasc Surg 17:682–685
10. Matsumura K, Hirano T, Takeda K, Matsuda A, Nakagawa T, Yamaguchi N et al (1991) Incidence of aneurysms in Takayasu's arteritis. Angiology 42:308–315
11. Towbin JA, Casey B, Belmont J (1999) The molecular basis of vascular disorders. Am J Hum Genet 64:678–684
12. Johnston KW, Rutherford RB, Tilson MD, Shah DM, Hollier L, Stanley JC (1991) Suggested standards for reporting on arterial aneurysms. Subcommittee on reporting standards for arterial aneurysms, ad hoc committee on reporting standards, society for vascular surgery and North American chapter, international society for cardiovascular surgery. J Vasc Surg 13:452–458
13. Reed D, Reed C, Stemmermann G, Hayashi T (1992) Are aortic aneurysms caused by atherosclerosis? Circulation 85:205–211
14. Tilson MD (1992) Aortic aneurysms and atherosclerosis. Circulation 85:378–379
15. Xu C, Zarins CK, Glagov S (2001) Aneurysmal and occlusive atherosclerosis of the human abdominal aorta. J Vasc Surg 33:91–96
16. Vardulaki KA, Walker NM, Day NE, Duffy SW, Ashton HA, Scott RA (2000) Quantifying the risks of hypertension, age, sex and smoking in patients with abdominal aortic aneurysm. Br J Surg 87:195–200
17. Lederle FA, Johnson GR, Wilson SE (2001) Abdominal aortic aneurysm in women. J Vasc Surg 34:122–126
18. Brady AR, Thompson SG, Fowkes FG, Greenhalgh RM, Powell JT (2004) Abdominal aortic aneurysm expansion: risk factors and time intervals for surveillance. Circulation 110:16–21
19. Thompson RW, Geraghty PJ, Lee JK (2002) Abdominal aortic aneurysms: basic mechanisms and clinical implications. Curr Probl Surg 39:110–230
20. Melrose J, Whitelock J, Xu Q, Ghosh P (1998) Pathogenesis of abdominal aortic aneurysms: possible role of differential production of proteoglycans by smooth muscle cells. J Vasc Surg 28:676–686
21. Dobrin PB, Mrkvicka R (1994) Failure of elastin or collagen as possible critical connective tissue alterations underlying aneurysmal dilatation. Cardiovasc Surg 2:484–488
22. Satta J, Juvonen T, Haukipuro K, Juvonen M, Kairaluoma MI (1995) Increased turnover of collagen in abdominal aortic aneurysms, demonstrated by measuring the concentration of the aminoterminal propeptide of type III procollagen in peripheral and aortal blood samples. J Vasc Surg 22:155–160
23. Traub O, Berk BC (1998) Laminar shear stress: mechanisms by which endothelial cells transduce an atheroprotective force. Arterioscler Thromb Vasc Biol 18:677–685
24. Visse R, Nagase H (2003) Matrix metalloproteinases and tissue inhibitors of metalloproteinases: structure, function, and biochemistry. Circ Res 92:827–839
25. Kazi M, Thyberg J, Religa P, Roy J, Eriksson P, Hedin U et al (2003) Influence of intraluminal thrombus on structural and cellular composition of abdominal aortic aneurysm wall. J Vasc Surg 38:1283–1292
26. Fontaine V, Touat Z, el Mtairag M, Vranckx R, Louedec L, Houard X et al (2004) Role of leukocyte elastase in preventing cellular re-colonization of the mural thrombus. Am J Pathol 164:2077–2087

27. Sakalihasan N, Limet R, Defawe OD (2005) Abdominal aortic aneurysm. Lancet 365:1577–1589
28. Sakalihasan N, Delvenne P, Nusgens BV, Limet R, Lapiere CM (1996) Activated forms of MMP2 and MMP9 in abdominal aortic aneurysms. J Vasc Surg 24:127–133
29. Curci JA, Liao S, Huffman MD, Shapiro SD, Thompson RW (1998) Expression and localization of macrophage elastase (matrix metalloproteinase-12) in abdominal aortic aneurysms. J Clin Invest 102:1900–1910
30. Davis V, Persidskaia R, Baca-Regen L, Itoh Y, Nagase H, Persidsky Y et al (1998) Matrix metalloproteinase-2 production and its binding to the matrix are increased in abdominal aortic aneurysms. Arterioscler Thromb Vasc Biol 18:1625–1633
31. Mao D, Lee JK, VanVickle SJ, Thompson RW (1999) Expression of collagenase-3 (MMP-13) in human abdominal aortic aneurysms and vascular smooth muscle cells in culture. Biochem Biophys Res Commun 261:904–910
32. Longo GM, Xiong W, Greiner TC, Zhao Y, Fiotti N, Baxter BT (2002) Matrix metalloproteinases 2 and 9 work in concert to produce aortic aneurysms. J Clin Invest 110:625–632
33. Tromp G, Gatalica Z, Skunca M, Berguer R, Siegel T, Kline RA et al (2004) Elevated expression of matrix metalloproteinase-13 in abdominal aortic aneurysms. Ann Vasc Surg 18:414–420
34. Thompson RW, Parks WC (1996) Role of matrix metalloproteinases in abdominal aortic aneurysms. Ann N Y Acad Sci 800:157–174
35. Tamarina NA, McMillan WD, Shively VP, Pearce WH (1997) Expression of matrix metalloproteinases and their inhibitors in aneurysms and normal aorta. Surgery 122:264–271 discussion 71-2
36. Knox JB, Sukhova GK, Whittemore AD, Libby P (1997) Evidence for altered balance between matrix metalloproteinases and their inhibitors in human aortic diseases. Circulation 95:205–212
37. Middleton RK, Lloyd GM, Bown MJ, Cooper NJ, London NJ, Sayers RD (2007) The pro-inflammatory and chemotactic cytokine microenvironment of the abdominal aortic aneurysm wall: a protein array study. J Vasc Surg 45:574–580
38. Juvonen J, Surcel HM, Satta J, Teppo AM, Bloigu A, Syrjala H et al (1997) Elevated circulating levels of inflammatory cytokines in patients with abdominal aortic aneurysm. Arterioscler Thromb Vasc Biol 17:2843–2847
39. Sprague AH, Khalil RA (2009) Inflammatory cytokines in vascular dysfunction and vascular disease. Biochem Pharmacol 78:539–552
40. Charo IF, Taubman MB (2004) Chemokines in the pathogenesis of vascular disease. Circ Res 95:858–866
41. Lopez-Candales A, Holmes DR, Liao S, Scott MJ, Wickline SA, Thompson RW (1997) Decreased vascular smooth muscle cell density in medial degeneration of human abdominal aortic aneurysms. Am J Pathol 150:993–1007
42. Uitto J, Christiano AM, Kahari VM, Bashir MM, Rosenbloom J (1991) Molecular biology and pathology of human elastin. Biochem Soc Trans 19:824–829
43. Liu JM, Davidson JM (1988) The elastogenic effect of recombinant transforming growth factor-beta on porcine aortic smooth muscle cells. Biochem Biophys Res Commun 154:895–901
44. Foster J, Rich CB, Florini JR (1987) Insulin-like growth factor I, somatomedin C, induces the synthesis of tropoelastin in aortic tissue. Coll Relat Res 7:161–169
45. Allaire E, Muscatelli-Groux B, Mandet C, Guinault AM, Bruneval P, Desgranges P et al (2002) Paracrine effect of vascular smooth muscle cells in the prevention of aortic aneurysm formation. J Vasc Surg 36:1018–1026
46. Bombeli T, Karsan A, Tait JF, Harlan JM (1997) Apoptotic vascular endothelial cells become procoagulant. Blood 89:2429–2442
47. Bombeli T, Schwartz BR, Harlan JM (1999) Endothelial cells undergoing apoptosis become proadhesive for nonactivated platelets. Blood 93:3831–3838

48. Durand E, Scoazec A, Lafont A, Boddaert J, Al Hajzen A, Addad F et al (2004) In vivo induction of endothelial apoptosis leads to vessel thrombosis and endothelial denudation: a clue to the understanding of the mechanisms of thrombotic plaque erosion. Circulation 109:2503–2506
49. Franck G, Dai J, Fifre A, Ngo S, Justine C, Michineau S et al (2013) Reestablishment of the endothelial lining by endothelial cell therapy stabilizes experimental abdominal aortic aneurysms. Circulation 127:1877–1887
50. Hackam DG, Thiruchelvam D, Redelmeier DA (2006) Angiotensin-converting enzyme inhibitors and aortic rupture: a population-based case-control study. Lancet 368:659–665
51. Schouten O, van Laanen JH, Boersma E, Vidakovic R, Feringa HH, Dunkelgrun M et al (2006) Statins are associated with a reduced infrarenal abdominal aortic aneurysm growth. Eur J Vasc Endovasc Surg 32:21–26
52. Yamawaki-Ogata A, Hashizume R, Satake M, Kaneko H, Mizutani S, Moritan T et al (2010) A doxycycline loaded, controlled-release, biodegradable fiber for the treatment of aortic aneurysms. Biomaterials 31:9554–9564
53. Baxter BT, Pearce WH, Waltke EA, Littooy FN, Hallett JW Jr, Kent KC et al (2002) Prolonged administration of doxycycline in patients with small asymptomatic abdominal aortic aneurysms: report of a prospective (Phase II) multicenter study. J Vasc Surg 36:1–12
54. Walton LJ, Franklin IJ, Bayston T, Brown LC, Greenhalgh RM, Taylor GW et al (1999) Inhibition of prostaglandin E2 synthesis in abdominal aortic aneurysms: implications for smooth muscle cell viability, inflammatory processes, and the expansion of abdominal aortic aneurysms. Circulation 100:48–54
55. Yoshimura K, Aoki H, Ikeda Y, Fujii K, Akiyama N, Furutani A et al (2005) Regression of abdominal aortic aneurysm by inhibition of c-Jun N-terminal kinase. Nat Med 11:1330–1338
56. Allaire E, Muscatelli-Groux B, Guinault AM, Pages C, Goussard A, Mandet C et al (2004) Vascular smooth muscle cell endovascular therapy stabilizes already developed aneurysms in a model of aortic injury elicited by inflammation and proteolysis. Ann Surg 239:417–427
57. Losy F, Dai J, Pages C, Ginat M, Muscatelli-Groux B, Guinault AM et al (2003) Paracrine secretion of transforming growth factor-beta1 in aneurysm healing and stabilization with endovascular smooth muscle cell therapy. J Vasc Surg 37:1301–1309
58. Hashizume R, Yamawaki-Ogata A, Ueda Y, Wagner WR, Narita Y (2011) Mesenchymal stem cells attenuate angiotensin II-induced aortic aneurysm growth in apolipoprotein E-deficient mice. J Vasc Surg 54:1743–1752
59. Schneider F, Saucy F, de Blic R, Dai J, Mohand F, Rouard H et al (2013) Bone marrow mesenchymal stem cells stabilize already-formed aortic aneurysms more efficiently than vascular smooth muscle cells in a rat model. Eur J Vasc Endovasc Surg 45:666–672
60. Fu XM, Yamawaki-Ogata A, Oshima H, Ueda Y, Usui A, Narita Y (2013) Intravenous administration of mesenchymal stem cells prevents angiotensin II-induced aortic aneurysm formation in apolipoprotein E-deficient mouse. J Transl Med 11:175
61. Yamawaki-Ogata A, Fu X, Hashizume R, Fujimoto KL, Araki Y, Oshima H et al (2014) Therapeutic potential of bone marrow-derived mesenchymal stem cells in formed aortic aneurysms of a mouse model. Eur J Cardiothorac Surg 45:e156–e165
62. Sharma AK, Lu G, Jester A, Johnston WF, Zhao Y, Hajzus VA et al (2012) Experimental abdominal aortic aneurysm formation is mediated by IL-17 and attenuated by mesenchymal stem cell treatment. Circulation 126:S38–S45
63. Riera Del Moral L, Aramburu CL, Garcia JR, de Cubas LR, Garcia-Olmo D, Garcia-Arranz M (2012) Experimental model for coadjuvant treatment with mesenchymal stem cells for aortic aneurysm. Am J Stem Cells 1:174–181
64. Tian X, Fan J, Yu M, Zhao Y, Fang Y, Bai S et al (2014) Adipose stem cells promote smooth muscle cells to secrete elastin in rat abdominal aortic aneurysm. PLoS One 9:e108105

65. Park HS, Choi GH, Hahn S, Yoo YS, Lee JY, Lee T (2013) Potential role of vascular smooth muscle cell-like progenitor cell therapy in the suppression of experimental abdominal aortic aneurysms. Biochem Biophys Res Commun 431:326–331
66. Kuroda Y, Kitada M, Wakao S, Nishikawa K, Tanimura Y, Makinoshima H et al (2010) Unique multipotent cells in adult human mesenchymal cell populations. Proc Natl Acad Sci U S A 107:8639–8643
67. Hori E, Hayakawa Y, Hayashi T, Hori S, Okamoto S, Shibata T et al (2016) Mobilization of pluripotent multilineage-differentiating stress-enduring cells in ischemic stroke. J Stroke Cerebrovasc Dis 25:1473–1481
68. Dezawa M (2016) Muse cells provide the pluripotency of mesenchymal stem cells: direct contribution of Muse cells to tissue regeneration. Cell Transplant 25:849–861
69. Kuroda Y, Wakao S, Kitada M, Murakami T, Nojima M, Dezawa M (2013) Isolation, culture and evaluation of multilineage-differentiating stress-enduring (Muse) cells. Nat Protoc 8:1391–1415
70. Gimeno ML, Fuertes F, Barcala Tabarrozzi AE, Attorressi AI, Cucchiani R, Corrales L et al (2017) Pluripotent nontumorigenic adipose tissue-derived Muse cells have immunomodulatory capacity mediated by transforming growth factor-beta1. Stem Cells Transl Med 6:161–173
71. Alessio N, Ozcan S, Tatsumi K, Murat A, Peluso G, Dezawa M et al (2017) The secretome of MUSE cells contains factors that may play a role in regulation of stemness, apoptosis and immunomodulation. Cell Cycle 16:33–44
72. Hosoyama K, Wakao S, Kushida Y, Ogura F, Maeda K, Adachi O et al (2018) Intravenously injected human Muse cells selectively engraft into mouse aortic aneurysms and attenuate dilatation by differentiating into multiple cell types. https://doi.org/10.1016/j.jtcvs.2018.01.098
73. Asahara T, Murohara T, Sullivan A, Silver M, van der Zee R, Li T et al (1997) Isolation of putative progenitor endothelial cells for angiogenesis. Science 275:964–967
74. Lu X, Dunn J, Dickinson AM, Gillespie JI, Baudouin SV (2004) Smooth muscle alpha-actin expression in endothelial cells derived from CD34+ human cord blood cells. Stem Cells Dev 13:521–527
75. Daugherty A, Manning MW, Cassis LA (2000) Angiotensin II promotes atherosclerotic lesions and aneurysms in apolipoprotein E-deficient mice. J Clin Invest 105:1605–1612
76. Bi Y, Zhong H, Xu K, Zhang Z, Qi X, Xia Y et al (2013) Development of a novel rabbit model of abdominal aortic aneurysm via a combination of periaortic calcium chloride and elastase incubation. PLoS One 8:e68476
77. Katagiri H, Kushida Y, Nojima M, Kuroda Y, Wakao S, Ishida K et al (2016) A distinct subpopulation of bone marrow mesenchymal stem cells, Muse cells, directly commit to the replacement of liver components. Am J Transplant 16:468–483
78. Yamauchi T, Kuroda Y, Morita T, Shichinohe H, Houkin K, Dezawa M et al (2015) Therapeutic effects of human multilineage-differentiating stress enduring (MUSE) cell transplantation into infarct brain of mice. PLoS One 10:e0116009
79. Uchida H, Morita T, Niizuma K, Kushida Y, Kuroda Y, Wakao S et al (2016) Transplantation of unique subpopulation of fibroblasts, Muse cells, ameliorates experimental stroke possibly via robust neuronal differentiation. Stem Cells 34:160–173
80. Kinoshita K, Kuno S, Ishimine H, Aoi N, Mineda K, Kato H et al (2015) Therapeutic potential of adipose-derived SSEA-3-positive Muse cells for treating diabetic skin ulcers. Stem Cells Transl Med 4:146–155
81. Iseki M, Kushida Y, Wakao S, Akimoto T, Mizuma M, Motoi F et al (2017) Muse cells, nontumorigenic pluripotent-like stem cells, have liver regeneration capacity through specific homing and cell replacement in a mouse model of liver fibrosis. Cell Transplant 26:821–840
82. Lewellis SW, Knaut H (2012) Attractive guidance: how the chemokine SDF1/CXCL12 guides different cells to different locations. Semin Cell Dev Biol 23:333–340

83. Vessey DA, Li L, Honbo N, Karliner JS (2009) Sphingosine 1-phosphate is an important endogenous cardioprotectant released by ischemic pre- and postconditioning. Am J Physiol Heart Circ Physiol 297:H1429–H1435
84. Ratajczak MZ, Suszynska M, Borkowska S, Ratajczak J, Schneider G (2014) The role of sphingosine-1 phosphate and ceramide-1 phosphate in trafficking of normal stem cells and cancer cells. Expert Opin Ther Targets 18:95–107
85. Tanaka T, Nishigaki K, Minatoguchi S, Nawa T, Yamada Y, Kanamori H et al (2017) Mobilized Muse cells after acute myocardial infarction predict cardiac function and remodeling in the chronic phase. Circ J 82:561–571
86. Diaz-Flores L Jr, Madrid JF, Gutierrez R, Varela H, Valladares F, Diaz-Flores L (2007) Cell contribution of vasa-vasorum to early arterial intimal thickening formation. Histol Histopathol 22:1379–1386
87. Daugherty A, Cassis LA (2002) Mechanisms of abdominal aortic aneurysm formation. Curr Atheroscler Rep 4:222–227
88. Diez-Tejedor E, Gutierrez-Fernandez M, Martinez-Sanchez P, Rodriguez-Frutos B, Ruiz-Ares G, Lara ML et al (2014) Reparative therapy for acute ischemic stroke with allogeneic mesenchymal stem cells from adipose tissue: a safety assessment: a phase II randomized, double-blind, placebo-controlled, single-center, pilot clinical trial. J Stroke Cerebrovasc Dis 23:2694–2700

Chapter 16
Muse Cells and Ischemia-Reperfusion Lung Injury

Hiroshi Yabuki, Tatsuaki Watanabe, Hisashi Oishi, Masato Katahira, Masahiko Kanehira, and Yoshinori Okada

Abstract Ischemia-reperfusion injury (IRI) is one of the main causes of primary graft dysfunction that accounts for 25% of mortality after lung transplantation. Disruption of blood supply and subsequent reperfusion result in organ damage with activating innate and adaptive immune response, leading to inflammatory insults. The IRI after lung transplantation is primarily manifested by permeability pulmonary edema on the basis of pulmonary vascular endothelial cell injury as seen in acute respiratory distress syndrome (ARDS). Stem cells have potent anti-inflammatory and immunomodulatory properties through local paracrine mechanisms. The application of mesenchymal stem cells (MSCs) for ARDS as well as IRI in various organs, therefore, has been interested and extensively investigated in animal models with promising results. Furthermore, two recent clinical randomized, placebo-controlled pilot studies demonstrated that treatment of ARDS with MSCs appears to be safe and feasible.

Muse cells are stress-tolerant and non-tumorigenic endogenous pluripotent-like stem cells. They comprise small proportions of cultured fibroblasts and MSCs and can be isolated from these populations. Muse cells are known to migrate to the damaged tissue after local or systemic administration, spontaneously differentiate into the tissue-compatible cells, and also secrete factors related to immunomodulation and tissue repair. We have recently shown the effect of Muse cells on ameliorating lung IRI in a rat model. With 2 h of warm ischemia and subsequent reperfusion on the left lung, the lung showed severe pulmonary edema. Administration of Muse cell through the left pulmonary artery immediately after reperfusion more significantly improved lung oxygenation capacity, compliance, and histological damage on days 1 and 3 after reperfusion compared with MSCs, and this was associated with higher expression levels of proteins related with anti-inflammation and tissue repair in the lung. Encouraging results of this study advocate further investigation of the ability of Muse cells to prevent and treat IRI after lung transplantation.

H. Yabuki · T. Watanabe · H. Oishi · M. Katahira · M. Kanehira · Y. Okada (✉)
Department of Thoracic Surgery, Institute of Development, Aging and Cancer, Tohoku University, Sendai, Japan
e-mail: yoshinori.okada.a1@tohoku.ac.jp

M. Dezawa (ed.), *Muse Cells*, Advances in Experimental Medicine and Biology 1103, https://doi.org/10.1007/978-4-431-56847-6_16

Keywords Muse cells · Mesenchymal stem cells · Lung transplantation · Ischemia-reperfusion injury · Acute respiratory distress syndrome · Rats

16.1 Introduction

Despite success of lung transplantation as a viable therapeutic modality for patients with end-stage lung diseases, 1-year survival rate of lung transplantation remains 80%, and primary graft dysfunction (PGD) accounts for 25% of mortality [1]. PGD is a form of acute lung injury that typically occurs within the first 72 h after transplantation. This syndrome is clinically characterized by hypoxemia, elevation of pulmonary arterial pressure, and infiltrative shadows on the chest X-ray. The pathogenesis of PGD after lung transplantation is primarily attributed to permeability pulmonary edema on the basis of vascular endothelial cell injury as seen in acute respiratory distress syndrome (ARDS), and pathological landmark of PGD in severe cases is diffuse alveolar damage. A variety of events occurring in the unavoidable process for lung transplantation can potentially be a cause of PGD. These include the inflammation associated with brain death, denervation, disruption of lymphatic tracts, ischemia, and reperfusion after transplantation. Of these, ischemia-reperfusion injury (IRI) has been identified as one of the main causes of PGD. Disruption of blood supply to the organ during the retrieval, transportation, and transplantation results in an imbalance between metabolic demand and supply in the organ. When the blood supply is subsequently restored, the sudden reperfusion and reoxygenation paradoxically aggravate the organ damage with activating innate and adaptive immune response, leading to inflammatory insults. These include upregulation of inflammatory cytokines, production of reactive oxygen species, influx of neutrophil into the alveolar space, and platelet aggregation in the pulmonary capillaries. All these reactions lead to the structural damage of the graft with the development of interstitial and alveolar edema [2]. In most patients, PGD is of mild to moderate severity with presenting mild hypoxemia and few infiltrates on chest X-ray, and these cases can be managed with standard supportive therapy in the intensive care unit. In some patients, however, PGD can be severe and is extended to show a picture of full brown lung edema occasionally requiring extracorporeal membrane oxygenation. There are no established therapies for this condition.

Stem cells are considered to have potent anti-inflammatory and immunomodulatory properties through local paracrine mechanisms and potential for tissue repair by proangiogenic, antifibrotic, and antiapoptotic actions [3]. Stem cells also have the ability to transfer functioning mitochondria to injured cells [4]. On the basis of these functions, the application of stem cells for ARDS as well as IRI in various organs has extensively been investigated. In this chapter, we will briefly outline the current topics regarding utilization of mesenchymal stem cells (MSCs) in an attempt to prevent and treat lung IRI and ARDS in experimental and clinical settings. Moreover, we will summarize the result of our recent experimental investigation on the effect of human multilineage-differentiating stress enduring (Muse) cells on ameliorating acute lung IRI in a rat model [5].

16.2 MSCs and Ischemia-Reperfusion Lung Injury

MSCs are adherent fibroblast-like cells that are isolated from the bone marrow and connective tissue of almost all organs. MSCs are characterized by their ability to propagate in vitro and differentiate into several cellular phenotypes, including bone, cartilage, and adipose tissue. In addition, MSCs have been found to possess anti-inflammatory and immunomodulatory functions. The interest in MSCs as a potential therapeutic modality for IRI results largely from their ability to modulate immune response and to promote tissue repair following injury [6]. MSCs promote regulatory T cell expansion and can modify immune cells including T cells and decrease their release of pro-inflammatory cytokines and increase their release of anti-inflammatory cytokines [7]. MSCs also secrete anti-inflammatory factors of several kinds, such as indoleamine 2, 3-dioxygenase (IDO), prostaglandin E2 (PGE2), and nitric oxide (NO). MSCs also act to support repair process required for regeneration of damaged tissue with secreting growth factors such as vascular endothelial growth actor (VEGF), insulin-like growth factor (IGF), hepatocyte growth factor (HGF), keratinocyte growth factor (KGF), and fibroblast growth factor (FGF) [8]. In rodent models of endotoxin-induced acute lung injury, infused MSCs have been shown to decrease levels of pro-inflammatory cytokines, increase levels of anti-inflammatory cytokines, and improve survival rates [9]. MSCs have also been shown to ameliorate pulmonary IRI in animal models. Chen and coauthors demonstrated that ischemia post-conditioning combined with MSC treatment synergistically protects the lung against IRI injury in a rat model [10]. Watanabe and coauthors showed that pretransplant administration of MSCs via the pulmonary artery of the transplanted lung attenuated IRI after 18 h of cold ischemia and 6 h of reperfusion in a mouse model of lung transplantation, as evidenced by reduced protein concentrations and cell counts in bronchoalveolar lavage fluid and less pathological scores for intra-alveolar hemorrhage and capillary congestion of the grafts treated with MSCs compared with the control animals [11]. A study with ex vivo perfused human lungs from four discarded donor lungs in a protracted ischemic model suggested improvement in the inflammatory profile and histology after intratracheal administration of multipotent adult progenitor cells [12]. Furthermore, a recent clinical randomized, placebo-controlled pilot study (NCT01902082) demonstrated that treatment of ARDS with allogeneic adipose-derived MSCs appears to be safe and feasible [13]. However, the clinical effect with the doses of MSCs used in this study was weak, and the authors suggested the need for further optimization of this strategy to reach the goal of reduced alveolar epithelial injury in ARDS. Another clinical phase I study (NCT01775774) with single intravenous infusion of allogeneic, bone marrow-derived human MSCs was well tolerated in nine patients with moderate to severe ARDS. Based on this experience, the authors had proceeded to phase II testing of MSCs for moderate to severe ARDS [14].

16.3 Muse Cells and Ischemia-Reperfusion Lung Injury

Muse cells are stress-tolerant and non-tumorigenic endogenous pluripotent-like stem cells. They comprise small proportions of cultured fibroblasts and MSCs and can be isolated from these populations [15]. Muse cells are known to migrate to the damaged tissue after local or systemic administration, spontaneously differentiate into the tissue-compatible cells, and achieve the recovery of the organ function in the kidney, muscle, brain, liver damage animal models [15–20]. Muse cells also express factors related to immunomodulation and tissue repair [21]. However, the reparative role of Muse cells on pulmonary disorders had been poorly investigated. We have recently shown the effect of Muse cells on ameliorating lung IRI in a rat model [5]. We will describe a brief summary of our experiment here and discuss a potential of Muse cells as a therapeutic tool for IRI after lung transplantation.

In our animal experiment, we induced IRI in the left lung of rats by clamping the left bronchus, pulmonary artery, and vein and subsequently reperfusing the left lung after 2 h of warm ischemia. Human Muse cells were collected from human bone marrow-MSCs with 7th to 8th subcultures as stage-specific embryonic antigen-3 (SSEA-3)-positive cells by magnetic-activated cell sorting. The collected cells contained 70% or more of SSEA-3-positive Muse cells in the study. Phosphate-buffered saline (PBS) (vehicle group; 200 μl PBS), human MSCs (MSC group; 1.5×10^5 cells/200 μl PBS), or human Muse cells (Muse group; 1.5×10^5 cells/200 μl PBS) were administered through the left pulmonary artery immediately after reperfusion. The functional and histological assessments were conducted on day 3 and 5 after reperfusion (Fig. 16.1).

The function of the left lung was evaluated by collecting arterial blood from the ascending aorta and measuring the static lung compliance, while the contralateral right hilum was cross-clamped. The Muse group showed a significantly better ratio

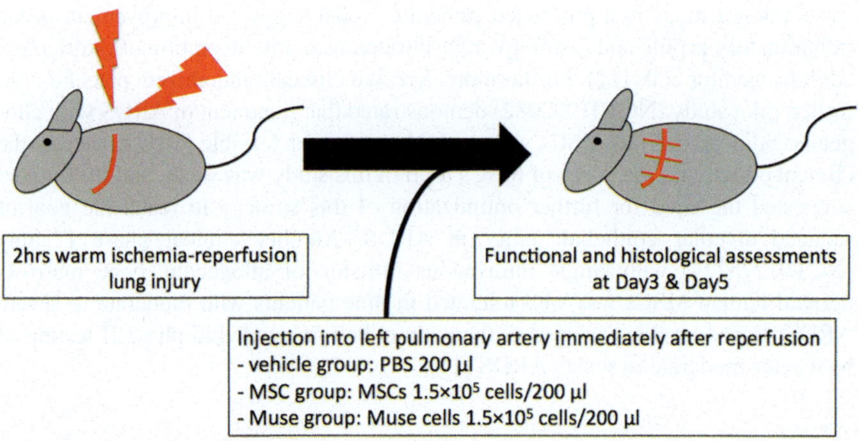

Fig. 16.1 Scheme of the experimental model of ischemia-reperfusion injury of the left lung in rats

of arterial oxygen partial pressure to fractional inspired oxygen, alveolar-arterial gradient, and static lung compliance on both days 3 and 5 compared with those in other groups (Fig. 16.2). On histological assessment of the left lung, the Muse group showed less intensity of intra-alveolar edema, intra-alveolar hemorrhage, capillary congestion, and neutrophil infiltration on day 3 after reperfusion (Fig. 16.3). Terminal deoxynucleotidyl transferase deoxyuridine triphosphate nick end labeling (TUNEL) assay also revealed significantly reduced TUNEL-positive cells in the Muse group compared with the other groups on both day 3 and 5.

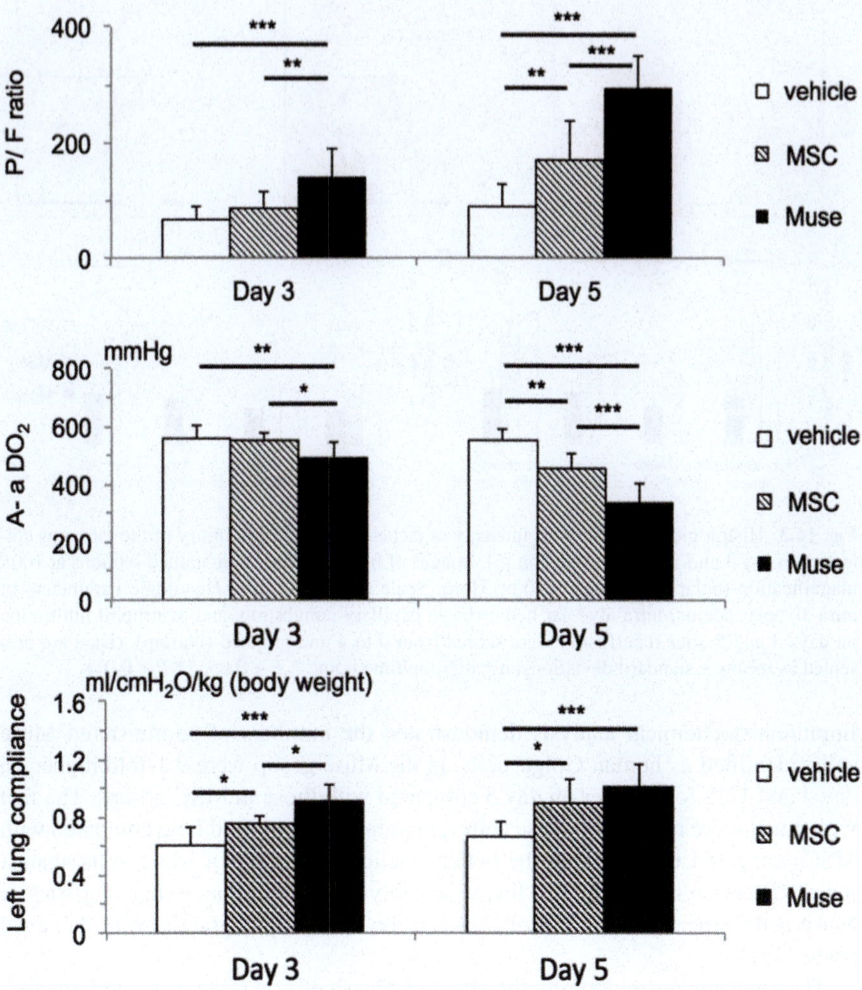

Fig. 16.2 Arterial oxygen partial pressure to fractional inspired oxygen ratios, alveolar-arterial oxygen gradients and static lung compliance of the control, the MSC, and the Muse groups following cross-clamp of the right hilum on days 3 and 5 after reperfusion [5]. (Data are presented as means ± standard deviations; n = 8/group/time point. * $P < 0.05$; ** $P < 0.01$; *** $P < 0.001$)

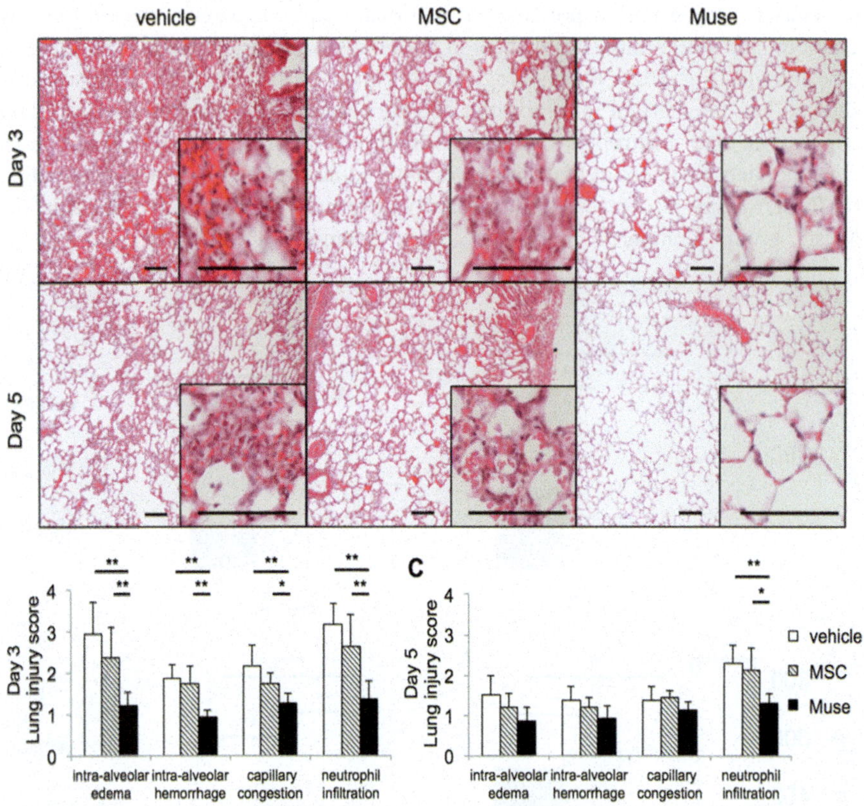

Fig. 16.3 Histological evaluation of intensity of ischemia-reperfusion injury of the rat lungs collected on day 3 and 5 after reperfusion [5]. Images of hematoxylin-eosin-stained sections at 100× magnification (enlarged images = 200×) (top). Scale bars = 100 μm. Histologic parameters of intra-alveolar edema, intra-alveolar hemorrhage, capillary congestion, and neutrophil infiltration on days 3 and 5 after reperfusion were scored from 0 to 4 and graphed (bottom). (Data are presented as means ± standard deviations; n = 8/group/time point. * $P < 0.05$; ** $P < 0.01$)

Immunohistochemical analysis demonstrated the numbers of administered Muse cells, identified as human Golgi$^+$ cells, in the Muse group were 9.1-fold higher on day 3 and 12.5-fold higher on day 5 compared with those in MSC groups. The fact with much more number of Muse cells remaining in the injured lung compared with MSCs appeared to contribute the better ameliorative effect of Muse cells against lung IRI seen in this study, and this is possibly due to two major characteristics of Muse cells: stress-tolerant nature [22] and the ability to migrate toward damaged tissue [19].

The mechanisms underlying the effect of Muse cells on protecting lung function and histology after IRI are attributed, at least in part, to the higher protein expressions of keratinocyte growth factor (KGF), interleukin-6 (IL-6), phospho-protein

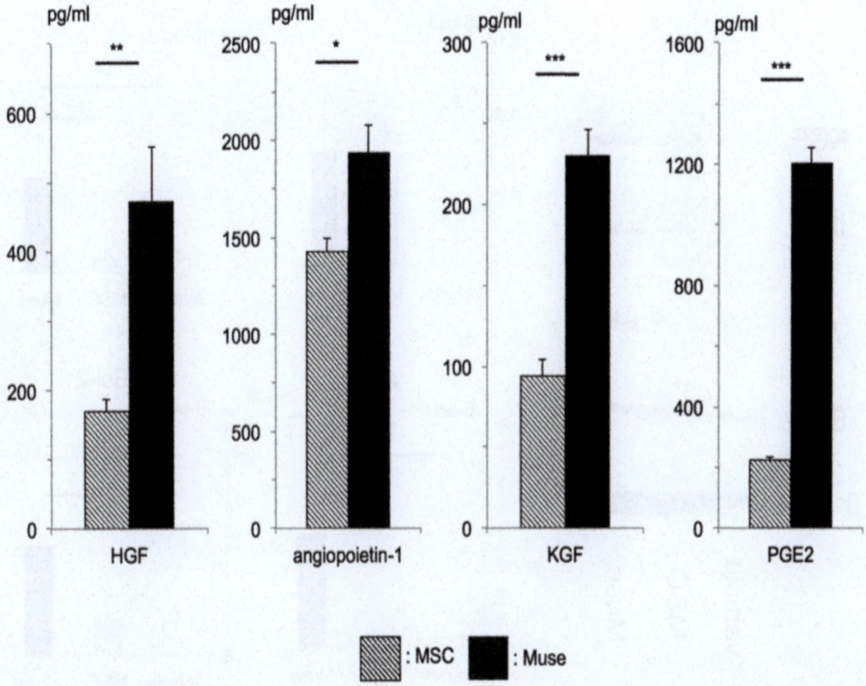

Fig. 16.4 Representative Western blots of the lungs subjected to ischemia-reperfusion injury and harvested on day 3 [5]. Each protein expression level was normalized to β-actin. *KGF* keratinocyte growth factor, *IL-6* interleukin-6, *Akt* protein kinase B, and *Bcl-2* B-cell lymphoma 2. (Data are presented as means ± standard deviations; n = 3/group. * $P < 0.05$; ** $P < 0.01$)

kinase B (phospho-Akt), and B-cell lymphoma-2 (Bcl-2) in the injured lung in the Muse group compared with the other groups (Fig. 16.4). KGF [23] has the effect to improve vascular permeability. KGF [24, 25] and IL-6 [26] have been shown to contribute to proliferation of alveolar epithelial cells and also suppress the apoptosis through Bcl-2/Akt signaling [27]. To better understand the paracrine ability of Muse cells compared with MSCs, we further conducted in vitro study. One million Muse cells or MSCs were cultured in 6 mL of DMEM without FBS for 72 h at 37 °C under an atmosphere of 5% CO_2 in a 6-cm dish. The culture supernatant was collected, and concentration of several substances related to tissue repair was measured using an enzyme-linked immunosorbent assay (ELISA). The concentrations of hepatocyte growth factor (HGF), angiopoietin-1, KGF, and PGE2 in the culture media supernatant of human Muse cells were higher than those of human MSCs (Fig. 16.5). Accordingly, Muse cells were demonstrated to have superior secretory ability of factors related to anti-inflammation and anti-apoptosis, ameliorating pulmonary vascular permeability, and the proliferation of alveolar cells compared with MSCs. Figure 16.6 shows hypothetical therapeutic mechanisms of Muse cells against lung IRI.

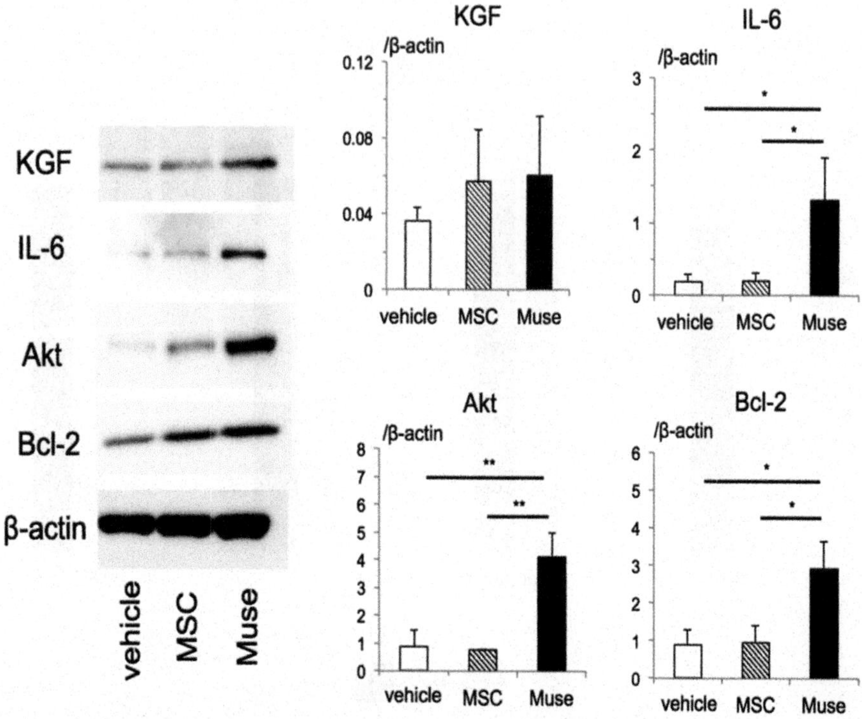

Fig. 16.5 Analysis of protective factor production in human Muse cells and human MSCs [5]. Human cytokine levels in the supernatants from 72 h cultures containing 10^6 cells were measured via enzyme-linked immunosorbent assay. *HGF* hepatocyte growth factor, *KGF* keratinocyte growth factor, *PGE2* prostaglandin E2. (Data are presented as means ± standard deviations. * $P < 0.05$; ** $P < 0.01$; *** $P < 0.001$)

Our experimental studies in vitro and those in vivo with a rodent lung IRI model clearly showed promising effect of Muse cells on protecting the lung against IRI through the ability of Muse cells to exert anti-inflammatory and reparative effects. Based on the encouraging results of our study, feasibility, effect, and the optimal procedures of intervention with Muse cells for prevention and treatment for IRI after lung transplantation and possibly for ARDS will merit further investigation.

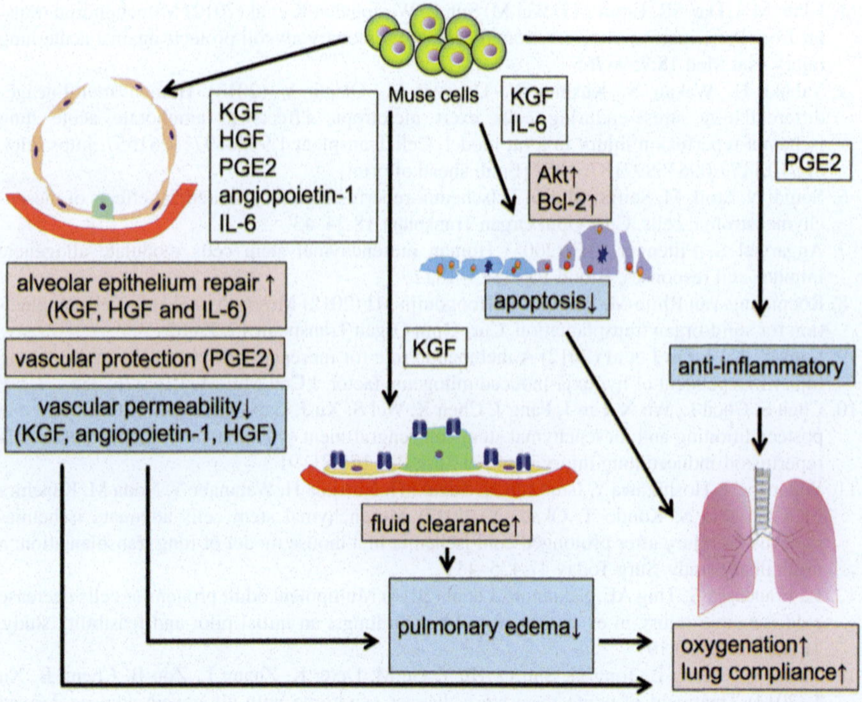

Fig. 16.6 Hypothetical schematic diagram of the mechanism by which the pleiotropic effects of human Muse cells promote structural and functional recovery of the lung rendered to ischemia-reperfusion injury [5]. The numbers of viable type I and type II alveolar epithelial cells increased in response to anti-apoptotic effects mediated by Bcl-2 and Akt and type II cell proliferation mediated by IL-6, KGF, and HGF. Both processes contributed directly toward improved lung function. KGF accelerates alveolar fluid clearance and improves vascular permeability together with angiopoietin-1 and HGF to ameliorate pulmonary edema. PGE2 induces vascular protective effects and mediates anti-inflammatory effects to attenuate inflammation in the injured lung. In our study, administered human Muse cells conferred all of these effects, leading to improved lung functionality in the ischemia-reperfusion injury model rat

References

1. Chambers DC, Yusen RD, Cherikh WS, Goldfarb SB, Kucheryavaya AY, Khusch K, Levvey BJ, Lund LH, Meiser B, Rossano JW, Stehlik J (2017) International society for heart and lung transplantation. The registry of the international society for heart and lung transplantation: thirty-fourth adult lung and heart-lung transplantation report-2017; focus theme: allograft ischemic time. J Heart Lung Transplant 36:1047–1059
2. Gelman AE, Fisher AJ, Huang HJ, Baz MA, Shaver CM, Egan TM, Mulligan MS (2017) Report of the ISHLT working group on primary lung graft dysfunction part III: mechanisms: a 2016 consensus group statement of the international society for heart and lung transplantation. J Heart Lung Transplant 36:1114–1120
3. Singer NG, Caplan AI (2011) Mesenchymal stem cells: mechanisms of inflammation. Annu Rev Pathol 6:457–478

4. Islam MN, Das SR, Emin MT, Wei M, Sun L, Westphalen K et al (2012) Mitochondrial transfer from bone-marrow-derived stromal cells to pulmonary alveoli protects against acute lung injury. Nat Med 18:759–765
5. Yabuki H, Wakao S, Kushida Y, Dezawa M, Okada Y (2018) Human multilineage-differentiating stress-enduring cells exert pleiotropic effects to ameliorate acute lung ischemia-reperfusion injury in a rat model. Cell Transplant 1:963689718761657. https://doi.org/10.1177/0963689718761657 [Epub ahead of print]
6. Souidi N, Stolk M, Seifert M (2013) Ischemia-reperfusion injury: beneficial effects of mesenchymal stromal cells. Curr Opin Organ Transplant 18:34–43
7. Aggarwal S, Pittenger MF (2005) Human mesenchymal stem cells modulate allogeneic immune cell responses. Blood 105:1815–1822
8. Roemeling-van Rhijn M, Weimar W, Hoogduijn MJ (2012) Mesenchymal stem cells: application for solid-organ transplantation. Curr Opin Organ Transplant 17:55–62
9. Song L, Xu J, Qu J et al (2012) A therapeutic role for mesenchymal stem cells in acute lung injury independent of hypoxia-induced mitogenic factor. J Cell Mol Med 16:376–385
10. Chen S, Chen L, Wu X, Lin J, Fang J, Chen X, Wei S, Xu J, Gao Q, Kang M (2012) Ischemia postconditioning and mesenchymal stem cells engraftment synergistically attenuate ischemia reperfusion-induced lung injury in rats. J Surg Res 178:81–91
11. Watanabe T, Hoshikawa Y, Ishibashi N, Suzuki H, Notsuda H, Watanabe Y, Noda M, Kanehira M, Ohkouchi S, Kondo T, Okada Y (2017) Mesenchymal stem cells attenuate ischemia-reperfusion injury after prolonged cold ischemia in a mouse model of lung transplantation: a preliminary study. Surg Today 47:425–431
12. La Francesca S, Ting AE, Sakamoto J et al (2014) Multipotent adult progenitor cells decrease cold ischemic injury in ex vivo perfused human lungs: an initial pilot and feasibility study. Transplant Res 3:19
13. Zheng G, Huang L, Tong H, Shu Q, Hu Y, Ge M, Deng K, Zhang L, Zou B, Cheng B, Xu J (2014) Treatment of acute respiratory distress syndrome with allogeneic adipose-derived mesenchymal stem cells: a randomized, placebo-controlled pilot study. Respir Res 15:39
14. Wilson JG, Liu KD, Zhuo H, Caballero L, McMillan M, Fang X, Cosgrove K, Vojnik R, Calfee CS, Lee JW, Rogers AJ, Levitt J, Wiener-Kronish J, Bajwa EK, Leavitt A, McKenna D, Thompson BT, Matthay MA (2015) Mesenchymal stem (stromal) cells for treatment of ARDS: a phase 1 clinical trial. Lancet Respir Med 3:24–32
15. Kuroda Y, Kitada M, Wakao NK, Tanimura Y, Makinoshima H, Goda M, Akashi H, Inutsuka A, Niwa A, Shigemoto T, Nabeshima Y, Nakahata T, Nabeshima Y, Fujiyoshi Y, Dezawa M (2010) Unique multipotent cells in adult human mesenchymal cell populations. Proc Natl Acad Sci U S A 107:8639–8643
16. Uchida H, Niizuma K, Kushida Y, Wakao S, Tominaga T, Borlongan CV, Dezawa M (2017) Human muse cells reconstruct neuronal circuitry in subacute lacunar stroke model. Stroke 48:428–435
17. Shimamura N, Kakuta K, Wang L, Naraoka M, Uchida H, Wakao S, Dezawa M, Ohkuma H (2017) Neuro-regeneration therapy using human muse cells is highly effective in a mouse intracerebral hemorrhage model. Exp Brain Res 235:565–572
18. Uchida N, Kushida Y, Kitada M, Wakao S, Kumagai N, Kuroda Y, Kondo Y, Hirohara Y, Kure S, Chazenbalk G, Dezawa M (2017) Beneficial effects of systemically administered human muse cells in adriamycin nephropathy. J Am Soc Nephrol 28:2946–2960
19. Uchida H, Morita T, Niizuma K, Kushida Y, Kuroda Y, Wakao S, Sakata H, Matsuzaka Y, Mushiake H, Tominaga T, Borlongan CV, Dezawa M (2016) Transplantation of unique subpopulation of fibroblasts, muse cells, ameliorates experimental stroke possibly via robust neuronal differentiation. Stem Cells 34:160–173
20. Iseki M, Kushida Y, Wakao S, Akimoto T, Mizuma M, Motoi F, Asada R, Shimizu S, Unno M, Chazenbalk G, Dezawa M (2017) Use cells, nontumorigenic pluripotent-like stem cells, have liver regeneration capacity through specific homing and cell replacement in a mouse model of liver fibrosis. Cell Transplant 26:821–840

21. Gimeno ML, Fuertes F, Barcala Tabarrozzi AE, Attorressi AI, Cucchiani R, Corrales L, Oliveira TC, Sogayar MC, Labriola L, Dewey RA, Perone MJ (2017) Pluripotent nontumorigenic adipose tissue-derived muse cells have immunomodulatory capacity mediated by transforming growth factor-beta1. Stem Cells Transl Med 6:161–173
22. Alessio N, Özcan S, Tatsumi K, Murat A, Peluso G, Dezawa M, Galderisi U (2017) The secretome of muse cells contains factors that may play a role in regulation of stemness, apoptosis and immunomodulation. Cell Cycle 16(1):33–44
23. Lee JW, Fang X, Gupta N, Serikov V, Matthay MA (2009) Allogeneic human mesenchymal stem cells for treatment of E. coli endotoxin-induced acute lung injury in the ex vivo perfused human lung. Proc Natl Acad Sci U S A 106:16357–16362
24. Chen J, Li C, Gao X, Li C, Liang Z, Yu L, Li Y, Xiao X, Chen L (2013) Keratinocyte growth factor gene delivery via mesenchymal stem cells protects against lipopolysaccharide-induced acute lung injury in mice. PLoS One 8:e83303
25. Plantier L, Marchand-Adam S, Antico Arciuch VG, Boyer L, De Coster C, Marchal J, Bachoual R, Mailleux A, Boczkowski J, Crestani B (2007) Keratinocyte growth factor protects against elastase-induced pulmonary emphysema in mice. Am J Physiol Lung Cell Mol Physiol 293:L1230–L1239
26. Tadokoro T, Wang Y, Barak LS, Bai Y, Randell SH, Hogan BL (2014) Il-6/stat3 promotes regeneration of airway ciliated cells from basal stem cells. Proc Natl Acad Sci U S A 111:E3641–E3649
27. Kuwano K (2007) Epithelial cell apoptosis and lung remodeling. Cell Mol Immunol 4:419–429

Chapter 17
Clinical Trials of Muse Cells

Mari Dezawa

Abstract Among many kinds of somatic stem cells, hematopoietic stem cells are the cells that have been successfully applied to treating leukemia patients as forms of bone marrow and cord blood transplantation. Mesenchymal stem cells, collectable from several sources including the bone marrow and adipose tissue, are also widely applied to clinical trials for their easy accessibility and low risks of tumorigenesis, while their outcomes were shown to be not clinically relevant in several target diseases. The most important issue for the stem cells is whether the cells are safe and effective for curing diseases. In this chapter, the outline of the clinical trial in Muse cells is discussed.

Keywords ES cells · iPS cells · Mesenchymal stem cells · Non-tumorigenicity · Autologous transplantation · Allogenic transplantation · Donor · Intravenous injection

Because Muse cells are non-tumorigenic and comprise a part of MSCs, which are already widely applied to clinical trials, they are feasible for clinical trials from the viewpoint of safety. Compared with ES and iPS cells, Muse cells have strong unique advantages. First, Muse cells do not require genetic manipulations to newly acquire pluripotency or to differentiate into purposive cells because they are already pluripotent and able to spontaneously differentiate into tissue-compatible cells after homing. Thus, naïve Muse cells collected from tissue sources can be directly applied for patient treatment. Second, ES and iPS cells, after differentiation in vitro, must be delivered directly to the target site, either by surgical operation or topical injection. These cells, irrespective of their differentiation state, do not home to the target tissue when administered intravenously. Muse cells, on the other hand, are able to preferentially home to the target tissue following intravenous injection. Based on these unique characteristics, Muse cells are applicable for clinical treatment by

M. Dezawa (✉)
Department of Stem Cell Biology and Histology, Tohoku University Graduate School of Medicine, Sendai, Japan
e-mail: mdezawa@med.tohoku.ac.jp

M. Dezawa (ed.), *Muse Cells*, Advances in Experimental Medicine and Biology 1103, https://doi.org/10.1007/978-4-431-56847-6_17

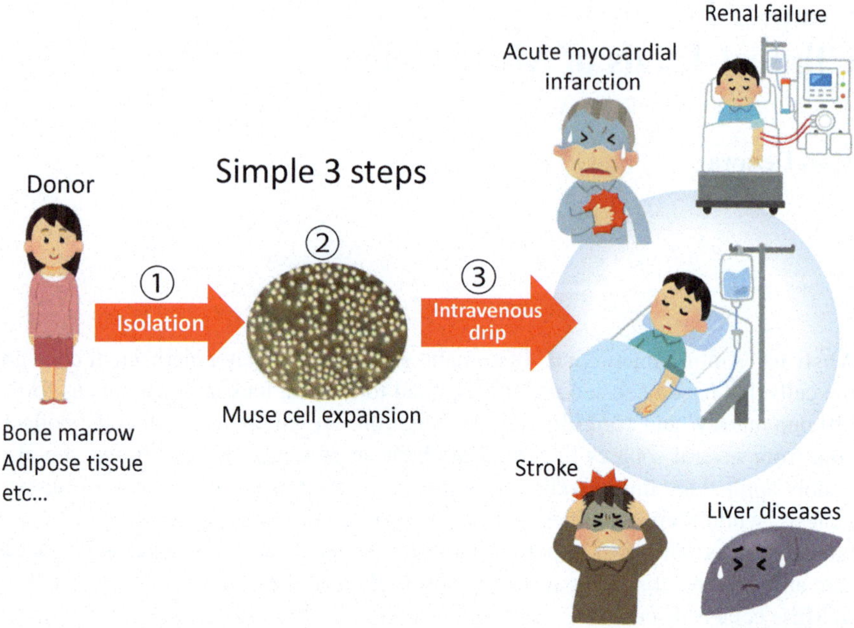

Fig. 17.1 The concept of clinical trial of Muse cells

simple steps: collection of Muse cells from the tissue source, expansion, and intravenous administration to patients (Fig. 17.1).

Another important point for clinical application is the usability of allogeneic cells. If stem cells are applicable for only autologous transplantation, their applicability will be substantially limited despite the many advantages of an autologous system. First, an autologous system is not applicable in the acute phase. A certain time period is required to establish patient-derived stem cell collection and expansion to clinical scale. Second, collection and expansion of patient-derived stem cells are not practical for patients whose health is already compromised or who are aged. From the viewpoint of clinical application, allogeneic transplantation is considered feasible as allogeneic Muse cell effects compare favorably with those of autologous Muse cells in animal models (Fig. 17.1).

Regenerative medicine is a promising new strategy for curing intractable diseases, but it is costly and requires multiple step manipulations with a long preparatory period. Furthermore, invasive routes of administration, such as by surgery, are not available in general clinics. The possibility of supplying regenerative medicine to patients in outpatient clinics by intravenous drip, however, could transform current medicine. Therefore, Muse cells have the potential to innovate medicine.

A phase I clinical trial to evaluate Muse cell application for the treatment of acute myocardial infarction has been initiated in Japan by a group of companies funded by Mitsubishi Chemical Holdings Company. Currently, Life Science Institute Ltd. is conducting the clinical trial. After verification of the safety of Muse cell application, the target diseases will be expanded. Estimation of the clinical safety and effectiveness of Muse cells is eagerly anticipated.

Chapter 18
Future of Muse Cells

Wise Young

Abstract Discovered nearly 10 years ago by Professor Mari Dezawa and her colleagues, Muse cells are entering clinical trials faster than any other stem cell for three reasons. First, Muse cells have multiple fail-safe mechanisms to keep themselves from growing out of control and do not form tumors. In contrast, embryonic stem cells and induced pluripotent stem cells form tumors and must be differentiated before transplantation. Second, Muse cells possess potent anti-immune mechanisms, including human leukocyte antigen G and indoleamine 2,3-dioxygenase that prevent both cellular and humoral immunity. Muse cells engraft even though they do not match HLA antigens with the host. Third, Muse cells are able to determine what kind and how many cells they need to make for tissue repair. While the mechanisms responsible for these traits are not well understood, Muse cells are able to enter severely injured tissues of all kinds and repair them. Study of mechanisms underlying these traits of Muse cells is likely to yield new therapies for cancer prevention, autoimmune diseases, and repair of injured tissues. The future is bright for Muse cells.

Keywords Muse · HLA-G · IDO · Tumor · Anti-immune · Tissue repair

18.1 Introduction

Muse cells were discovered by Professor Mari Dezawa and colleagues 10 years ago and are already entering clinical trials. They have been used to treat acute myocardial infarcts [1] and animal models of chronic myocardial infarction [2], liver [3, 4], kidney [5], brain [6–9], and other organs. Muse cells are moving into clinical trials faster and earlier than embryonic stem cells (ESCs) or induced pluripotent stem cells (iPSCs) because they have three unusual traits.

First, Muse cells do not form tumors. Muse cells apparently have evolved multiple fail-safe mechanisms to keep themselves from growing out of control. In

W. Young (✉)
W. M. Keck Center for Collaborative Neuroscience, Rutgers, State University of New Jersey, Piscataway, NJ, USA

M. Dezawa (ed.), *Muse Cells*, Advances in Experimental Medicine and Biology 1103, https://doi.org/10.1007/978-4-431-56847-6_18

contrast, ESCs evolved to make babies. Failing that, they make teratomas. The FDA requires that ESC must be extensively differentiated before transplanted [10]. Unless differentiated or suppressed in some way, iPSCs can also cause teratomas [11].

Second, Muse cells possess potent anti-immune mechanisms, including high expression of the human leukocyte antigen G (HLA-G) and the immune modulatory enzyme indoleamine 2,3-dioxygenase (IDO), to suppress both cellular and humoral immunities. HLA-G is the mechanism evolved by placental mammals to prevent immune rejection of the placenta and the fetus that the placenta nurtures. Allogeneic Muse cells engraft without immune rejection and do not have to be autologous or isogenic to achieve immune tolerance.

Third, Muse cells are clearly endowed with mechanisms to recognize what kind and how many cells they need to make for tissue repair. Injured tissues are often severely damaged and no longer possess the cellular niches necessary to tell stem cells what to do or make. Muse cells must have special mechanisms to determine what cells to make. While these mechanisms are not well understood, Muse cells repair tissue damage in many organs.

These characteristics of Muse cells represent important new opportunities for stem cell biology and therapies. Learning why Muse cells are not tumorigenic will provide important clues concerning how to prevent tumors. Understanding the anti-immune mechanisms of Muse cells will provide insights into how to avoid immune rejection. Finding out how stem cells detect and make the right type and number of cells is the central question of stem cell biology.

The following sections will discuss these three characteristics of Muse cells, their implications for the future of Muse and other stem cell therapies, and possible approaches to solving the problems of tumorigenesis, immune rejection, and decision-making by Muse and other stem cells. In addition, of course, understanding how Muse cells work and repair different tissues will allow improvement of Muse cell therapy.

The future is bright for Muse cells. The discovery of Muse cells heralds a new era of tissue repair by a unique cell type evolved to protect and restore injured tissues. The Muse cell is creating a brave new world of cellular repair and replacement that we have not had access to before. In the coming decades, we will be able to heal faster and better than before. Muse cells provide answers to not only trauma and disease-induced cell damage but also aging as well.

18.2 Non-tumorigenic Cells

Professor Dezawa and her group have greatly advanced our understanding of why and how Muse cells do not form tumors. While Muse cells can proliferate at a rapid rate when needed, they engage in less self-renewal than asymmetric reproduction. The number of muse cells in cultures diminishes over time, and the cultures do not cause tumors. In vivo, Muse cells do not inflame cells around them or stimulate growth factor production.

Pluripotency and anti-immunity are dangerous properties, especially when combined. Pluripotent stem cells make teratomas, tumors that include all three lineages of epidermal, mesodermal, and endodermal cells. Anti-immunity would prevent the immune system from detecting and eliminating the cells. Muse cells clearly possess very effective mechanisms to prevent tumors because there are no reports of teratomas after Muse cell transplantations.

The difficulty of growing pure Muse cultures is evidence of tumor suppression mechanisms. However, the mechanisms that prevent tumor formation are not yet well understood. Recently, Hanna et al. [12] identified MBD3 as the first of many pluripotency suppressor genes. Silencing MBD3 dramatically increased the efficacy of Yamanaka genes to induce pluripotency [13]. Other pluripotency suppressor genes have been discovered [14]. Study of Muse cells will likely yield new mechanisms of pluripotency and tumor regulation.

18.3 Anti-immunity

Mesenchymal stem cells express HLA-G [15–18]. In Chap. 4, Kushida, Wakao, and Dezawa (Fig. 4.13) show that 87.5% of the Muse cells entering infarcted rabbit hearts express HLA-G [19], much higher than the usual 20% expression rate of HLA-G expression by mesenchymal stem cells. Allograft Muse cells that engrafted at the infarct border expressed HLA-G on day 3 after intravenous injections.

HLA-G is the most powerful anti-immune mechanism in the body. This non-classical HLA allele is much more strongly expressed on MSCs of human umbilical cord (hUC) than MSC from other sources. Ding et al. [20] reported that 90.8% of hUCMSC express HLA-G1, HLA-G5, and HLA-G7. The cells inhibited proliferation of lymphocytes by 35 ± 3% and could be blocked by HLA-G antibodies. Trophoblasts at the placental-maternal border express HLA-G.

Cancers that express HLA-G form aggressive tumors [21]. Placental mammals evolved HLA-G to protect the placenta and fetus. Many human pluripotent stem cell lines created by transient exposure of placental cells to BMP4 strongly express HLA-G [18, 22]. Placental trophoblasts isolated by Percoll gradient and negative antibody sorting express high levels of HLA-G [23].

Ability to evade the immune system is a dangerous property. Cancer cells that have this ability tend to form aggressive tumors. Muse cells must have evolved effective mechanisms to prevent loss of growth control. These mechanisms likely include not only very effective single- and double-strand DNA repair mechanisms but also high expression of genes [24–26] that suppress retrotransposon activity that may mutate growth control genes [27, 28] and increase the risk of cancer [29]. Such mechanisms also increase the ability of Muse cells to tolerate stress.

18.4 What Cells to Make?

Until recently, the only known mechanism by which tissues can tell stem cells what type and how many cells to make is through "niches." In injured tissues, "niches" are unlikely to survive or to know what to instruct Muse cells. Muse cells must be able to determine what kind and how many cells to make when it arrives injured tissues. It would not be acceptable, for example, for Muse cells to arrive in the heart and start making skin cells.

One mechanism, suggested by Mari Dezawa in her presentations, is that Muse cells phagocytose dead or dying apoptotic cells and then utilize signals from the phagocytosed cells to decide what cells to make. If so, this would be an elegant mechanism. However, when Muse cells are grown in culture, they often make a variety of cell types without necessarily having been exposed to these cell types in culture.

Muse cells may obtain their cues for which cells to produce from antigen-presenting cells that phagocytose dead cells and present their antigens. For example, monocyte-derived dendritic cells (DC) present tumor antigens [30], a process that depends on NF-kappa-B and is inhibited by IkappaB [31, 32]. Interferon-producing natural killer (NK) DC cells [33, 34] or other NK cells may also play a role.

In addition to not knowing how Muse cells are exposed to signals regarding what cells to make, we do not know the nature of the signal. In theory, the signal is likely to be complex (there are many cell types), specific (it would not do if the wrong signal is given), and located on cells that have died (so that their antigens would be acquired and presented). Potential candidates for such signaling include lectins and cell surface glycan signatures [35].

18.5 Master of Tissue Repair

We have much to learn from Muse cells. As more evidence accumulate showing the diverse armamentarium of pleotropic, paracrine, cell replacement, anti-inflammatory, and anti-immune tools that Muse cells can bring to bear in injured tissues, these individual tools can be tested as therapeutic modalities or by facilitating these capabilities in Muse cells.

Activating Muse cells to repair specific tissues before transplantation may allow them to hit the ground running. For example, rather than waiting for naive Muse cells to identify the cell types that they are supposed to replace, preexposure of Muse cells to target cells may increase their potency and efficacy. Finally, Muse cells likely work with other cells to repair tissues. Muse cells are likely to work with other cells to repair tissues, e.g., mesenchymal cells to rebuild tissue structure, pericytes to rebuild vascular networks, and neural stem cells to rebuild blood-brain barriers.

Finally, while Muse cells possess potent anti-inflammatory and anti-immune mechanisms, their progeny may not retain these mechanisms for long. It would not be good, for example, if cardiomyocytes made by Muse cells incorporated into the heart and are then immune-rejected some months later. Recruitment of autologous

Muse cells or use of HLA-matched Muse cells may be better for long-term engraftment. Muse cells can be collected and stored from many tissues, including adipose, bone marrow, umbilical cord, and placenta.

18.6 Conclusions

The discovery of Muse cells presents exciting new research opportunities for the field of stem cell research. Adult Muse cells evolved to repair injured tissues. They do not form tumors, resist immune rejection, and can detect and decide what type and number of cells to produce. Understanding the mechanisms of these traits will provide many insights concerning how stem cells repair tissues.

Unlike ESC and iPSC, Muse cells do not form teratomas. While Muse cells can proliferate rapidly if necessary, they engage in more asymmetric divisions than self-renewal. The progeny of Muse cells tends to grow faster than the Muse cells themselves. As a consequence, the percentage of Muse cell in culture falls as the culture grows, declining to only a few percentage of the cells in a culture over time.

Like MSC, Muse cells express anti-immune mechanisms. In particular, they express HLA-G, which turns off immune cells, including cellular (NK cells) and humoral (lymphocytes) immunity. However, nearly 90% of Muse cells express HLA-G compared to only about 20% of MSC. Because they are anti-immune, Muse cell is useful not only for cell replacement but also to suppress autoimmune disease.

Muse cells can detect and decide which cells have died to make the appropriate type and number of cells to replace them. Neither the detection mechanism nor signals are known. Whatever the signal is, it must be complex to represent hundreds of cell types, specific to identify the cell type that is needed, and present on the cell. Lectins and glycans have sufficient complexity and specificity to serve this purpose.

Muse cells are phagocytes, and it is possible that they eat dead and dying cells to determine which cells to make. Many inflammatory cells, however, phagocytose cells and present their antigens, including dendritic cells, NK cells, and others. The first laboratory to decode this signal will be able to prime the Muse cells before transplantation so that they can hit the ground running.

Muse cells also secrete general and tissue-specific factors that have paracrine and pleiotropic effects on tissues. Studying these factors and how Muse cells use these factors will teach us much about their roles and effects. Likewise, Muse cells are likely to work with other cells to rebuild tissues, including mesenchymal to restore tissue structure, pericytes and endothelial cells to rebuild the vasculature, and neural stem cells to close the blood-brain barrier.

Finally, while muse cells possess potent anti-inflammatory and anti-immune mechanisms to prevent rejection, it is not clear that their progeny will retain these mechanisms for the long term. For treatments with the intent for long-term engraftment, therapies that emphasize recruitment of endogenous Muse cells or use of HLA-matched Muse cells would be desirable. Muse cells can be collected from many tissues, including adipose, bone marrow, umbilical cord, and placenta.

References

1. Amin M et al (2018) Cardiotrophic growth factor-driven induction of human Muse cells into cardiomyocyte-like phenotype. Cell Transplant 27(2):285–298
2. Tanaka T et al (2018) Mobilized Muse cells after acute myocardial infarction predict cardiac function and remodeling in the chronic phase. Circ J 82(2):561–571
3. Iseki M et al (2017) Muse cells, nontumorigenic pluripotent-like stem cells, have liver regeneration capacity through specific homing and cell replacement in a mouse model of liver fibrosis. Cell Transplant 26(5):821–840
4. Katagiri H et al (2016) A distinct subpopulation of bone marrow mesenchymal stem cells, Muse cells, directly commit to the replacement of liver components. Am J Transplant 16(2):468–483
5. Uchida N et al (2017) Beneficial effects of systemically administered human Muse cells in adriamycin nephropathy. J Am Soc Nephrol 28(10):2946–2960
6. Uchida H et al (2017) Human Muse cells reconstruct neuronal circuitry in subacute lacunar stroke model. Stroke 48(2):428–435
7. Shimamura N et al (2017) Neuro-regeneration therapy using human Muse cells is highly effective in a mouse intracerebral hemorrhage model. Exp Brain Res 235(2):565–572
8. Uchida H et al (2016) Transplantation of unique subpopulation of fibroblasts, Muse cells, ameliorates experimental stroke possibly via robust neuronal differentiation. Stem Cells 34(1):160–173
9. Yamauchi T et al (2015) Therapeutic effects of human multilineage-differentiating stress enduring (MUSE) cell transplantation into infarct brain of mice. PLoS One 10(3):e0116009
10. Vogel G (2005) Cell biology. Ready or not? Human ES cells head toward the clinic. Science 308(5728):1534–1538
11. He J et al (2017) A safety checkpoint to eliminate cancer risk of the immune evasive cells derived from human embryonic stem cells. Stem Cells 35(5):1154–1161
12. Rais Y et al (2013) Deterministic direct reprogramming of somatic cells to pluripotency. Nature 502(7469):65–70
13. Yamanaka S (2008) Induction of pluripotent stem cells from mouse fibroblasts by four transcription factors. Cell Prolif 41(Suppl 1):51–56
14. Kareta MS, Sage J, Wernig M (2015) Crosstalk between stem cell and cell cycle machineries. Curr Opin Cell Biol 37:68–74
15. Yen BL et al (2009) Brief report – human embryonic stem cell-derived mesenchymal progenitors possess strong immunosuppressive effects toward natural killer cells as well as T lymphocytes. Stem Cells 27(2):451–456
16. Selmani Z et al (2009) HLA-G is a crucial immunosuppressive molecule secreted by adult human mesenchymal stem cells. Transplantation 87(9 Suppl):S62–S66
17. Siegel G, Schafer R, Dazzi F (2009) The immunosuppressive properties of mesenchymal stem cells. Transplantation 87(9 Suppl):S45–S49
18. Liu KJ et al (2011) Surface expression of HLA-G is involved in mediating immunomodulatory effects of placenta-derived multipotent cells (PDMCs) towards natural killer lymphocytes. Cell Transplant 20(11–12):1721–1730
19. Yamada Y et al (2018) S1P-S1PR2 axis mediates homing of Muse cells into damaged heart for long-lasting tissue repair and functional recovery after acute myocardial infarction. Circ Res 122(8):1069–1083
20. Ding DC et al (2016) Characterization of HLA-G and related immunosuppressive effects in human umbilical cord stroma-derived stem cells. Cell Transplant 25(2):217–228
21. Rouas-Freiss N et al (2014) The dual role of HLA-G in cancer. J Immunol Res 2014:359748
22. Yang Y et al (2015) Heightened potency of human pluripotent stem cell lines created by transient BMP4 exposure. Proc Natl Acad Sci U S A 112(18):E2337–E2346
23. Li L, Schust DJ (2015) Isolation, purification and in vitro differentiation of cytotrophoblast cells from human term placenta. Reprod Biol Endocrinol 13:71

24. Orecchini E et al (2017) ADAR1 restricts LINE-1 retrotransposition. Nucleic Acids Res 45(1):155–168
25. Wylie A et al (2016) p53 genes function to restrain mobile elements. Genes Dev 30(1):64–77
26. Schulz WA (2006) L1 retrotransposons in human cancers. J Biomed Biotechnol 2006(1):83672
27. Miki Y (1998) Retrotransposal integration of mobile genetic elements in human diseases. J Hum Genet 43(2):77–84
28. Gasior SL et al (2006) The human LINE-1 retrotransposon creates DNA double-strand breaks. J Mol Biol 357(5):1383–1393
29. Helman E et al (2014) Somatic retrotransposition in human cancer revealed by whole-genome and exome sequencing. Genome Res 24(7):1053–1063
30. Bonaccorsi I et al (2015) Acquisition and presentation of tumor antigens by dendritic cells. Crit Rev Immunol 35(5):349–364
31. Yoshimura S et al (2001) Role of NFkappaB in antigen presentation and development of regulatory T cells elucidated by treatment of dendritic cells with the proteasome inhibitor PSI. Eur J Immunol 31(6):1883–1893
32. Yoshimura S et al (2003) Antigen presentation by murine dendritic cells is nuclear factor-kappa B dependent both in vitro and in vivo. Scand J Immunol 58(2):165–172
33. Chan CW et al (2006) Interferon-producing killer dendritic cells provide a link between innate and adaptive immunity. Nat Med 12(2):207–213
34. Lee SW et al (2013) Natural killer dendritic cells enhance immune responses elicited by alpha-galactosylceramide-stimulated natural killer T cells. Biomed Res Int 2013:460706
35. Tao SC et al (2008) Lectin microarrays identify cell-specific and functionally significant cell surface glycan markers. Glycobiology 18(10):761–769

Correction to: Basic Characteristics of Muse Cells

Shohei Wakao, Yoshihiro Kushida, and Mari Dezawa

Correction to:
Chapter 2 in: M. Dezawa (ed.), *Muse Cells*, Advances in Experimental Medicine and Biology 1103, https://doi.org/10.1007/978-4-431-56847-6_2

The below listed corrections have been carried out in the following pages of the current version:

1. In page 14, the legend of Fig. 2.1 "Trilineage differentiation ability of single Muse cell-derived cluster. The central phase-contrast microscopic image shows expanded cells emerging from the single Muse cell-derived cluster in gelatin-coated culture dish. In expanded cells, cells positive for neurofilament, cytokeratin 7, smooth muscle actin, alpha-fetoprotein, and desmin were detected" is corrected to "SSEA-3(+) Muse cells differentiate into various kinds of ectodermal-, endodermal- and mesodermal-lineage cells in vitro and in vivo, either spontaneously or by cytokine induction."
2. In page 16, section 2.2, sub-section 2.2.1, the third paragraph, "Trilineage differentiation" was duplicated in the last sentence which is now corrected and the updated sentence now reads "Because single cell-derived pluripotent-like clusters are not formed, trilineage differentiation and self-renewal are"
3. In page 23, section 2.5, sub-section 2.5.1, the fourth paragraph, the word "uick" is incorrect. It is corrected to "quick".
4. In page 24, section 2.6, sub-section 2.6.1, Figure 2.8, the word "chorangiocytes" is incorrect. It is corrected to "Cholangiocytes".
5. In page 30, section 2.8, the ratio "0.01~0.03%" is incorrect. It is corrected to "0.1~0.03%".

The updated online version of this chapter can be found at
https://doi.org/10.1007/978-4-431-56847-6_2

M. Dezawa (ed.), *Muse Cells*, Advances in Experimental Medicine and Biology 1103, https://doi.org/10.1007/978-4-431-56847-6_19

The manufacturer's authorised representative in the EU is Springer Nature Customer Service Centre GmbH, Europaplatz 3, 69115 Heidelberg, Germany. If you have any concerns regarding our products, please contact ProductSafety@springernature.com

Printed and bound by CPI Group (UK) Ltd, Croydon, CR0 4YY
01/10/2026
02225866-0012